Raising Poultry on Pasture:
Ten Years of Success

The American Pastured
Poultry Producers Association
Compilation

Edited by
Jody Padgham

Raising Poultry on Pasture
Ten years of Success
First Edition 2006

Copyright 2006 by the
American Pastured Poultry Producers Association
www.apppa.org

Edited by Jody Padgham

Thanks to all who have
contributed their wisdom to this book.

All rights reserved. No part of this book may be reproduced without written permission from the publisher (APPPA). No part of this book may be reproduced, stored in a retrivial system of transmitted in any form by any means-- electronic, mechanical, photocopying, recording or other-- without written permission from the American Pastured Poultry Producers Association.

The information in this book is presented by independent authors and believed by them to be true and complete. All recommendations are made without guarantee on the part of the American Pastured Poultry Producers Association. The editor and publisher disclaim any liability in connection with the use of this information.

Photos have been contributed by the authors.

Any questions may be directed to Jody Padgham, 2240 310th St, Boyd, WI 54726

Printed in the United States.

Library of Congress Cataloging-in-Publication Data

Includes Bibliographical References and Index

ISBN 0-9721770-4-3

Raising Poultry on Pasture
Table of Contents

Introduction

1. Overview of Pastured Poultry
    Introduction to Pastured Poultry — 1
    Pastured Poultry Producers Speak Out! — 5
    Research Study of Large Pastured Poultry Farms — 7
    Pastured Poultry System Comparisons — 9

2. Brooding
    Brooding Basics — 12
    Guidelines for Brooding Chicks — 13
    The Brooder Nightmare — 15
    Bennett Ranch Brooder Checklist — 16
    A Few Ideas to Supplement Dan Bennett's "Brooder Nightmare" — 18
    Producer Profile: Charles and Laura Ritch — 20

3. Pastured Poultry Genetics
    Pastured Peepers Symposium — 22
    Business Profile: Moyer's Chicks — 25
    What's Happening at the Hatcheries — 26
    Raising Your Own Meat Flock? — 27
    The Cornish Cross: What is Wrong With This Picture? — 30
    Producer Profile: The Chicken Man of Hume — 33

4. Shelter Designs
    Creative Pen Designs — 36
    Portable Pens Compared — 38
    Inexpensive Mobile Pens, Storage etc. — 41
    Business Profile: Brower, A Division of Hawkeye Steel — 42
    Producer Profile: Mike and Deb Hansen — 43
    Mike Hansen Cattle Panel Pen — 46
    Pen Designs—2004 — 48
    Wood Pens, PVC Pens, Hog Panel Pens, and Now Rebar Yurts! — 50

5. Day-Range Systems
    Shelter Design Considerations — 51
    Producer Profile: Aaron and Kelly Silverman — 54
    Day-Ranging — 57
    Stationary Netting Model — 58
    Business Profile: Kencove — 59
    Typical Day-Ranging Model — 60
    Prairie Schooner Chickens — 61

Raising Poultry on Pasture
Table of Contents

6. Equipment
Watering System Comparisons	64
Scavenging for Fun and Profit	66
Business Profile: Gillis	67
Reefer Madness	68

7. Eggs on Pasture
What About Eggs?	71
More Pastured Egg Information	72
Hoophouses for Hens	74
Marketing Pastured Eggs	77
Eggs on Pasture? A Kansas Story	79
Business Profile: EggCartons.Com	81
Lightweight Nest Boxes	82
Egg Processing by Hand	82
Producer Profile: Springfield Farm	85
Keeping Eggs Clean	88
Environment Stresses on Laying Hens	88
Winterizing Laying Hens	90

8. Turkeys, Ducks and other Poultry
Turkeys	92
Turkeys Have Feelings Too!	96
Producer Profile: Alexander Family Farm	98
It's Turkey Time Again	100
Research Adventure	103
Raising Historical Turkeys	105
Raising Heritage Turkeys	107
Ducks and Pheasants: Taking the Salatin Model One Step Further	109
Producer Profile: Geese in the Pasture	111

9. Poultry Nutrition and Health
The "Pasture" in Pastured Poultry: An Oregon View	115
The "Pasture" in Pastured Poultry, Continued...	116
Grass Conversion Rates by Poultry	118
On Farm Research and Snake Oil	119
Why Do You Mix Your Own Feed?	121
Nutritional Tips	122
Poultry Feed Ration Update	123
Nutrition News For 2002	124
Looking for a Feed Mill?	125
Summer Ration to Winter Ration Changes for Pastured Layers	125
Business Profile: The Fertrell Company, Inc.	126

Raising Poultry on Pasture
Table of Contents

Poultry Nutrition and Health, continued
    Little Lessons Learned   127
    Producer Profile: Jonathon and Ellie Coulimore   128
    Mortality: Common Causes   130
    Business Profile: Acadian Seaplants Limited   133
    Dealing with Coccidiosis in Pastured Poultry   134
    Business Profile: Dotson Farm and Feed   136
    Introduction to Tenosynovitis (Viral Arthritis)   137
    What is Blackhead?   138
    Roost Mite Control 101   138
    Avian Influenza in Poultry   139
    Bird Flu Editorial   141
    Predators: Thieves in the Night   143
    You Are the Key to Preventing Avian Disease   144

10. Processing
    How to Process a Chicken at Home   145
    Small-Scale Poultry Processing Equipment   148
    Build your Own: Killing Cones and Transport Boxes   150
    Eviscerating Poultry the Easy and Sanitary Way   152
    Just 100 Chicks!   153
    Business Profile: The Poultry Man—Eli Reiff   154
    Business Profile: Schafer Farms Natural Meats   155
    Business Profile: Pickwick Zesco   156
    Cadillac Mobile Processing Plant   157
    WI Mobile Processing Unit   158
    Mobile Processing Units (MPUs)   159
    Revisiting the MPU   161

11. Marketing
    Direct Marketing Isn't About Selling, It's About Dreaming   165
    Steps to Successful Marketing   167
    Direct Marketing: Seven Basic Roles   169
    Marketing Meats and Poultry at Farmers Markets   170
    Marketing Options   173
    From Customers Into Converts   174
    Price Check/ Reality Check   177
    Developing A Brochure For Your Farm   178
    Business Profile: Growers Discount Labels   180
    Pastured Poultry on the Internet   181
    Web Pages for Your Farm   182
    Measuring the Interest for Marketing   184
    Give Them What They Want!   187

Raising Poultry on Pasture
Table of Contents

Marketing, continued
    Almost Organic?   188
    Producer Profile: Elmwood Stock Farms   191
    Mega-Farms Make Your Marketing Easier   193
    Business Profile: PASA   195
    Why Grassfed is Best   196
    Nutritional Analysis of Pastured Poultry Products   197
    How to Sell Grassfed Science   198

12. Record-Keeping and Insurance
    Tracking and Using Information About Your Farm   200
    Successful Poultry Production Begins With Record-Keeping   201
    Record-Keeping   204
    Producer Profile: Dan Bennett   206
    Profitability—Will it Make Money?   208
    What Will I Do Next Year?   210
    Creating a Business Plan   212
    How Much Insurance is Enough?   214
    How to Protect Your Business though Legal Structuring   217
    On Coyotes and Lawyers   219
    Rest Assured, You Can Find Insurance   222

13. The Good, the Bad and the Ugly: things to keep you thinking
    What If   224
    Producer Profile: Kip and Jackie Glass   226
    Zero Mortality: A Compelling Design Condition   228
    Producer Profile: Karla Tschoepe   231
    The Biological Integrity of Pastured Poultry   233
    It's All in the Vernacular   234
    Which Would You Choose?   235
    Alternative Poultry Production in France   236
    Lessons from a French System of Production- Label Rouge   240
    Poultry Research Review   242
    Composting Dead Livestock   245

14. Resources   247

Index   248

Introduction
By Jody Padgham Summer 2005

Enjoying the privilege of editing the APPPA GRIT! newsletter for a few years made me very aware of two things- first, how knowledgeable, creative and skilled a group of people our pastured poultry membership is, and second, what a lack of written material there is for striving pastured poultry producers to gather information from. Those thoughts merged into the germ of this book, which the American Pastured Poultry Producers Association (APPPA) is pleased to present to you. We offer here a compilation of articles from the APPPA GRIT newsletter, from its beginnings in January 1997 through issue number 37, which came out in the summer of 2005.

Within these pages we have collected the best of the experimentation of many creative poultry-focused minds over the years. From the true pioneers, such as Joel Salatin of Polyface Farm, to the voices of those on the beginning "make all the mistakes you can" edge of things, we present you with a multitude of ideas and numerous creative approaches to raising poultry on pasture.

We offer this information with a few disclaimers. The first is that some of these articles, although full of relevant information, were written several years ago, in the beginning stages of a developing industry. For example, you will read in " Turkeys" by Joel Salatin (page 92), Joel's first thoughts after two years of turkey production in 1998. This is a great article that brings a lot of basic information to the beginning turkey producer, but much has been learned about raising turkeys on grass since 1998, and subsequent articles will add to the base that Joel lays. We recommend that you read an entire chapter (well, really, the whole book!) to collect a diversity of information and "lessons learned" before you forge off to make decisions and start your own trial and error process. We present the articles here in topical order, not in historical order. Note the original publication date (in each heading) to see what place in the timeline of learning the particular article may represent. Many early ideas have held up through the years, others have been expanded and refined. Some have been discarded and replaced.

We also, through this compilation, bring you theories, successes and experiments from many different farmers. The APPPA membership ranges over most of the 50 U.S. states and a few foreign countries. Member operations produce from zero birds (just learning...) to the mid range of a few thousand, to the high end of 50,000 broilers on pasture per year. Each producer brings their own background, personality and resources to the pastured poultry industry. The diversity of our authorship will hopefully reflect the diversity of you, our readers. We hope that the one primary lesson that you bring home from reading this book is that although there is a lot to learn from those who have gone before you in pastured poultry production, and definitely some right and wrong things to do, there is no "one way" to succeed at pastured poultry production. Each bioregion, each market center, each family farm will have its own challenges, its own successes. As you will read in one of the articles in this book: the way to succeed at any small farming practice is to "read every book you can find and then throw all the books away." Take what you learn here, mull it over, and set up systems that reflect and support your personal needs, values and resources. We won't encourage you to throw this book away, but do invite you to soak in all of the wisdom it has to offer, and then take it to you neighbor's house and pass it on. (Perhaps after you make copies of the chapter on "nutrition and health"!)

Our third disclaimer follows from the second: none of our authors proclaim themselves as "experts." For the most part they have learned what they offer in these pages through trial and error. We take no responsibility for any information that may seem conflicting or incorrect, and can guarantee no particular results if you follow the recommendations offered here.

Many began their journey into pastured poultry after reading the ground breaking *Pastured Poultry Profit$* written by Joel Salatin in 1993. Salatin was one of the originators and initial board members of APPPA. Throughout this book you will see references to "the Salatin method" and "Salatin pens." Salatin, a true master of ingenuity, did a wonderful job of creating and then sharing his ideas of how to succeed at raising poultry on grass. APPPA, and thus this book, takes up the story from there. We do not repeat much of what Salatin lays out in his book. Those interested in learning more about Salatin's original pen design (which he still uses) and other inspirations will want to read his book, which is often available from your local public library or may be purchased from several mail-order or online book stores. For details, see the resources section, Chapter Fourteen.

This book presents information, from overviews to details, in thirteen chapters plus a resource section and index. Scattered throughout the book we offer two "in-depth" views—the *APPPA Producer Profiles* and *APPPA Business Member Profiles*. Producer Profiles give you an inside look at the choices producers of all sizes and types around the country have made in their operations. These will be very helpful to anyone just starting out in understanding what a successful pastured poultry operation looks like. The APPPA Business Member Profiles highlight businesses around the country that support and offer services to pastured poultry producers. This series was started in 2002 of the APPPA GRIT! Contact information for Business Members is current as of the first printing of this book, but not guaranteed forever. Hopefully a web search or call to the APPPA office will bring you up-to-date in years to come if these contacts become incorrect.

It is unfortunate that many of the early articles, including those describing pen designs, did not contain duplicatable photos we could include here. All photos in the book were contributed by the authors or those being described in each article.

We wish you the best in succeeding in this dynamic, consumer-driven, innovative industry. We also invite you to join the American Pastured Poultry Producers Association to share what YOU have learned and continue learning from those who are passionate about raising birds on pasture. For more information about APPPA, see www.apppa.org.

About the American Pastured Poultry Producers Association

The American Pastured Poultry Producers' Association (APPPA) was established in 1997 to assist all pastured poultry producers in North America. APPPA is a clearinghouse for innovations and refinements to the pastured poultry model. APPPA offers the opportunity for people to learn and exchange information about raising poultry on pasture.

Offerings from APPPA include
¤ An information-filled newsletter
¤ Website with resources for consumers and producers
¤ Three membership levels, including Producer, Producer-Plus and Business
¤ Active electronic "listserve" for Producer Plus members

Visit the APPPA website today for more information. www.apppa.org

CHAPTER ONE
Overview of Pastured Poultry

Introduction to Pastured Poultry
By Jody Padgham Fall 2005

Many of us remember going to "Grandma's House" and having a tasty chicken dinner. Or, maybe we are young enough that we just hear other people talking about how wonderful, and easy to take for granted, that experience was.

Anyone you ask who has memories like this will at some point say "you just can't get chicken like that anymore." Although many country people still "keep a few hens" or have 25 roasters running around to catch for Sunday dinner, poultry production has in general gone the way of all things agricultural in the United States - bigger is better, efficiencies reign, and animal lives are compromised and speeded up - all for the sake of keeping food prices low and profits high. The average supermarket chicken has never seen the light of day, may be only four to five weeks old when butchered, and is grown in very close quarters with thousands of other birds inside huge buildings. Their beaks have been cut off so that the stress of being in uncomfortable living conditions doesn't lead to pecking their fellows to death. Hens producing eggs are in very small cages, one on top of another, never to see the sun or grass. Is it no wonder that chicken and eggs today don't taste like they used to?

The Beginnings of Pastured Poultry
Enter the new world of pastured poultry production! Early in the 1960s, a pioneer named Joel Salatin of Swope, Virginia, started experimenting with growing chickens outside, so they could live more "normal" poultry lives, eat bugs and grass and enjoy the rain and sun. Salatin is passionate about efficiency, and he knew that to make his poultry dreams work in his farm system, the chicken production needed to be cost effective, efficient, good for the birds and good for the land they were living on. The bonus is that the system produces very high quality food.

After several years of experimentation, Salatin developed a "Pastured Poultry Model" which he captured in his book *Pastured Poultry Profit$*, first published in 1993. Over the years thousands of people have visited Polyface Farm in Virginia to learn about growing poultry on pasture. In 1997, an organization to bring all of those people together, the American Pastured Poultry Producers Association (APPPA), was formed, with Joel Salatin as one of the charter board members.

APPPA has thrived over the years, with its keystone being the "APPPA GRIT!," a now bi-monthly newsletter offering features and innovations from poultry producers from all over the country. Producers have studied the concepts that Joel Salatin pulled together into his pastured poultry model, and developed tangents of their own. Because of this, it is not as easy as it was ten years ago to describe what pastured poultry is. We offer the general concepts here, which will be elaborated in the remainder of this book. In this introduction we will give you an overview of chicken broiler production. Specific details on laying hen, turkey, goose and duck production can be found in later chapters.

Is Pastured Poultry for You?
Perhaps you own a few acres of land that you'd like to "just do something with," or you have a family farm that you would like to diversify. Pastured poultry can be raised on very small acreage, either as an independent operation or as an integrated piece with other farming ventures, such as a market garden or grazing ruminants. Poultry can be raised for family consumption only, or as a profit-making venture. The size of your operation, efficiencies, and time put into marketing are the primary things that will distinguish a commercial pastured poultry operation from a home operation, but much else will remain the same. Whether you raise 50 or 50,000 birds a year, many of the same concepts hold.

Almost anyone can do the work involved with raising poultry on pasture. It is an excellent project for school age children or as a side-business for any farm. In general, there is a very low overhead to get started (it can be done without buildings or large mechanical equipment, for example) and the turn around time of the product is quick (the chicks you buy today will be ready to sell as ready-to-eat broilers in eight weeks).

Find Your Market First!

Those wishing to get into pastured poultry to make money must first explore potential markets in the area. Are there others selling pastured poultry in your region? Do they fill the market or is there room for more? Will the local farmers market let you sell chicken? If so, what are the regulations for licensing or handling? Most states limit the number of birds you can process and sell from your farm. How many birds do you want to try to raise and sell? Who will those be sold to?

Although in most areas the market for pastured poultry is greater than the supply, it can take a few years to both develop your production systems so that you can produce consistent quality and not suffer a lot of loss, and to get the word out that you have pastured poultry available to sell. The best way to get started is to grow poultry for a few years for family and friends to "get the bugs out" of your production and slowly grow a business with neighbors and friends. Most successful growers say they started small and grew their businesses, mainly through word-of-mouth expansion of their sales base. Don't make the mistake of starting out with 300 birds ready to sell and no customers in sight.

Make a Plan

We recommend that you do some planning before you jump right into poultry production. First, read this book(!). Join APPPA, so that you can keep up with innovations and get answers to questions that invariably come up. Visit other farms that are already raising poultry on pasture so that you can see some of the equipment first hand and learn about decisions and challenges those producers have experienced.

Once you have done those things, you are ready to make a plan. What kind of production system will you try first: pen or day-range? How many birds will you raise the first year, in five years, or ten? Where will the pens go? Who will do the work? When do you want your first birds to arrive? How many batches per season? Who will process the broilers? What are the laws of your state that regulate poultry production and processing? Where will you market the birds, and what do you need to do to make that happen—apply to be at a farmers market, take out an ad in the paper, put up posters at the grocery? Write your thoughts down so that you have a general idea of what you are up against.

Decide on a Production System

As you will read later in this book, pastured poultry producers have in the past ten years taken two main paths: the first group raises birds in small, enclosed shelters (anywhere from 8 x 10 ft. to 12 x 14 ft., of varying heights). These shelters are placed out on a pasture, with chicks, feed and water and moved every day by pulling or using a dolly ("moveable pens"). The other group of producers builds a shelter that may be bigger and may have a floor, but certainly has doors which allow the birds to come in and out. This shelter will be surrounded by electric fencing (either netting or strands of electric fence) and moved only occasionally. The electric fence is moved regularly to allow the birds access to new pasture ("day-range").

There are different advantages and challenges to each of these systems, as shown by research shared by Kip Glass in his article "Pastured Poultry Systems Comparisons" (page 9). Once you decide on a system and pen design, you will need to build or purchase the type of pen you'd like to start with. This must be fitted with a watering system and feed trays. Details on shelter systems and equipment can be found in Chapters Four, Five and Six.

Starting the Chicks

Most pastured poultry producers use day old "broiler" chicks purchased from a local or regional hatchery. There is a list of hatcheries at the end of this book. They can also be found with an internet search. The most common broiler chick is the "Cornish Cross." Some producers are working with hatcheries to develop genetics more suited to pasture production--the Cornish Cross has been developed for the confinement poultry industry and is not ideal for pasture production. You can read more about this in Chapter Three: Genetics.

Day-old chicks will generally be shipped in the mail, unless you can drive to get them yourself. You will need to get the young chicks set up immediately in a brooder, where they will stay for two to three weeks, depending on conditions. You can learn about the keys to brooder management in Chapter Two. Key issues in the brooder are bedding, space, heat, access to water and feed and predator control. Loss in the brooder is not uncommon and can occur suddenly. Some losses are due to poor conditions during shipping and can not be predicted or avoided on your end. If you see more than 8% loss during brooding, you must trouble shoot your brooder situation.

Feed

Feed may be purchased as ready-made "chick starter" from the local farm store or mixed with a ration from this book (Chapter Nine: Poultry Nutrition and Health) by your local feed mill. Note that a generic chick starter in a bag will probably have antibiotics and other medications in it. If you'd like to raise your birds "chemical free" you will have to avoid medicated feeds and use alternatives (such as vinegar in the water) to keep the chicks thriving. A clean brooder set-up is the key to keeping little chicks healthy; medications only mask poor management. Game bird or waterfowl feed generally is not medicated and can be used if desired.

Some like to feed baby chicks more finely ground feed, but this is not essential. Poultry feed generally contains corn, roasted soymeal, sometimes oats or wheat, another protein source such as fish or crab meal, calcium and minerals. Soybeans MUST be roasted, as raw soy contains an enzyme that inhibits protein absorption and will kill chickens.

Those desiring to sell organic poultry must feed certified organic feed. Check with regional feed mills for availability of organic poultry feed, which will not contain any medications and is made with organically grown grains.

A broiler will eat 9-15 lbs. of feed during its lifetime. Although broilers love to eat bugs and grass while on pasture, the amount of each of these they eat does not significantly affect the volume of feed they consume.

When out on pasture, it has been found that if feed is allowed to run out during the evening hours (vs. keeping a full feed tray 24-hours per day) the birds will have fewer problems with legs and heart attacks (common in confinement systems).

There are a variety of feeder designs for pastured birds. Some use traditional feeders that can be purchased at farm supply stores. Inexpensive feeders can be constructed by cutting lengths of six-inch PVC in half lengthwise and stabilizing with end caps or attaching together into "feed-rafts" for large groups of birds.

On the Pasture

Chicks are moved out to pasture pens anywhere from two to three weeks of age. The exact timing depends on the time of year and weather conditions. Young chicks can do very well if protected, but will die from exposure if wet and cold. Some producers wrap pasture pens with tarps or plastic early and late in the year in order to extend the season.

It is critical to keep water in front of the birds at all times, but especially in hot weather. Some producers put two watering systems into each pen in case of failure. Others check pens every hour, or several times a day, to be sure waterers are working properly in extreme heat. There is nothing worse than to put a lot of time and care into birds and then find that one broken hose on a hot day has caused several of them to expire. You can read more about how individual producers have dealt with this in the Producer Profiles found throughout the book.

A general rule of thumb is that each bird will need 1.5 sq. ft. of space in a pen as it matures. This means that you can put 100 three-week-old chicks into a 10 x 10 ft. pen, but when they are bigger, at five to six weeks of age, you will want to move 25-50 birds into another pen so that they have room to grow.

A day-range system can plan on about .66 sq. ft. of housing per bird, as the birds generally only come into the pen in the heat of the day and at night. In a day-range the birds will not move far to find food and water; for this reason many recommend keeping nourishment inside the day-range shelter.

Most producers will run numerous batches of birds per season, allowing a succession of chicks to run through the brooder and then out into the pens. You can have chicks growing up in the brooder while your previous batch is finishing out on the pasture. Once the pasture birds go to processing, chicks can move right in from the brooder. Depending on the area of the country you live in, the season starts anywhere from February to May, with those lucky producers in Texas, New Mexico and other far southern states able to run batches year round.

Predators

One reason large-scale commercial poultry producers have moved their birds inside is to avoid problems with predators. It seems that just about every predator on earth loves chicken. In the brooder, rats can be a problem; out on pasture, everything from skunks to raccoons to weasels and dogs to hawks, eagles and owls will try to eat your flock. Throughout the book you will read suggestions in dealing with many of these. A bad predator problem has shut many a pastured poultry producer down. You must learn to act decisively and quickly when dealing with predators.

Processing

Most broilers are ready to process at eight weeks of age, and will be four to six pounds processed weight. In ideal situations they will grow faster, and some producers regularly process at seven weeks. Others like slower growing or larger birds and will wait till birds are nine to ten weeks old. Once you explore your market and know the size bird you are aiming for, you can plan accordingly. Remember that these birds are growing fast. A week in a broilers life is a long time, so you must plan your chick order dates carefully to be sure that you have time to deal with processing on the exact dates the chickens are ready, seven to nine weeks later.

It is important to set up processing arrangements before you order your first birds. Finding someone to process your chickens (or especially ducks or geese) can be very challenging in some areas, as most of the small processing plants that were plentiful in days gone by are now shut down. You may have luck finding a list of processing plants at your State Department of Agriculture; otherwise, you will have to ask around to find where you can take birds to be processed.

If you plan to take birds across state lines to be sold, they must be processed in a USDA-inspected processing plant. In general, those selling at a farmers market or to stores or restaurants must have birds processed at a state (or federal) inspected plant. Some states allow limited on-farm processing for direct-to-consumer sales. Check with others in your area or your state Department of Agriculture to find out the details of your state laws. For more on how to process birds at home and other processing alternatives, see Chapter Ten.

Marketing

As stated earlier, by the time you get to this point you should already know who is buying your chickens. Many established pastured poultry producers ask customers to pre-order birds at the beginning of the season; some even ask for a deposit with each order. This helps a producer plan on the number of birds to raise in a given year and relieves market pressure later in the season when in the thick of production.

If raised correctly, the fine taste of a pasture-raised bird will "sell itself." Successful producers create brochures highlighting their farm story and the qualities of their poultry products. Many find that web sites are helpful to encourage sales. Some producers ship poultry nationwide to take advantage of distant markets. You can explore numerous marketing options in Chapter Eleven.

One key to a successful business is the proper pricing for your product. It is important to learn this lesson early. Understanding the issue of pricing may present a huge learning curve for anyone that has not approached it before- and it can easily destroy any unaware entrepreneur. It is a basic truth that if you sell something for less than it costs you to raise it, you will LOOSE MORE MONEY WITH EACH ADDITIONAL PIECE YOU SELL. This is counter-intuitive to some, they think selling more of something is always a good thing. If you have not figured how much it costs to raise up your birds, and added in amounts to cover your labor and a little profit, it is not worth raising poultry to sell unless you are just doing it for fun and don't care if you loose money. Chapter Twelve, on Record-keeping, offers some simple examples of how to keep track of costs and suggestions on how to set prices.

One of the reasons many have found success with pastured poultry is that they enjoy the opportunity to grow high-quality food and to talk to their neighbors and members of their community about how that food is grown. Your relationships with your customers are key and should not be taken for granted. In this world where fewer and fewer have direct connections to farms, the option of buying from you can be a very important experience for your customers. Make the connection a strong one and you will never regret it.

Conclusion

Does this short outline of pastured poultry intrigue you? Many have found raising poultry on pasture to be a very satisfying experience. We offer in this book plenty of food for thought to get you started on a successful pastured poultry adventure, whether it involves 50 or 5,000 birds.

Pastured Poultry Producers Speak Out!
Summary of a Questionnaire
By Anne Fanatico Summer 2000

APPPA Grit! and Free-Range Poultry Forum included a questionnaire for range poultry producers to which 100 producers responded several months ago. The survey was designed to assess needs for a three-year poultry project sponsored by Heifer Project International and the National Center for Appropriate Technology (ATTRA) to help pastured poultry farmers interested in expanding operations and sales. Many thanks to the producers who took the time to share their opinions for this survey!

Producer Profile

48% of respondents farm full-time. 59% use a computer for farm operations. Farms tend to be diversified: 87% list other enterprises such as beef, lambs, goats, dairy heifers, pigs, vegetables, flowers, grains, hay, etc. 69% also raise layers for egg production and 31% raise turkeys. Other poultry products include stewing hens, guinea fowl, ducks, and pheasants. Some use poultry as part of a fertility program for vegetable production or to diversify products available to customers. 82% want to expand but only modestly—only 5% plan to expand beyond 10,000 broilers per year (see Question 6). Those who do not want to expand often cite time for processing as a limiting factor.

1: Broilers produced per year

0 broilers	10%
0-500 broilers	48%
500-1,000 broilers	19%
> 1,000 broilers	21%

2: Egg production per year

0 dozen	31%
0-500 dozen	22%
500-1000 dozen	16%
>1000 dozen	33%

3: Yearly gross sales

	Broilers	Eggs	Turkeys
$0-5,000	63%	65%	44%
$5,000-10,000	13%	3%	4%
$10,000-15,000	6%	2%	2%
$15,000+	5%	2%	0%

4: Price per pound of meat

$0.50-1.00	1%
$1.00-1.50	22%
$1.50-2.00	49%
$2.00-2.50	10%
$2.50+	6%

5: Price per dozen eggs

$1.00-1.50	34%
$1.50-2.00	28%
$2.00-2.50	9%
$3.00-3.50	3 %

6: Broiler expansion (growth to ideal numbers of birds as expressed by respondents)

1-1000	26%
1,000-5,000	30%
5,000-10,000	14%
10,000+	5%

Others want to expand "as much as possible to support my family" which is hard to quantify; others only plan to expand egg production.

Processing and Regulations:
7: Type of processing

On-farm non-inspected	73%
On-farm state-inspected	1 %
On-farm fed-inspected	1%
Custom	13%
Other	1 %

62% are dissatisfied with their current processing labor requirements and 64% are dissatisfied with regulations. Regulations are often ambiguous or too confining. Without government inspection, sales are limited—some states only allowing 1000 birds per farm per year to be sold non-inspected.

Processing is too labor-intensive—it is inefficient due to low-volume equipment, insufficient labor or lack of technical experience. It is difficult to find "on-call" labor—some producers ask their customers for help in processing. Equipment is expensive and upgrades have to be carefully thought out.

13% of respondents have their birds custom-processed. Birds are hauled in pick-ups or trailers to a custom processing facility and brought home in coolers or tubs on ice. Producers who have access to custom-processing consider factors such as cost, shipping, distance to processor, regulations, and organic certification against factors involved in setting up their own processing operation.

Egg processing is also problematic. "We are a beginning egg operation and have yet to find the necessary equipment for efficient egg processing."

38% indicate they would consider building a government-inspected processing plant. Others indicate they may be interested "down the road." One producer comments "(not) unless it could be done cooperatively or with state and/or federal assistance." Many producers see a mobile processing unit (MPU) as a good option.

Marketing
80% list direct marketing as their number one method of marketing. Farmers markets, stores, and restaurants are also listed.

Business Planning and Growth
If expanding, 40% would do a feasibility study and 48% would do a marketing study. Many do not consider a feasibility study to be important because they will only expand in response to customer demand and will remain relatively small. Most comment on the high demand for pastured poultry; in fact they cannot keep up with it.

Some indicate a need for measuring profitability, especially before expanding. Cost of production and processing at a higher volume are needed in particular.

Stock
65% of respondents indicate they are satisfied with the breeds they use. Cornish Cross are the main genetics used for meat. Layer breeds include Rhode Island Red, Barred Rocks, Golden Buff, Orpingtons. However, producers appear barely satisfied with Cornish Cross because they have many negative qualities (weak legs, heart attacks, ascites, poor foraging ability, poor heat tolerance, etc.) While some value the short grow-out period, others find it "unnatural."

Hatcheries got a good report from producers: 83% are satisfied with the hatcheries they use. Producers weigh factors such as distance, quality, consistency, price, availability, service, technical support, and interest in pastured poultry.

Feed
Only 38% use commercial feed with 32% indicating the feed is non-medicated. Availability of non-medicated feed can be a problem, especially for chicks. 72% are interested in certified organic feed. 70% use a custom feed, but it can be problematic and time-consuming.

Collaboration
85% are interested in collaborative work or purchases with other range poultry producers. "A co-op type organization to buy chicks, feed, and equipment would be helpful. A co-op marketing venture would help also."

Respondents are interested in sharing knowledge for production, processing, and marketing, especially feed formulation and sourcing feed ingredients.

Research Study of Large Pastured Poultry Farms
By Jody Padgham Summer 2002

A study undertaken in the summer of 2001 at the University of Wisconsin by Steve Stevenson and Don Schuster of the University of Wisconsin Center for Integrated Agricultural Studies looks at the history and economics of large pastured poultry operations in the U.S. Though the sample size is small, and thus results statistically insignificant, there are several interesting pieces of information in the data.

In the summer of 2001 a national search was done to identify pastured poultry producers who processed 4,000 or more birds per year. 12 producers were identified in the U.S. Nine agreed to an in-depth phone survey about the processes and economics of their businesses.

Of the nine, two had raised birds for over 14 years and seven were in the business less than six years. The average number of years of involvement was seven and a half years. Four of the nine identified prior farming experience; the other five came from other occupations (some of which were ag-related).

Annual production ranged from 4,000 to 50,000 meat birds, with an average of 14,500. The total production from these nine operations was 130,500 birds, which represents only .00016% of the total broiler industry in the U.S. Average production was 600 birds per batch.

Production
At one point in their history, all the producers raised birds in moveable 10 x 12 ft. pens "because they read Joel Salatin's book." Five of the nine have since switched to day-range systems, most citing labor efficiencies as the reason.

Farm size ranged from 4 to 312 acres, with an average of about 625 birds per acre. All were raising Cornish Rock type cross. Four of the producers hire labor outside the family.

Costs for chicks varied between $0.53 and $0.75 per chick, with an average of $0.61. Death rate was about 10%, with brooder death ranging from 2-10%, field loss from 1-10%. Weather and predator loss were the largest causes of death. Most used existing facilities to brood chicks, four heated with electric heat, four with gas brooders and one with wood heat. One producer calculated he spent $1,283 on heat for the season; others did not have numbers to share.

All used shovel and wheelbarrow to clean out the brooders; 55% cleaned after every batch. One producer cleaned once a week, one once a year and one after every third batch.

The cost of the day-range houses ran between $187 and $800 per house, which included building labor, waterers, feeders and electronet. 10 x 12 ft. pens cost $200-365 to build, with an average of $285 (figures include labor, waterers, feeders and some electric wire for predator control). Two thirds of the producers use some sort of automatic watering system, either nipple, gasum or homemade system with black pipe. The other third use 5-gallon waterers filled with pails.

All systems fed feed by hand using 5-gallon pails, with each feeding system being different.

Marketing and Finances
All producers interviewed relied on marketing their birds as "tasting better" and "better for you" than "store bought" birds. All advertised no antibiotics or growth hormones. 33% were certified organic, 67% were not organic but do not use antibiotics or growth hormones. All producers were able to sell all the birds they raised.

Two of the producers sell all 4,000 birds direct to retail customers. Six sell both retail and wholesale. One also sells through a cooperative, delivering the birds live to the co-op. 31% sell retail from the farm, fresh or frozen, whole or cut. 26% sell direct to groceries. 18% sell to restaurants, most fresh sales. 11% sell at farmers markets, mostly whole birds and 50% fresh. Most all producers say their first sales were at farmers markets. One producer sold 40% of their birds from an on-farm store.

Prices ranged from $1.15 (for live birds) to 2.75 per pound. On-farm sales ranged from $1.65-2.25 per pound, retail ranged from $1.50-2.75 per pound, sales to restaurants ranged from $1.65-2.75 per pound, and sales at farmers markets ranged from $2.25-2.50 per pound. Sale weights ranged from 3.5-4.25 pounds, with the average 3.91 pounds.

Advertising was in the form of fliers, newsletters and ads in papers, with 67% of the farms putting money into advertising. Ad budgets ranged from $200-2,000 per year, with an average of $725.

Half of the producers said they were planning on expanding production in the following year.

Four had birds processed off-farm and traveled between five and 150 miles to get the birds processed. All used pickup trucks with trailers to transport the birds. Costs for off-site processing ranged from $1.00 to $4.00 per bird. (Average $1.41, excluding the $4.00, which was $2.08 higher than the next lower cost). Four of the five plants used were privately owned; one was a co-op. Two were "very satisfied" with their processors, two were "satisfied" and one producer was "not at all satisfied".

Five of the producers process on-farm. All use family to help. Hours spent processing ranged from 155-468 hours per year. On-farm facilities processed between 100 and 1,000 birds per day, with the average 385. The range of costs estimated to process (including labor) was $0.85-1.50 per bird (average $1.17). This figure is $0.24 less than the average cost of processing off-site.

These nine farms produce 90% of their family income from the farm. The pastured poultry part of the operation ranged from 0-90%, with an average of 32%. (The 0% had high farm debt and worked off-farm.) Other farm income came from other meat sales, CSA and vegetables. All felt that chicken was a good way to get new customer interest in the farm for their other farm products. Only three of the nine producers had off-farm income.

Overall financials showed that the average investment before processing (not including land, capitalization, management time, insurance or interest on investment) was $4.58 per bird. Average gross income per bird was $8.12. This leaves a pre-processing income of $3.54. Average processing on farm was $1.17, so average gross profit per bird was $2.37. Off-farm processing averaged $1.41 per bird, leaving an average gross profit of $2.13 per bird.

Lessons Learned and Advice to Others
- Slow growth - start with a small customer base, ensure quality and expand slowly.
- Start with retail direct to customer and move to wholesale. The farmers market really helps to expand the demand.
- Growth of the market and thus production expansion was faster than expected.
- Restaurants really like the product.
- Spend time to educate the public about the quality.
- Attention to detail in production is extremely important.
- Stay close to home and watch the birds closely.
- Learn from others - and their mistakes!
- Read books and then "throw them away."
- Jump in if you feel the market is there.
- Marketing should be easy- the birds sell themselves

Challenges and Hard Spots:
- Government regulations can be a huge challenge.
- Finding a good processor can be an obstacle.
- Keeping up with the time demands.
- Challenges of marketing.
- Finding quality equipment.

Pastured Poultry System Comparisons
By Kip Glass Spring 2002

A major debate in the pastured poultry industry is whether the moveable pen or the day-ranging system is better. I decided to do a study to compare the two systems to help fuel the debate.

Under a Missouri Sustainable Agriculture grant, research was done in the spring and fall months of 2001. The situations we wanted to compare between the two systems were:

¤ Labor input: is there any time and work savings between the two production systems?
¤ Feed conversion: is one system better than the other in the total weight gain of the bird?
¤ Pasture considerations: does the pasture respond differently between the two models?
¤ Bird health: how does the condition of the birds compare, mortality rates etc?

Let me begin with a description of what the two production models are.

"Moveable pen" is where the birds are enclosed in a floorless structure and not allowed to range from it to forage. The only forage is what is available underneath them in the pen. Food and water is placed in the pen where it is constantly available to the birds. The pen is moved once or twice daily to allow the birds to forage and for manure dispersal.

"Day-range" is where a semi-permanent shelter with bedding houses the birds, but the birds are let out to range. An electrified netting, rotated on a regular basis, contains the birds in different paddocks around this shelter. The netting protects against ground predators. Water and food is supplied to the birds in the paddocks and in the shelter.

Some of the parameters followed were:
Shelters: shelter size was 12 x 16 ft. with 160 birds stocked into each pen during each production cycle. This allowed an average of 1.2 sq. ft. per bird.

The moveable pen was moved once a day until the birds reached five weeks of age and then twice a day until the seventh week. Half the birds (80) were harvested at seven weeks in each shelter. The moveable pen was then moved only once a day because of low pasture impact.

The day-range pen had a door on each end, and the electrified netting was rotated around the four different quadrants of the five-week cycle of the birds. The first quadrant had birds on it for 14 days, second quadrant 10 days, third quadrant 8 days and fourth seven days. The netting was 165 ft. long, making each quadrant approximately 40 x 40 ft. Again, half the birds were harvested at 7 weeks of age, the other half at eight weeks. After the harvest of the last birds the day-range shelter is skidded to the next fresh spot in the pasture to start the next batch.

Pasture coverage by the moveable pens, moved on the above schedule, covered 9,024 sq. ft. during the five weeks, the day-range pen covers 7,300 sq. ft. per five week schedule.

Feed: each pen was fed the same pounds of feed each day. The feed mix used was the Fertrell Broiler Grower formula (see pages 123-125). Feeders in the day-range areas were moved to different locations in the paddocks daily.

Water: the moveable pen was supplied with 5-gallon buckets attached to Plasson bell-type waterers. The day-range pen had a trough float valve waterer in the pasture attached to a 250-gallon water tank with supply line. Because of daytime heat, I incorporated a bell waterer in the shelter attached to a 55-gallon supply barrel.

Weather and field conditions: two comparisons were run from early spring to early summer, and two were run from late summer to early fall. This was to balance the study over carrying pasture and temperature conditions. The pasture was mainly fescue, with thinning brome, numerous broadleaf weeds and, in the summer through late fall, very thick lespedeza.

Observations
Labor (see time chart, next page)
As you can see from the chart, there is really not significant labor time savings between the moveable pen and the day-range system. Actually, if the moveable pens are designed thoughtfully and light enough to be easy to move, then there is no real strenuous labor involved with the moveable pen.

LABOR COMPARISON	
Moveable pen: chores	Time (min)
Clean up waterers and set up	5
Feed & water, 2 x/day, 35 day @ 6 min	210
Move pen 1 x/day, 35 day @ 3 min	105
Move pen 2 x/day, 14 day @ 3 min	42
Total minutes	362
Day range pen: chores	Time (min)
Skid shelter to new location	10
Clean and set up waterers and fill	40
Move in bedding, load and unload	30
Set up fence 4x @ 15 min	60
Feed & let out (35 days @ 2 min)	70
Put up at night (35 days @ 2 min)	70
Scrape out bedding, scatter or move to compost	30
Total minutes	310
RESULTS: The moveable pen takes 52 more minutes per batch	

With the day-range pen, daily chores are easier. All you have to do is let the birds out, feed them and put them in at night. Moving the electrified netting is very simple and is only done four times during each batch. Skidding the heavy shelter is not as time consuming, but it takes mechanical or animal means to do it. The more labor intensive task is the hauling in and out of the bedding. If you don't save it, you have to spread it on the pasture either by hand or mechanically. And if you save it to use for compost, you have to haul it away. Of course, there is a great value in the compost, especially if you have a market garden or other use for it.

With the moveable pen there is a quicker turn around time for restocking the birds after the last harvest. Just move the shelter forward one spot, throw in feed and water and add the birds.

In comparing labor, I think the main perception is that there are a lot of days in the day-range where we don't really do a lot, and it makes the days in which we have to do the hard work not seem so bad. More on the labor issue in the closing statements.

Feed and Water Issues

With the day-range, watering is a lot easier, using a large supply tank and automatic waterers. You only have to fill the main supply tank a few times each batch.

With the moveable pen this could be done, but would be a little more difficult as the pens are moving all the time and the supply lines would be snagging on the grass, etc. But it would be doable. We elected in this study to use buckets that have to be filled every day as most people that use moveable pens do it this way.

One major issue with water in the day-range group is that you have to have water in the shelter during the hot days, as the birds will not even venture three feet out of the shelter into the sun to drink. They would rather perish in the shade then go outside to drink.

As far as feed, the birds in the moveable pen consume their feed a lot quicker and I'm sure they would have eaten a lot more, in a shorter time span, if they had the feed in front of them at all times. With the day-range pen, if it is hot and sunny out, they would not go out very often to eat and would eat only in the mornings and then wait till the end of the day to eat as it got cooler. I'm sure this limited their total consumption, but this could be remedied by putting feeders inside.

Bird Condition

As far as the birds' condition, the moveable pen birds were a little more soiled because part of the day they were in their own manure. The day-range birds were noticeably cleaner. If it rained, the moveable pens could get pretty wet if water ran under the pens and soaked the birds. This can be a major problem during extended rainy spells. Aesthetically, the birds roaming in the grass just looked happier and anyone else that saw the two models made that comment too.

Field Conditions

Even though the moveable pen covered more total area during each batch, the effect on the pasture was significantly more noticeable. Even though in the day-range pen we moved the feeders around every day, manure dispersal was very sporadic. Since the birds spend most of the hot day and all the night in the shelter, I would estimate that 50% of the total manure produced goes into the bedding. With the moveable pen the pasture is more evenly coated with manure. One negative factor of the day-range shelter sitting five weeks in one spot: there is an area of pasture killed where it sat.

Mortality

A majority of my mortality problems, I feel, are because of my brooding facility. My conditions are too cramped, and the light level is lower than I would like. This, I think, leads to more than normal leg problems which I will correct next year.

There really weren't any noticeable problems except for batch One. The moveable pen birds got pretty wet one chilly night and that led to later ascites problems.

Hawks were a small problem for batch Three, but not that bad. I know a lot of other pastured poultry producers who have significant hawk problems with the day-range system. This is an area that needs to be addressed for many.

Costs

The addition of a floor added a lot to the cost of the day-range pens. Also, the use of the netting added over $150 to the day-range costs per pen. These two factors, of course, can be amortized over several batches for several years. My bedding source was free, but in most areas that is another expense.

Weight Comparisons (see table)

For the most part, the day-range pen lagged behind the moveable pen in bird harvest weights.

In batch Two, I can't explain why they were so close in weight. Batch Three, I harvested two thirds of the birds in the moveable pen at the seventh week because they were cockerels and would have been too big by the eighth week. This skewed the results for that batch and would have made a larger total weight for the moveable pen. I think batch Four moveable pen weights were lighter because of cooler temperatures. These birds had to lie on the cool ground, using up more of their feed to produce heat energy, thus the lighter weights.

I think that the larger weight difference between the moveable pen and the day-range pen can be attributed to the feed being right in front of the birds at all times in the moveable pens. The birds in moveable pens could have consumed a lot more feed and probably reached market weight quicker if I would have given them more feed than the day-range. That would be another good study to do.

Summary

As with any extensive study, these results bring up more questions. I think the pastured poultry industry needs to do more comparisons, look at more possible models. There are really good factors with each production model that I wish could be incorporated into another study.

Every individual needs to assess the good and bad points of each system, and see how well it will fit into their situation of production. Maybe for those who raise under 1,000 birds a year the moveable pen is more feasible, but above 1,000 the day-range system might be better. Do some number crunching and see if that a third of a pound lighter weight will affect you very much with all your fixed costs.

Pen Comparisons	MP1	DR1	MP2	DR2	MP3	DR3	MP4	DR4
Chick start date	3/14/01	3/14/01	4/18/01	4/18/01	7/25/01	7/25/01	8/29/01	8/29/01
Harvest dates	5/05, 5/12	5/5, 5/12	6/9, 6/16	6/9, 6/16	9/15, 9/22	9/15, 9/22	10/20, 10/27	10/20, 10/27
# birds in pen	160	160	160	160	160	160	160	160
# birds lost per pen	13	3	1	10	7	8	3	2
# birds processed	147	157	159	150	153	152	157	158
Total live weight	927	923.5	835.5	785.5	992	926	775.5	878.5
Feed per bird in field (lbs)	11.5	11.5	10.3	10.3	15.6	15.6	14.68	14.68
Avg live weight (lbs)	6.31	5.68	5.25	5.23	6.48	6.09	4.93	5.56
Avg dressed weight (lbs)	4.73	4.41	3.94	3.92	4.86	4.56	3.70	4.17
MP= Moveable Pen DR= Day-Range								

CHAPTER TWO
Brooding

The first two-to-four weeks of a small birds' life is the most critical. The largest losses, and the largest impact on success, can occur in the brooder. Brooding isn't complicated, but you must pay close attention to a few critical things. We outline those from several perspectives here.

Brooding Basics
By Robert Plamondon Spring 2004

Whether you are an experienced producer or just starting out, it is a good idea to remind yourself of brooding basics each year. In this article, I'll cover what I consider to be the five biggest issues in brooding: Heat, water, feed, coccidiosis, and predators (rats!).

Heat
Use electric heat if you can, propane if you must. Stay away from coal, kerosene, and wood-fired brooders.

Overhead Heat Lamps
Overhead heat bulbs can give excellent results in cold weather if you stop any hint of a floor draft and understand that the number of chicks that can be brooded per bulb varies with the temperature. The old rule of thumb is that a 250-watt heat lamp can brood a number of chicks equal to the average temperature plus 25. For example, in 50° F weather, you can brood 75 chicks per heat lamp. At zero degrees, you can brood 25 chicks.

Any floor draft ruins the effectiveness of heat lamps for cold-weather brooding, so a draft shield, such as a circle of cardboard, is essential.

With overhead lamps, always use a real brooder fixture from the feed store, not a clamp light from the hardware store. A real brooder fixture hangs from the ceiling and has two semicircular wires that cross under the bulb to make a guard. If it falls to the floor, the bulb not only won't make contact with the litter, the fixture rolls onto its side, preventing fires.

The heat lamp must be high enough that all the chicks can easily fit inside the circle of heat. If you set it too low, they'll fight to get into the too-small heated area. Always turn the heat on well before the chicks arrive. One of the essential rules of brooding is that the area under the brooder must be warm and dry to the touch before the chicks arrive. I also like the rule, "Use two bulbs or new bulbs" to prevent disaster when an old bulb burns out. Just use two 125-watt heat lamps instead of a 250-watt one.

Insulated Electric Brooders
You can provide more reliable heat at less expense by using a homemade insulated electric brooder. A 250-chick brooder can be built in about two hours for $20 or so. For more information, see my Web site at http://www.plamondon.com/brooder.shtml, or my book, *Success With Baby Chicks*.

Water
Chicks need to be provided with warm water from the start. It needs to be in a well-lit, easy-to-find place. With electric lamp brooding, all these requirements tend to be met if you put the waterers inside the area that's lit and warmed by the lamps.

I use quart glass canning jars on screw-on plastic bases. I've had a lot of trouble with day-old chicks getting soaked in larger waterers, an experience they generally don't survive. The glint of the glass jars seems to attract the chicks. Real canning jars are cheaper and better than the plastic jars they sell at the feed store.

I use ordinary sugar in the first water at the rate of one pound per gallon, which works out to about one cup of sugar per gallon. When you refill the waterers, rinse them out and don't use any more sugar. Sugar has been demonstrated to reduce early chick mortality in controlled tests. You can give electrolytes to the chicks if you want to, but I find them to be unnecessary. I put the chick waterers on 2 x 4 in. scraps to keep them from sinking down into the litter.

Some hatcheries use a gel compound in their chick boxes. It's solid, but contains 90% water. The chicks peck at it during shipment, and arrive much less thirsty than usual. This lack of thirstiness can be a little unnerving, since one of the ways I judge the condition of chicks is how readily they take to food and drink!

For a longer-term water setup, I use Little Giant Poultry Waterers with the normal bowls (not the shallow chick bowls). The chick bowls are only useful for the first few days, when I'm using the quart-jar waterers, and are a nuisance after that because they spill more water.

Feed

Commercial Chicken Production Manual recommends that feed not be given until three hours after the chicks have been installed. Apparently, it helps to get the chicks drinking before they start eating. I've also seen recommendations that you give feed immediately, or wait just one hour. Take your pick. If there's any chance that you won't make it back to the brooder house at the appointed time, it's probably better to put the feed down right away.

Chicks want to scratch at their feed with their feet. It's best to indulge them for the first few days by placing feed in some kind of tray they can stand in. Some chick box lids are suitable for this. Egg flats are also good; one flat per fifty chicks. I have also used plastic cafeteria trays. The lazy way is to spread a single sheet of newspaper on the floor and pour feed onto it. Thick wads of newspaper become slick and nasty, but single sheets fall apart and become incorporated with the litter instead. No need to remove it.

I like to have tube feeders installed from day one, sitting flat on the floor. This tends to lead to spillage, so a sheet of newspaper under each feeder will prevent waste. Keep the waterers far enough away from the feed that the feed doesn't get wet.

When chicks are chilled, whether in shipping or in the brooder house, they may suffer from paste-up, where lumps of feces adhere to their rear ends. The traditional preventative is to feed chick scratch instead of chick starter for the first 48-hours. This works like a charm. I like this practice because I never know if my chicks have been chilled during shipment.

Guidelines for Brooding Chicks
Heifer Project International Fall 997

Because nothing can decimate a flock of chicks faster than chills coming from drafts or dampness, new chicks need an area that is predator proof, free of drafts, dry, and has proper bedding materials. In addition, the chicks should also have as much natural light as possible to stimulate the pituitary gland for proper growth and development.

Some guidelines include:

¤ An area of at least 5 x 5 ft. should be provided for 100 chicks.

¤ This enclosure should have rounded corners to prevent the chicks from piling up on each other.

¤ About six inches of bedding material such as wood shavings should be provided (avoid fine sawdust because the chicks may eat it.)

¤ Sawdust or shavings from treated lumber should not be used.

¤ A heat lamp or other heat source will be needed to keep the chicks warm to 90° F during the first two or three days. Raising the heat lamp can then gradually reduce the temperature. If the chicks are huddled under the lamp, it is too cold. If they are huddled against the walls, it is too hot. If they are moving, feeding, and eating it is okay.

¤ Chicks should stay in the brooder two to three weeks depending on the weather.

Mortality is lower when the bedding material is aerated and small amounts of new shavings are added between batches of chicks.

Coccidiosis

There's a tendency among alternative producers to assume that disease is something that happens to other people. This is true in many cases, but not with coccidiosis.

If you have a significant number of stunted, runty chicks, or if your chicks seem healthier on pasture than they did in the brooder house, or if brooder house mortality doesn't fall to zero and stay there after the first few days, your chicks are probably suffering from coccidiosis. I suspect that many problems that are blamed on feed or "bad genetics" are really the result of undiagnosed coccidiosis outbreaks.

Coccidiosis is usually 100% preventable; there's no reason why any chick should ever show symptoms of it. Because it's preventable, I feel that it's bad management to allow it to show up at all. Treating a coccidiosis outbreak is like shutting the barn door after the horses have escaped.

The simplest way of preventing coccidiosis is to use medicated chick starter. I recommend that you do this unless you are taking definite steps to prevent coccidiosis by other means. Medicated starter may be considered "politically incorrect" but I find it highly effective. (Ed note: anyone raising birds they wish to market as "organic" cannot use medicated feeds.)

It takes several days for the coccidia in chicken manure to undergo the life-cycle change that makes them infectious (they have to go from being "oocysts" to "sporulated oocysts"). Daily moves of pasture pens move the chickens away from the manure before it becomes infectious. Unfortunately, brooding in daily-move pens is difficult. But it means that coccidiosis is not an issue once chicks are moved from the brooder house to daily-move pens.

Getting back to the brooder house, raising the chicks on wire floors will prevent coccidiosis, since the manure drops through the floor to where the chicks can't get at it.

Old litter that's been used for at least six months has anticoccidial properties (presumably because it acquires a population of organisms that eat coccidia), but not if it's caked over with a layer of manure. Thus, old-time poultrykeepers who relied on the anticoccidial properties of deep litter skimmed the surface and stirred what remained every day. If you simply heave the wet, caked-over part into a corner of the brooder house, it will heat up and miraculously turn back into litter in a few days. Stirring in ten pounds of hydrated lime per 100 sq. ft. of floor space between batches will help keep the litter friable. I've found this technique to be difficult to put into practice with modern broilers, who grow fast and generate so much manure that it's hard to keep up with them. To do this you'll need to be able to work with tools in your brooder, and you'll want a minimum of 6 in. of litter, or ideally 12-18 in.

Deep litter can produce enormous amounts of ammonia, so you want to be sure your brooder house is well ventilated, ideally with vents or windows high up in the house that don't contribute to floor drafts.

Rats

My first few years were rat-free. Beginner's luck, I suppose. Rats will stay away from larger chickens, but will kill all your baby chicks if you let them, killing them much faster than they can eat them and filling tunnels with the bodies.

I have had no luck at all with spring-style rat traps and mixed success in excluding rats from my brooder houses, though all of them have concrete floors. What really works for me is rat poison. This wasn't my first choice, but you gotta do what works.

The most important thing is to set down rat poison before the brooding season starts and clean it up before the chicks go in the brooder. I like one-ounce "nibbler blocks" that have a hole in the middle. These are reasonably waterproof and can be nailed to boards so the rats can't fill tunnels with rat poison instead of eating it. Before the chick season starts, I put bait blocks inside each brooder house, one in each corner, and set up bait stations outside. My current bait station is a length of 2 x 8 in. board with a couple of bait blocks nailed to it. This is leaned against the side of a brooder house so a rat will have a nice secluded space to nibble the bait in. The bait is also protected from the weather and hidden from the eyes of wild birds and larger critters who might otherwise mess with it.

The bait inside the brooder houses is very important. If it shows any signs of nibbling, you have a rodent problem inside the brooder house, which must be solved before the chicks arrive! Not surprisingly, the brooder house next to my barn, which houses half a dozen cats, has less rodent activity than the ones further away. But we've lost chicks to rats in both locations.

The Brooder Nightmare
By Dan Bennett   Winter 2003

Who has experienced "The Brooder Nightmare?" You know, you raise a couple of batches per year, a few hundred birds total, with no real brooder problems--2-3% death loss--you watch the birds pretty close and nothing really out of the ordinary happens. Then you make the big decision to raise your numbers to a little over a thousand and sure, you have plenty of weather problems out on the pasture but this brooder thing is no big deal. Next, you get really serious and you begin bringing in 10 to 20 batches per year--problems you have never seen begin to pop up in the most unlikely of places--the brooder for example. Death loss creeps up over 5%. The next year when you increase numbers again, death loss keeps increasing. You have just experienced "The Brooder Nightmare."

Joel Salatin discussed something he called "slippage" in his "The Pastoralist" column in the September 2003 issue of *The Stockman Grass Farmer*. His point is that as you grow, your profitability tends to slip. "The Brooder Nightmare" is definitely in the category of "slippage." I quote Joel's final sentence here as our reason for this discussion. "Dealing with slippage aggressively is the only way to make your business successful."

I'm writing this article for the intermediate-stage brooder person. I'm assuming you have the basics of heat, food, water and draft down pat. Look to past Grit! articles or Joel Salatin's book *Pasture Poultry Profit$* for the basics.

By way of introduction, we at the Bennett Ranch have been brooding chicks since 1995. Our basic approach is the classic barn style brooder with natural lighting where possible. We use pine shavings as floor bedding in the brooder. We have used the deep bedding approach utilized by Salatin in the past and currently are utilizing the technique of cleaning out the brooders every three weeks as a new batch of chicks arrive. We like the clean out approach better because it's less work. Both approaches work and they both have advantages.

The brooder problem discussed earlier is the reality we've experienced on the Bennett Ranch over the past couple of years. We began to have brooder problems during the 2000 season as our numbers went from 25 batches of 200 chicks per week in '99 to 27 batches of 400 per week. We tried a number of things during that season which helped some but never got to the real problem. Then in 2001, we went to 600 per week and this made the problem worse. At the end of the 2001 season, we knew we had to do things differently as our death loss percentages of 8% were unacceptable. I talked to Jeff Mattocks at The Fertrell Company a number of times and everyone else I could think of to try to solve the problem. We decided to implement a number of new ideas for the 2002 season, but as the season got underway, we knew fairly soon thereafter that the problem was not yet solved! We studied more, prayed more and finally came up with the solution we will discuss in this article.

The sad fact is that the solution was very simple. Although we had implemented a number of brooder improvements over the years, and even some new ones during the 2002 season, by far the single biggest issue has alluded us. Surprise... it was cleanliness. If I had to do one thing in my brooder to reduce death loss and improve the health of my chicks, I'd walk out each morning and spread a thin layer of fresh bedding. This may sound wasteful to some and expensive to others, but in our situation, fresh bedding each morning is our best investment! This obviously is not a new idea, it's just one that is hard to bite the bullet on; especially when you get to watch the amount of money that flows out of your checking account for brooder chips each year. My wife Jenny was actually surfing the web one night last summer in search of organic food information. At one of these sites, she ran across a discussion of brooder care and the topic of fresh bedding was discussed in detail.

As I said, we've implemented a number of improvements in our brooders over the years and I will discuss all of these through the use of our "Brooder Checklist." We developed the checklist to aid our helpers this summer who had to pick up the ball when our primary brooder queen (my daughter Jamen) was off gallivanting around the country.

Bennett Ranch Brooder Checklist

Morning:
1) Turn on full-spectrum lights.
2) Check each drink cup and floor waterer to ensure cleanliness and proper operation.
3) Check for dead chicks--remember to write down on the death loss chart in the kitchen.
4) Put fresh grass and grit in each brooder--use clean side of paper bag in the brooder and if dirty, get new bag.
5) Check for soiled areas in each brooder and add fresh pine shavings from the open bag using the scoop and/or bucket.

Afternoon:
1) Supply each brooder with beef liver--get from freezer prior to going out--use clean side of paper bag in the brooder and if dirty, get new bag.
2) Check each drink cup and floor waterer to ensure cleanliness and proper operation.
3) Check for dead chicks--remember to write down on the death loss chart in the kitchen.
4) Fill all feeders--top off even if ½ full.
5) Turn off full spectrum lights.

Wed. and Sat.:
1) Put ½ cup of apple cider vinegar in 5-gallon brooder gravity flow water tank.

Sat.:
1) Take floor waterers out of 1-1/2 week old chicks.

Ongoing as needed:
1) Adjust height of waterers for size of birds.
2) Turn off heat lamps as chicks get older - adjust with weather.
3) Put 1-gallon of raw milk into brooders using 1-gallon plastic waterers when chicks show signs of unkeptness or a pattern of death occurs.
4) Adjust doors according to weather and corresponding air flow needs.

The highlights of the brooder checklist are as follows in the order they are listed in the checklist:

Morning:
1) Full spectrum lights are used in each brooder to provide the effect of sunlight, as our barn has no windows.
2) We have utilized cup drinkers available from GQF in all our brooders. This allows us an automatic water source in all seven brooders within our barn fed by one gravity flow tank located within the barn. The cup waterers also allow us to raise the waterers as the chicks grow to improve cleanliness.
3) We track death loss on the daily basis by batch of chicks so we have the information needed to make decisions.
4) We supply fresh grass and grit to each brooder each morning on a folded-over paper bag. This allows us to have a clean surface morning and night to put our grass, grit and beef liver on. We simply fold the bag over to the clean side, when all four are soiled, we put in a new one. We got a few thousand of these at an auction a few years ago for next to nothing. The focus is cleanliness!
5) Fresh pine shavings each morning are a must. As I mentioned earlier, this improvement is the single most important thing!
6) Depending on weather, open or close brooder doors for air flow adjustment.

Bennett Ranch Brooder Checklist

Evening:
1) We supplement with beef liver because Joel Salatin said we should and it seems to help!
2) We use 60 in. turkey feeders from Brower after the chicks are one-week-old and therefore we can get away with feeding only once a day in the evening.
3) Depending on weather, open or close brooder doors for air flow adjustment.

Biweekly, Weekly and Ongoing:
1) We treat the water with apple cider vinegar twice a week to encourage the good bugs in the gut to fight coccidiosis.
2) We use a GQF circular gravity-fed waterer during the first week and a half because the chicks can't seem to figure out the cups until then. The cups are better for cleanliness and we use them for the rest of the three week brooder stay.
3) For heat, we use standard heat lamps in addition to a GQF electric floor hover/brooder for the first week or two. Adjusting the heat for the weather is obviously crucial.
4) Treating the symptom of coccidiosis with raw milk is a technique I read about last year in an APPPA Grit! article by Tim Shell. The idea is that raw milk provides the chicks with the right bugs in the gut to fight the disease. This approach certainly seems to slow things down, but as we all know, prevention is better than treating a symptom and a clean environment for the chicks is our best prevention!

Additional Thoughts:
Our chick starter aids: When the chicks arrive from the post office, we fill 1-gallon plastic waterers with sugar water to get the chicks going quickly--1/2 cup per gallon. We also add a liquid probiotic to the water as well to promote a healthy digestive system.

On predators in the brooder: We have experienced the terror of rats. For my money this is the worst predator issue I've experienced. When you find hundreds of dead chicks stacked in hard to reach places around your brooder, you will agree. My suggestion: Keep your brooders as tight as possible and more importantly, encourage the growth and vitality of your cat population!

I hope this helps in your effort to raise the best chicken in the world. This has definitely been a learning process for us and I'm quite certain it will continue to be so. Happy Brooding!

A Few Ideas to Supplement Dan Bennett's Helpful Article, "Brooder Nightmare"
By Timothy Shell Spring 2003

Dan's basic outlook and commitment to fight "slippage" with a "zero tolerance" policy towards a dirty brooding environment is right on target. That commitment qualifies as the irreplaceable foundation for "near-zero mortality" brooding. The specific details, (drinker, feed spillage, poop zones, etc.), which need to be managed in each grower's situation are too variable to be answered specifically by an article of this nature. But a caregiver who stays alert to the need for "hyper-cleanliness" in the brooder will be tuned in and be able to bring corrective action prior to the development of problems. The following ideas may be helpful as well.

1. Observation: Zero tolerance for non-observant caregivers.

Only by careful and consistent observation was a solution discovered for the problem. It is so easy once a brooder is set up and functioning to reduce and eliminate the amount of time spent observing the chicks. Most brooder problems will be solved by spending 10 to 15 minutes after chores are done doing nothing but simply sitting and observing the chicks. Observation slippage precedes all other slippage. The lower the caregiver-to-chick ratio the more likely for observation to slip. This is why larger operations should consider adopting more efficient technologies, (nipple drinkers), and practices, (bulk feed/water infrastructure and logistics), which allow the caregiver to properly nourish more birds in less time and not compromise observation time which should remain a constant.

A Good Day in the Brooder

2. Drinker Wet Zones: Zero tolerance for wet "cake" zones around drinkers.

The area surrounding a drinker is the primary zone where fresh bedding is required on a daily basis. All types and styles of drinkers must be carefully managed for the elimination of wet zones in the brooder bedding. While 85 to 90% of the litter in a brooder is still relatively clean and functioning (dehydrating droppings), wet zones around the drinkers can become pathogenic. When droppings fall on litter that is unable to absorb moisture (due to saturation of the litter medium with water), those droppings become a food source for anaerobic, pathogenic bacteria. Only by dehydration of droppings in an aerobic environment is the growth of pathogens impeded. It is these wet zones that are the primary culprits in generating disease in a brooder. This condition in confinement poultry enterprises is managed with antibiotics. Many chicks destined for a pasture home don't "do drugs" and have no means of overcoming excessive pathogenic buildup. We have found correctly installed, properly functioning nipple drinkers in the brooder to be the most convenient way to combat wet zones. All reservoir-type drinkers, including cup drinkers, allow build up of litter, fecal dust, and feed residue in the drinker. Chicks always discover the water in the nipples drinkers within 60 seconds of exposure, eliminating the need for retraining once started on reservoirs. The gallon floor drinkers are especially bad for allowing the birds to actually step in the drinker. This guarantees rapid inoculation of pathogens to entire flock. The level of work needed to keep all the floor drinkers elevated on bricks or boards and refilled creates a big job that is easy to neglect. Birds that have run out of water for long enough will actually begin to eat wet litter.

3. Feed Spill Zones: Zero tolerance for spilled feed.

When feed is spilled over the edge of a feeder it falls in the exact zone where the birds are positioned while eating. Your birds which are health-challenged will often drop "exotic" forms of manure which are evidence of the battle for supremacy occurring in the gut of the bird between the "good" bugs and the "bad" bugs. This manure (diarrhea) is very sticky, much more so than normal bird droppings. It will adhere to the feet of the birds and end up, you guessed it, right back at the edge of the feeder. Growers must not allow birds to either spill feed or stand in the feeder.

Properly made reel feeders are designed to overcome this problem by allowing several adjustments to the height of the feeder and the reel. Feeders should never be filled over 3/4 full. This allows the birds to scratch in the feed without raking it up over the side. If you allow the birds to become very hungry they will sort through the litter looking for the spilled/soiled feed, increasing pathogen uptake.

4. Sleeping Quarters: Zero tolerance of "cake".

Pastured broiler production models follow confinement models rather closely from day one through day 21 to 28, or while in the brooder, about half of the birds' life cycle. Since most confinement operations use large houses with cardboard rings to begin brooding they can remove them and expand brooding to the whole house as the birds outgrow their rings. Confinement brooding as practiced by most pastured growers does not allow for this expansion of brooding area. Therefore, a pastured growers' late term brooding conditions tend to produce caked litter around week three to four rather than near maturity (week six+) as in confinement houses.

This expansion ability actually gives the confinement grower a more sanitary environment for the same growing period. A confinement house is still relatively clean at three to four weeks due to all the unsoiled litter available. In small, non-expandable brooders, the birds tend to nestle in the same preferred site each night to sleep. This area receives a bulk load of droppings each night while the rest of the litter is in nearly new condition. It is essential to inject air into this zone by stirring and/or covering with new litter. This renews the litters ability to dehydrate the droppings and give the air (oxygen) a chance to challenge pathogenic bacteria.

5. Feeding raw milk:

It is best to feed the milk by adding to the feed. Mix the feed with enough milk to achieve a semi-dry "cookie-dough" consistency. This ensures that all the birds receive uniform dosage in proportion to individual size and appetite (the bigger birds need more and eat more). Some individuals may not prefer the fluid milk if water is also available, and withholding water is not recommended. Feed only the amount of feed that will be consumed in 24-hours. Feed only for 48-hours; for severe problems use a two day on/five day off routine.

Just for fun, take a minute to review the following descriptions of brooding caregivers. See if you don't understand why some folks have high brooder mortality.

Diary of a Poor Caregiver:

Rushes in to brooder with feed pail, (just woke up), fills feeders to overflowing, (not sure when will return; not enough feeder space; just a little bit left in the bucket), sloshes water around in drinker, (eeh, it'll be enough till I get back), dumps debris into empty feed pail, (feed residue, bedding, feces, further contaminates next feed in pail), gives a quick look around, rushes out, flings dirty water into grass, tosses bucket toward feed barrel, splits the scene! Chances of achieving "near-zero" brooder mortality = not good.

Diary of a Good Caregiver:

Approaches brooder in alert and observant frame of mind. Notices predator droppings and scratching in grass alongside brooder, makes mental note to set foot trap. Steps slowly into brooder with feed pail so as not to frighten birds, notices smells in the air while looking around for anything out of order. Checks for "stuffiness" from poor ventilation, feels okay. Tips feed residue to one end of feeders and fills 3/4 full with fresh feed, adjusts legs and reel, notices several "exotic" droppings, makes mental note to add extra probiotics to water. Notices birds not well under brooder, raises height a little to reduce heat. Notices birds ducking some to drink from nipple drinker, raises one notch, checks water level and float in pail. Stirs up litter from sleeping site, mixing with some cleaner litter. Picks up two dead chicks, adds mentally to tally. Sits down on upturned pail with chin on hand and watches birds play, studies dead chicks for clues of demise; one has crop distended, tears open and finds filled with litter. Notices stronger than usual draft from windy day, lowers vent some. Notices one chick with closed, runny eye. Notices birds are out of grit, brings some in. Notices reel not spinning on one feeder, adjusts metal end, bending out slightly. Slowly exits brooder, returns with probiotic, doses water pail with amount. Sets foot trap outside brooder w/ dead chick. Returns pail to feed barrel, closes lid securely. Puts other dead chicks on compost pile in route to other chores. Chances of achieving "near-zero" brooder mortality = very good.

May we all overcome the weak links in our operations that prevent us from being good caregivers.

APPPA PRODUCER PROFILE

Charles and Laura Ritch

By Jody Padgham and Aaron Silverman   Summer 2003

Charles Ritch, his wife Laura, two teenage daughters and a few helpers raised a little over 11,000 birds in 2002 on their farm in northern Alabama. The Ritches do seven batches of broilers a year, with the first batch processed in April and the last in October. A typical batch is anywhere from 1,500 – 2,000+ birds.

Broiler Production System
Charles uses a mixture of the typical Salatin type pull pen and the day-range, or mini-barn system. The pull pens have anywhere from 90 birds to a pen in the spring and fall and 80 birds in the summer. The pens used now are constructed exactly like those Salatin uses, but Charles is experimenting with other designs. The mini-barns were built with floors, with plans to brood the chicks inside, but this did not work and the cleaning of the chip bedding has proved to be too much work. Charles is now working on removing the floors. The mini-barns are surrounded with poultry netting and moved as needed. The mini-barn has less death loss and produces a higher quality bird, and Charles thinks that it is more "free-range."

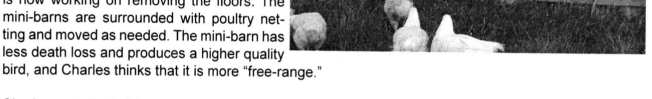

Charles uses the bell-type waters exclusively. He tried a nipple system, but did not like it, feeling that the birds were not getting enough to drink and the waterers were problematic.

Birds are purchased as straight runs (all male) and harvested at eight weeks. Overall death loss runs around 5% to 8%, but is improving.

Charles believes brooding is THE single most important aspect of this system. The Ritches currently place chicks in battery brooders for the first weeks and then move them onto fresh wood chips. They do not clean out the wood chips after each batch, but aerate and put a fresh layer on top. The further they go into the season the more aerobic the bedding becomes, and Charles notices a ton of bugs in the litter. The brooder heat sources are gas-fired pancake brooders. Again, only bell-type waters are used. Charles is currently building a hoophouse to do brooding in.

Feed
Charles uses Jeff Mattocks at The Fertrell Company exclusively for his ration mix, and says they "don't know how we would make it without Jeff and his vast knowledge of nutrition." Feed runs $.15 to $.16 per pound and is bought bulk and delivered to the silo. Feed is regularly tested for mold and toxins and overall feed quality.

Labor and Processing
Daily labor is family with some outside help from time to time. Daily chores take about two hours.

APPPA PRODUCER PROFILE *Ritch*

Charles spends about 10 more hours each week with miscellaneous chores.

The family processes the birds themselves with 8 people. Charles, Laura, daughters (ages 15 and 12) plus four other helpers (including three other young home-schooled boys and girls) do the processing work. Charles figures it cost about $1 per bird for processing, bagging and waiting on customers, cleanup, etc., and that they lose about two birds per 100 in the processing. The use of killing cones and a shackle system allows them to do about 200 birds per hour.

Marketing

All sales are on farm. State law does not allow sales at farmers markets, and the Ritches do not deliver. Word-of-mouth is the most effective marketing tool and they offer incentives for referrals. They also advertise in the local health magazine and once a year in the paper of the largest town nearby, although Charles adds that newspaper advertising is very hard to evaluate. The Ritches send out newsletters and notify email customers when there is something on the web that might be interesting. They also contact newspaper editors and request to be interviewed when an opportunity presents itself.

Record-Keeping

Charles keeps accurate records of all feed used, death loss and expenses on a daily basis per batch. At the end of each batch he does a spreadsheet analysis of every ratio he can think of: total sales, number harvested, number purchased, pounds harvested, number sold, number rejected, feed used, feed cost, chick cost, direct processing cost (labor, deprecation of equipment, etc), gross profit, profit per bird, percent mortality, feed per pound of birds sold, feed per bird and feed conversion ratio.

A spreadsheet is used to keep track of orders. Customers are tracked using a contact database. The Ritches know who is buying what and when they bought it.

The Bottom Line

Since pastured poultry is a superior product, Charles finds the concept easy to sell. But, the cost and the trip to the farm are prohibitive for the consumer. The Ritches have resisted offering delivery because they feel it will increase sales without increasing profits, and it will take them away from the farm. They already find they're so busy that it is hard to get everything done.

Words from Charles on challenges: "Niche markets are always tough. You run wide open to stay in place. Keeping customers excited and motivated is a constant job. Growth is hard to finance. New brooder space, new field pens, better processing equipment, crashes, and mistakes all eat into the profit."

CHAPTER THREE
Pastured Poultry Genetics

There is much conversation on the genetics of birds used in pastured systems-- many use the common Cornish Cross that has been bred to thrive in confinement systems. Read opinions on that cross and alternatives, old and new, for pastured systems.

Pastured Peepers Symposium
By Skip Polson Fall 1999

Thirty-one pastured poultry producers (PPPs) from 16 states gathered in Virginia in August 1999 to learn the current state of the art of producing hatching eggs and broiler chicks from parent stock raised on pasture. The first-ever Pastured Peepers Symposium was organized by the American Pastured Poultry Producers Association (APPPA) and made possible by financial support from Heifer Project International (HPI).

PPPs have long recognized that the broiler chicks most commonly used by the conventional poultry industry are not very well suited for the way we want to raise them on pasture. They have been bred to grow fast, and they do! So fast, in fact, that weak legs and hearts are common, they do not forage very well, and they are very susceptible to heat stress. All they want to do is eat, rest, and eat again. While this lazy couch-potato attitude toward life may be appropriate for a bird raised in confinement where there is not much else to do but eat and rest, it prevents these birds from enjoying life to the fullest when they are raised on pasture.

For pasture-based poultry production we definitely want a bird which will grow fast, but we don't want one whose heart cannot keep up with its body's growth rate. We want birds that are healthy, hardy, vigorous, and able to walk and forage well. We want birds that will thrive on pasture and produce a top quality meat carcass at the same time.

A few years ago the Salatin family and several other PPPs in Virginia began to wonder if some of the same positive benefits of raising poultry on green grass could be found in chicks whose parents were raised and kept on pasture. Compared to confinement rearing, they reasoned, we get healthier broilers on pasture, and better quality eggs with darker orange yolks from hens on pasture, so we just might get healthier chicks, too, if we raise their parents on pasture. Thus began the effort to produce and compare Pastured Peepers to the conventional industry chicks we've all been using.

The number of Pastured Peepers produced has grown from a very small beginning just three years ago. Most of the PPPs who have tried them have been favorably impressed by their vigor and survivability, but it is still too early to tell if they are significantly better than the conventional industry chicks. There is still a great deal to learn about this aspect of pastured poultry production. Timothy Shell and his wife, Naomi, of Mt. Solon, VA, are the current leaders in this field. Their 200-hen flock produced over 12,000 chicks in the 1999 season. The Shells were the primary hosts of our 1999 Pastured Peepers Symposium, as we focused on learning as much as possible from their experience. Our hope is that more PPPs around the country will raise some Pastured Peepers so we can quickly expand our collective knowledge and share this with each other through APPPA.

We are grateful to the Shells for being our hosts, for sharing their knowledge, and for continuing the spirit of sharing and cooperation we appreciate so much in APPPA. During our sojourn in Virginia we also spent an afternoon at Good Earth Farm with Andy Lee and Patricia Foreman, and one day at Polyface Farm with Joel and Teresa Salatin and family. We owe many thanks to them as well.

We all know that many things in life are more difficult and more complex than raising chickens. We also know, however, that raising chickens (just like any other enterprise) does have its own particular complexities and specialized knowledge. In this Symposium we learned of several important aspects of Pastured Peeper production which must be given special consideration and which are significantly different than raising broilers on pasture.

One of the most basic differences is the challenge of managing and controlling the grazing of a large group of adult birds in a manner consistent with our pasturing philosophy. Many people who wanted to produce eggs have added nest boxes to standard pasture pens and learned that this system requires a lot of labor compared to the relatively small number of hens in each pen. Portable hen houses ("eggmobiles") for day-ranging hens have worked well for some folks, but Tim Shell quickly points out this won't work so well for others who have smaller acreage and don't really want the hens roosting on their front porch or roaming uncontrolled through their garden.

Thus, we were thrilled to see the FeatherNet system of layer management now in use at both the Shell Farm and at Polyface Farm. This system uses poultry electronet to free-range laying hens around a portable hoophouse which provides shelter for the nest boxes and provides a place for the birds to sleep at night and get out of bad weather when necessary. Enough electronet is used to create a pasture large enough to accommodate the birds for three days. The fencing, hoophouse, birds and all are moved to a fresh spot of grass every three days. This system dramatically reduces the labor required for a large number of birds, while enabling the farm family to truly control and manage the grazing of their birds. Electronet fence is available from several suppliers, including Premier Fence Systems and Kencove Farm Fence.

A second major difference between raising broilers and brood hens is the restricted feeding program required for breeding stock. If these birds are fed free-choice (like the broilers) they will simply be too fat and overweight for good production. Both egg production and hatchability are adversely affected by obesity in replacement pullets and hens. And when I say restricted feeding, I mean RESTRICTED! Most of us could hardly believe that the birds would do well on the small amount of feed which was recommended. It makes sense when you think about it, but this requires a big mental adjustment for broiler producers: instead of producing a bird with a four pound carcass at eight weeks of age, you now want a bird that is only slightly larger than that at 24 weeks of age! This much slower growth rate translates to a huge difference in the amount of feed consumed by the birds each day.

The Shells have followed the feeding guide table contained in the *Hubbard Classic Breeder Management Guide*, published by the Research and Development Department of Hubbard Farms in New Hampshire. Hubbard Farms and several other poultry companies have their management guides posted on the worldwide web. A listing of some of those can be found at a site prepared by the University of Minnesota Extension Service (http://www.ansci.umn.edu/poultry/resources/hubbardfarms.htm).

We also observed that the physical act of feeding the birds requires some creative thinking in this restricted feeding regime. It is extremely difficult to walk among the ravenous birds if you're carrying their next meal in a bucket. You are almost certain to injure some of them because of their frenzied crowding around your feet. Tim Shell has solved this problem with lightweight feeder troughs made from PVC pipe which he is able to slide under the electronet fencing. This allows him to pull the empty troughs outside the fencing, fill them there without the birds getting in his way, and then pull them quickly back inside the fencing where the birds swarm over them to begin their meal. It is vital to have enough trough space for all the birds to eat at once, because all the feed will be consumed in about 15 minutes! If they cannot all get to the feed at the same time, some of the birds will not get their share.

The poultry company management guides also contain useful information about the best male to female ratio for maximum egg fertility. The Hubbard guide recommends eight to nine males per 100 females from the beginning of egg laying through peak production, and gradually reducing to seven or eight good males per 100 females by the end of the laying period. This ratio is important to maintain: too few males will result in infertile eggs, too many roosters will increase the physical stress on the females (even to the point of injuring them).

One suggestion which might help the hens maintain the feather covering on their backs and be more comfortable throughout the breeding period is to clip the tips of the toes of the males when they are only a few days old. Without toenails, the roosters will do a lot less damage to the hens. Some producers will not want to do this to their rooster chicks, and I understand that, but it is also pretty unpleasant when hens are severely scratched (and sometimes seriously wounded) by the roosters mounting them. Too many roosters can be a real problem.

Another critical aspect of this enterprise is egg handling. It is absolutely essential to manage your flock so the hens can lay their clean eggs in clean nests and you can keep them clean until they go into the incubator. No floor eggs or dirty eggs should be used for hatching. Egg cleanliness has a dramatic effect on hatchability.

Hatching eggs should also be selected for uniformity of size. Eggs with shell defects or double yolks should not be used for hatching. More detailed recommendations for egg handling can be found in the poultry company management guides.

During the Symposium we visited the local hatchery which does custom incubation and hatching for the Shell Family. There we learned about the many technicalities involved in this aspect of producing healthy chicks: cleanliness, temperature and humidity control throughout incubation, frequent egg turning, and adjusting the temperature and humidity during the final stages of hatching. I was amazed to learn how critical the final hatching stage is: everything can be fine up to that point, and then healthy embryos can die in the last day if the humidity in the hatcher is wrong. While some PPPs will want to acquire an incubator and hatch their own eggs, I believe it is highly advisable to find an experienced hatchery to handle the incubation and hatching of our pasture-produced eggs. I am sure they would do a better job of that than I would. It would take me a long time to learn the things they already know, so I would gladly pay a reasonable price for this service. (This year the Shells paid their custom hatchery eight cents per egg set.)

Tim Shell's Pastured Peepers

What does the future hold for Pastured Peepers? I think they have a lot of potential to strengthen pastured poultry enterprises by providing chicks more suitable for the pastured production model. We have to really prove that, however, by producing many more of them and seeing how they actually perform in different parts of the country. There is still a great deal to learn about how to better manage breeder and layer flocks on pasture. I'm sure that lots of innovation is still possible in the design of fencing, shelters, feeders and nest boxes for pastured poultry systems.

We also have an immense amount to learn about the genetics of our birds and which breeds and strains can perform best on pasture. This subject area holds a great deal of promise as well. The Pastured Peeper field is wide open for creative minds and energetic farm families who want to be part of healthy, wholesome agriculture in North America.

In addition to everything we learned about Pastured Peepers, we also learned a tremendous amount from the other Symposium participants about pastured poultry production and marketing in general in their part of the country. It was really uplifting to meet everyone and realize what a strong foundation of pastured poultry producers is growing throughout the country. This truly was a memorable and powerful gathering of the pioneers of pastured poultry. As one participant remarked, "This was a great learning experience. I'm very proud to be an APPPA member."

APPPA BUSINESS MEMBER PROFILE

Moyer's Chicks
By Jody Padgham Winter 2004

Moyer's Chicks, located in Quakertown, PA, has the goal of providing the best quality chicks and producing efficient results for their customers, whether they raise 50 or 50,000 chickens. Leon Moyer, co-owner, states: "We are committed to doing all we can to provide 100% customer satisfaction in what has been ordered from us. We are also available to provide assistance with any questions our customers may have about poultry raising." Moyer's Chicks serves the eastern U.S. and Canada, also sending mail orders throughout the U.S. and some international sales.

In 1946, Ernest Moyer began a small chick hatching business in Quakertown to supply area farmers with top quality baby chicks. Being service-oriented from the beginning, he chose to hatch chicks year round, not just seasonally like many other hatcheries in the area. During the first 12 months of operation he hatched a total of 60,000 baby chicks. His first delivery vehicle was the back seat of his 1942 Oldsmobile sedan. Over the years Moyer's reputation spread and they began supplying chicks throughout Pennsylvania and surrounding states. In the early 1960s, they even opened a second hatchery on the island of Puerto Rico, which they operated for the next 25 years. Today, after 15 expansions in 50 years, they are now hatching up to 16 million chicks annually. Their focus still remains offering top quality chicks coupled with the best customer service. Although they transport much of their production in climate-controlled trailer trucks, up to 130,000 per load, into the eastern provinces of Canada, they still have many long-time customers buying 500 or less. Many are sent by U.S. mail throughout the country. The business is owned and managed by the 2nd generation Moyer siblings: Ivan, Leon and Eileen.

Three types of baby chicks are currently being offered: the Cornish Cross for meat production, reaching four to five lbs. in five to six weeks; a brown egg layer, a cross between a Rhode Island Red male and White Rock female; and the White Leghorn pullet for white eggs. Moyer's also offers 17-week-old started pullets for those not choosing to raise their own replacement layers. These cannot be shipped in the U.S. mail. At their office they offer many types of small equipment, incubators, fertile eggs, and helpful books. The small incubators, books, baby chicks and fertile eggs can all be sent via U.S. mail.

Leon leaves us with this:
"Since being involved in American agriculture for over 50 years and seeing many changes over this half century, one of the most exciting developments of this decade is the privilege of supplying an increasing number of innovative, new generation farm families involved in multi-faceted, sustainable agricultural development. It is gratifying to see them building their growing poultry enterprises on the solid foundation of the baby chicks supplied by Moyer's for generations."

Contact Information: Moyer's Chicks, 266 E. Paletown Road, Quakertown, PA 18951 215-536-3155, www.moyerschicks.com

What's Happening at the Hatcheries
By Matt John Spring 2005

After a few years of working in commercial poultry right out of college, I became very familiar with the industry from the large corporations to the 'small-order' hatcheries that most pastured poultry growers buy from. I was surprised to find that a large number of chicks sold around the country are drop-shipped from one or two sources and nearly all of the broilers and sex-link layers are hatched from eggs purchased on the open market. In fact, at least one of the most popular 'hatcheries' has not owned chickens for many years, but drop-ships everything they sell.

My wife worked briefly for a company that sells broiler hatching eggs to corporate poultry companies as well as many of the small hatcheries across the U.S. The main lesson learned from that experience is that when Tyson, Purdue or one of the corporate companies wants a particular broiler cross, they get first choice. The smaller hatcheries that buy hatching eggs get whatever cross is left over. You may have noticed that batches of broilers from the same hatchery in the same year can vary greatly in size, growth rate, livability, etc. The only way to insure a consistent quality of chicks is to buy from a source that owns their breeding stock. Call and ask, if they refer to 'cooperating hatcheries,' 'affiliates,' 'associate breeders,' etc. then you will know they are not breeding their own chickens.

Week-Old Cornish Cross Chicks

When choosing a hatchery, the first consideration should be to buy local. Most of us who sell meat, eggs or other produce from our farms use the fact that it was grown locally as one of the main selling points. It seems logical that we would choose to support local businesses to provide our chicks or poults. I would try my best to find a hatchery that can deliver by the next day after hatching. For many readers in more remote parts of the country, this may be nearly impossible. However, the benefits of limiting shipping time and getting chicks into a stable environment and on feed and water as soon as possible have been known for many years.

If the hatchery closest to your farm does not meet your expectations for quality, price or any other reasons, then choose a hatchery that uses their own breeding stock. This said, it may be nearly impossible to find a hatchery near you that has its own Cornish Cross breeders. However, I want to suggest another option. Why not develop chickens that meet your specific needs? Until about 60 years ago, one could pick up any of the farm or poultry publications and find advertisements from poultry farmers who were selling pure breeds and first generation (F1) crosses of production chickens. The ads talked about high egg production numbers, fast growth, etc. While those advertised numbers pale in comparison to that of modern production layers and broilers of today's commercial industry, it is very difficult to find a New Hampshire, Rhode Island Red, Delaware or other dual purpose breed that can truly be considered a 'meat bird' as they were in the first half of the last century. Nearly all of the pure breeds offered in today's market have been selected for egg production and most private breeders have selected for exhibition qualities such as type, color, etc.

I have heard it said recently that "dual-purpose means no purpose." While it is true that many of the meat and egg traits are genetically negatively correlated, careful breeding can bring about the best of both worlds. It is unrealistic to expect any one breed to produce 300 eggs per year and a five pound broiler carcass at eight weeks, but it is a worthy goal to aim for economical production of your own meat and eggs from one breed of chicken. The sustainability of maintaining a particular flock that can meet the needs of your family's or farmers market customer for meat and eggs as well as reproduce itself year after year with little or no outside importation of new stock need not be further clarified!

According to a recent poultry genetics text, three multi-national poultry companies own the five primary layer breeder firms. It is estimated that in 1960, approximately 132 primary firms of egg breeder stock were participating in the USDA random-sample egg test. The control of broiler breeding stock is similar. The same text reported six major primary broiler breeders and most of the global market is supplied by two major multi-national companies. These products are hybrids based on crosses of several highly inbred pure lines. The genetic base of the vast majority of chickens in the U.S. is extremely narrow. These commercial birds are performing at a production rate unimagined even 20 years ago. However, nearly all commercial stock is completely dependent on constant use of sub-therapeutic antibiotics and I'm sure most of us are aware of the implications of antibiotic resistance pathogens in food to human health. A recent study completed in Georgia has indicated that antibiotic-resistant bacteria such as E. coli are surviving and being released into the environment through field application of litter and manure from factory-type poultry farms.

A gap in information exists about the meat qualities of the dual purpose breeds. The design of a study or studies to compare and contrast growth, carcass qualities, feed efficiency, etc., of these breeds and their F1 crosses is sorely needed to quantify where we are in the United States and give some direction to where small breeders and farmers need to go. One recent study from the University of Arkansas compared various traits among slower growth hybrids. The results of this study may serve as a baseline for comparison. In addition, recent conversations with one of the rare breed preservation organizations indicated that they agree there is merit in further study of this area.

Some pure breeds of chickens which could be readily crossed for a gourmet market broiler still exist. Some are in the hands of private breeders and, amazingly, some universities are maintaining lines with a lot of potential. These chicks will not grow as quickly as Cornish Cross. Educating your customers about the differences in appearance, taste and texture will be the key to marketing these gourmet broilers. For today's small farmers to become truly sustainable we need to have more control over the source of our own chickens.

Wouldn't it be great to go to a neighbor or someone just a couple of counties over for chicks or to purchase new stock for your own breeding program? Poultry production led the animal industry in consolidation, confinement and commercialization of agriculture starting in the 1960s. Pastured poultry farmers have been among the leaders in reminding Americans how high-quality food should taste. Isn't it time to stop depending on the commercial industry for our basic inputs and create our own breeding programs tailored to our own region, state or farm?

Raising Your Own Meat Flock?
By Tim Shell   Winter 2003

Would you like to be able to own and reproduce your own meat chicken flock? Would you like that flock to be growthy without the "couch potato" characteristics of modern Cornish Rock Cross chickens? Would you like to be able to produce 1,000 chicks a year for yourself and neighbors with only a dozen hens and a rooster? Would you like to be able to do it for less than $.20/chick in feed costs? Would you like to have a hen that would still be laying 50% after she finished doing all the above? And dressing out seven pound or more as a plump stewing hen when finished? Yes? Well...... now you can.

Hi, I'm Timothy Shell, who, with the help of my wife, Naomi, and three girls, Mary, (5) Ruth (3) and Miriam (5 months), have been building that breed of chicken just for homesteads. Amid the worms, pigs, goats, cows, sheep, and ducks on our 20-acre homestead in the Shenandoah Valley of Virginia, chickens have been the center of our agricultural focus, specifically, broiler breeders. Three years and six generations of linebreeding finds us closing in on our goal of having a standardized breed of broiler chicken. We are working to develop two strains of broiler chickens which will be able to reproduce true to type. We launched the breeding program in 2000 after discovering the process of linebreeding. Linebreeding is a mating program which mimics that which occurs in all of nature and has been used all through history by humans in the domestication of every animal species. We have closed the gene pool on our flocks and now have breeding stock to offer as well as chicks to grow for the table.

Folks can now simply order our day old chicks and grow them out, saving the peak performers for breeding, eating the rest and continuing the process year by year. It is the same chick grown for food or for new flocks just the way every other farmstead breed is used.

Our two strains are: the "Improved Corndel Cross," and the "Improved Pastured Peeper." The Improved Corndel Cross is a Cornish Rock x Delaware cross, a broiler that grows out in nine weeks to a four pound avgerage dress weight, a six pound dress weight in 12 weeks, and up to eight pounds in 15 weeks. The Improved Pastured Peeper is a standardized, commercial, Cornish Rock Cross, growing to four pounds in eight weeks, six pounds in ten weeks, and up to eight pounds in 12 weeks.

Why a New Breed?
A multitude of standardized egg-laying breeds, such as Rhode Island Reds, New Hampshires, etc., are readily available. There exists no standardized broiler chicken genetic package that can be reproduced true to type, available to the homesteader. All of the broiler chicks bought and sold in the U.S. are generated from hybridized parent stock that are under the proprietary control of large multinational corporations. These companies will not do business with small folks, requiring orders to be in the thousands of chicks, costing from $2.60 (females)/each to $4 (males)/each.

Most "independent" hatcheries around the country simply order eggs to hatch or just broker the chicks from these larger companies. Almost none of the small hatcheries today own breeder flocks. We desire to develop, (for the growing backyard poultry movement, otherwise described by terms such as "pastured poultry" or "chicken tractors"), a broiler genetic package from which anyone can retain breeding stock and propagate perpetually. This can help create independence from the industrial poultry production model so many of us deplore while enjoying the benefits of a true meat-type bird and including the harvest of many surplus eggs as well.

The beauty of our alternative broilers is that you can feed the same chick three different ways and get three different products: on full feed for maximum growth in limited time; on limited feed over a longer time period; or on a restricted feeding program to create new breeding stock. The growth data above is for birds on full feed with some fencing in to restrict movement which contributes to weight gain. One can feed much less feed per day over a longer period of time to encourage more natural scavenging by the birds themselves, yielding the same dress weight at a later date with lower feed costs. You get a bigger carcass in less time than with layer breeds, without the high mortality rates of regular broiler chicks. Then, you can feed a portion of the flock on restricted feed for breeding purposes.

Why Two Strains?
Our first thought was that by free-ranging the parent stock on pasture we could get a hardier broiler chick. Growing modern Cornish Rock chicks is like growing plants in a greenhouse; they are a "hot house" species that has to be gradually hardened off to the outdoor environment just like a greenhouse plant. This hardening off process yields up to 10% mortality in the shift from brooder to field conditions and up to another 5% in the field from heart attacks and ascites, a chronic heart failure disease.

Most folks familiar with raising laying hens from day-old stock are disgusted with the fragility and laziness of the commercial Cornish Rock in comparison. Most major producers now simply factor the mortality in as a cost of doing business. Our Pastured Peepers show a 10 to 20% improvement over the commercial birds, which is good, but we were looking for an 80 to 90% improvement. The Pastured Peeper is a good bird for large pastured broiler operations, such a Joel Salatin's Polyface Inc., which uses them.

We initiated the Corndel Cross using the Delaware, a heritage breed used in the mid 1900s for broiler production. The Corndel Cross is a 25% Delaware x 75% Cornish Rock Cross. These chicks show a 40 to 60% improvement over the commercial birds but do take a week longer to grow out.

The improvement is mainly seen in their ability to handle more weather-related stress in field conditions, such as extremes of cold, heat and moisture. The Corndel is a busier bird, a much better forager without the characteristic feet and leg trouble seen in the Cornish Cross. They tend to range further afield, covering more territory acquiring food. They still retain the Cornish Cross benefits of great feed conversion, the large, plump, double breast and the white feathering for a clean-looking, dressed carcass. Flavor is much improved. These hyperactive, pasture-hardy chicks are much more fun to raise, especially for young children who can be so easily disillusioned with death among the flock. The Corndel performs well in pasture pens/chicken tractors but really shines in the day-range model using electrified poultry netting, since they are willing to move around much more.

How Do You Grow Breeders?

Broiler breeders of any type must be grown out on a restricted feed program. This program is designed to grow the chick slowly to maturity rather than as fast as possible. If a full feed program is used the hens will develop so much fat around the reproductive organs, especially the egg canal, that they will prolapse their rectum when attempting to lay their first egg. The males will be clumsy and heavy, struggling to mount and breed the hens, falling off when mounted, and producing excessive wear and stress on the hen and even lacerations of the saddle. Amazingly, on a properly controlled diet, instead of looking ridiculously obese, waddling like a duck, etc., these birds are just like a normal egg-laying breed; aggressive, lithe and agile, able to run, fly and roost in trees (or on your tractor seat), dig out your wife's flower bed, etc. Yes, the roosters can and will "flop" you. How funny trying to imagine a normal broiler doing that!

To grow breeders you need a feed chart, which designates the appropriate amount of feed to be given for each week of age of the birds' life. It is a very challenging program to learn. Even though the birds on the program are in excellent health and act just like a normal chicken, they still have the ravenous appetite of a broiler. They truly believe they are starving.

This can create a problem, being stampeded with hungry birds every day at feeding time, and preparations must be made in advance for correct housing and control of the birds. It is easy to overfeed them, creating trouble later. We have developed a model that gives optimum control using a permanent shelter and paddocks rotated around it made with electrified poultry netting. Thus it is now possible for anyone to maintain a breeding flock that has most of the advantages of a meat-type bird with fewer of the drawbacks.

In Virginia, a broiler breeder hen will produce about 100 chicks over a 28-week period from March through September, with optimum management. Eggs can be set once a week The birds need about .33 to .36 pounds of feed per day when laying, so they eat a little more than a regular layer but the eggs are very large, yielding large chicks. The stewing hens when finished are enormous, over seven pounds dressed. The breeders can be used for two years, but it is more practical to renew them every year. To use them a second year a slim-down process must be enforced after the first cycle of laying to prevent over conditioning; this is difficult to do. It is 24 weeks from the time a baby broiler breeder hen hatches out to the laying of her first egg. Two generations can be produced in one year. A male can service about 10 to 12 hens. So, with a dozen birds you could produce up to 1000 chicks in a season, or about 30 to 35 a week. Or, you could set every other week, using eggs on odd weeks for table use if only 500 chicks are desired. These dozen birds will produce at this rate on under five pounds of feed per day, or about 1000 pounds for the production season. It takes about 20 pounds to get each bird from day old to first egg, 240 pounds for the 12-bird flock. At $.13/pound feed x 1240 pounds = $161.2 or $.16/chick in feed costs. That leaves a lot of room to account for a return to labor. They will still be laying about 50% at the end of 28 weeks of chick production. The flock can be dressed for stewing or allowed to continue into winter producing table eggs once the season for growing and processing broilers is over.

To choose breeding stock, simply harvest all birds with undesirable characteristics, (color, small size, disease, etc.) and mate the best with the best. This will continue the standardizing process, allowing you to set your own criteria and generate a population whose performance is adapted to your specific production model and environment.

The longer you take to grow birds the healthier they will be. [I grow breeders out to the same live weight (6 lbs.) in 24 weeks that broilers grow to in eight weeks. They act just like normal chickens except that they believe they are on the verge of starvation 24/7.] You only need to cut back on the feed a little along the way to do this. Of course, you also need enough feeder space for all to eat at once so the smaller ones don't get pushed back. I know this is a lot more attention to detail than most folks are willing to put forth. It is a lot easier to just keep the feed trough full. That is why alternative breeds of broilers will probably stay in the market for the near future.

Ed. Note: Tim Shell "moved on" in life a few years ago, to achieve his dream of helping people raise pastured poultry in China. His flock was dispersed. Keep on the lookout for "Corndel Cross" genetics as we watch those who are successfully continuing his work.

The Cornish Cross: What is Wrong With This Picture?
By Harvey Ussery Winter 2001

I stopped raising Cornish Cross chickens for several years due to their many weaknesses and flaws, known only too well to anyone who has worked with them. I returned to raising "barnyard chickens"—the old standard dual-purpose breeds—to put broilers on the table and give to family and friends. But my wife and I missed those plump roasters you can produce with the Cornish Cross but not, in my experience, with the standard breeds (without caponizing, at any rate), which are simply too tough at a dressed weight of five to seven pounds. So last year, when I learned that Tim Shell was producing Cornish chicks from parent stock on pasture, I decided to try again. When I called Tim, he confirmed that he and others who have been using his chicks have found them to be hardier and more robust than the ever-deteriorating equivalents from the large commercial hatcheries. I ordered a batch of 60.

All was well through the brooder phase—just a single loss—most impressive in comparison with past batches of Cornish. I moved them onto pasture at about four weeks, in a netted area along the lines of Andy Lee's "day-ranging model." The birds showed the usual Cornish lethargy about foraging; but weight gain was, as always, impressive, especially in comparison with a group of standard chicks hatched at the same time by natural mothers in my main flock. And then in June we had a spell of hot, humid weather. When I went out one afternoon to check on the birds, I found a number of Cornish—now right at broiler stage—either dead or seriously distressed with heat exhaustion. (I lost 22 of them over the next day or so.) Despite the fact they had been on that pasture more than two weeks, drinking from a float-operated waterer right outside their shade shelter, they sat on their rears inside the shelter and died rather than walk ten feet for a drink of water!

I turned 180 degrees from the sight of scattered bodies and looked at my young standard chickens in a separate netted pasture. They were bright and active, scooting about in the hot afternoon like water bugs. Whenever they felt the need for a drink, they would cross the entire area to the waterer.

I turned back to the appalling sight of dead and dying birds and asked: "What is wrong with this picture?!"

We in the pastured poultry movement have turned our rhetorical guns on the Tyson's and the Frank Perdue's of the broiler industry. We have blasted the waste, the pollution, the lack of sustainability, the inhumanity, and the contamination of both our groundwater and our food supply that flow from a debased production system. Striving for a model which protects natural and agricultural resources and offers our customers poultry fit to eat, we have rejected all that—all, that is, except the very heart of the industry's flawed system: The Cornish Cross chicken.

The Cornish Cross's greatest virtue is also its greatest vice: its phenomenal rate of growth. That growth is constantly outstripping all its bodily systems—its internal organs and nervous system as well as its skeletal structure. The inevitable results include not only the leg problems and tendency to heart failure; the digestive system clearly lags behind as well. Look at the droppings: they always contain a fair amount of undigested feed-indeed, sometimes they look like nothing more than a wet feed mash. Whatever the statistics about conversion of feed to flesh seem to imply, clearly there is a great deal of waste and inefficiency here. A standard chicken's droppings, in contrast, are usually firm, gray with a white coating, and show no trace of undigested feed.

The Cornish Cross—like the huge supermarket strawberry whose growth has been forced by over-fertilization and irrigation—is lower in flavor than a bird that has had a more natural growth curve. Of course, to folks whose only experience of chicken has been the supermarket version—or worse yet, Mega McNuggets—pastured poultry has been a revelation: "Man, chicken was never like this!" But I would be happy to put one of my "barnyard chickens" (slaughtered at about 12 weeks) up against any pastured Cornish broiler in the land in a taste test: they unquestionably have more flavor. And if flavor is a measure of nutritional value—as I believe in a natural, unprocessed food it is—then again we should be asking, "What is wrong with this picture?"

I may be accused of waxing "mystical" here, but I believe that when we eat another living thing, plant or animal, we are eating not only its physical nutrients but its vitality as well. We have quite rightly condemned the broiler industry for producing chickens that are sick, propped up by antibiotics, growth hormones, and other industrial voodoo. And yet we continue to offer the same bird—raised without those contaminants and in a far more sanitary manner, to be sure—but weak and low in vitality, propped up by high management inputs.

Let me emphasize as strongly as I can: these observations are not intended as a criticism of all those good folks who are working so hard to make pastured poultry a viable alternative. I know that the market has come to expect and demand that broad, plump Cornish breast. I know that the economic pressures on pastured growers are fierce and that most feel they must have that seven or eight week grow-out in order to stay in business. However, I believe that the pastured poultry movement has matured to the point that we should be setting as a goal the production of an even better product. We should start by finding a viable alternative to the Cornish Cross.

In the long run, of course, the solution is to breed a better bird. And since no corporate or governmental agency is doing any breeding research relevant to pastured poultry needs, we are going to have to do that job ourselves. We should all start learning about the genetics of breeding. Some of us can contribute by making experimental crosses of our own or working with one of the standard breeds that were the foundation of the broiler industry before the Cornish Cross by selecting for traits that will maximize both vitality and production on pasture. Perhaps APPPA could put together a working group and/or sponsor a conference to explore options in breeding a better pastured bird.

Recently I talked again with Tim Shell, who likewise is frustrated with the weaknesses of the Cornish. Excited about the prospects for "linebreeding," he has begun crossing cocks from one of the former broiler breeds onto hens of one of the Cornish Cross strains; he plans to continue crossing and selecting until he has a bird more suited to pastured production. He believes that careful selection through 10 to 14 generations will develop a genetically consistent chicken adapted to our model.

Such a bird would:
¤ Have enhanced viability—meaning that it would be sturdier, healthier, would have more "on reserve" to deal with episodes of stress such as changes in weather.
¤ Exhibit much of the plumpness and broadness of breast of the Cornish.
¤ Be superior in flavor and nutrition to the Cornish.
¤ Have a moderately fast growth rate—but a balanced growth in which not only muscle tissue but all systems are developing healthily and in sync.
¤ Have an efficient digestive system which converts feed to flesh with a minimum of loss and waste.
¤ Be vastly more proficient than the Cornish at "rustling its own grub," while ranging on pasture.

Tim's goal is to produce such a bird with an ideal grow-out of about nine and a half weeks. The costs of the increased "turn-around" would be offset by lower mortality, lower labor input, and reduced feed costs. Note that the latter two factors assume a model more along the lines of Andy Lee's day-ranging model than the now-classic Polyface mobile pens, developed before the availability of electric net fencing in this country. I agree with Andy that the future of pastured production lies more with net fencing than with the pens in any case, given the lower initial materials and equipment costs and the reduced labor. But netting is essential if we develop a less lethargic, more active bird, which will need more space to range and forage a greater percentage of its feed.

The expanded foraging range may be the key factor in the sustainable production of an improved broiler. I regret that, due to the mixed nature of my own flock, I cannot give accurate feed conversion comparisons. Years of experience has convinced me, however, that when standard chickens have a large enough pastured area in which to roam, their per-pound slaughter weight has a lower feed cost than the Cornish, even with the longer grow-out factored in.

In the short term, there are a couple of things that might be useful to try. I urge producers who have been working exclusively with Cornish to raise a few of the standard breeds for broilers as well. (The White and Barred Rock, Delaware, New Hampshire, Wyandottes and others, as well as crosses among them, have at various times been important in broiler production.) See if you don't agree that these birds "work with you" in a way which is encouraging, in contrast to the fragility of the Cornish. Put some on your own table and see if you too find the flavor superior. Provide some to your more long-term and/or discriminating customers and ask whether they prefer the flavor. Certainly there is now a large enough universe of "pastured poultry palates" to ensure there are many with the discrimination to recognize poultry that is even better.

For a larger fowl suitable for roasting, I would like to see poultry folks reviving the almost-lost art of caponizing. Capons of some of the standard breeds could give us roasting fowls even larger than a Cornish roaster. (I have begun caponizing some of my young cockerels—New Hampshire Red, Newcomer strain.)

We must of course keep Andy Lee's warning in mind: that the wing-walker makes sure of the new handhold before letting go the old! Certainly those who have worked so hard putting into place a model which works for them should not abandon any element of the system—including its foundation, the Cornish Cross, without due care, experiment, and thought. In the long run, however, we must adopt the goal of producing a better bird. Let Perdue and Tyson have the Cornish Cross—we can do much better than that!

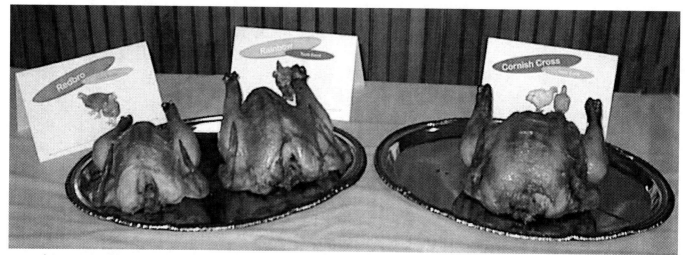

A taste test with slower grow out Canadian Redbro, (left), Rainbow from Arkansas (center) and Cornish Cross (right). Tasters chose Rainbow for best flavor and texture.

Genetics   33

APPPA PRODUCER PROFILE

The Chicken Man of Hume
By Jody Padgham Winter 2004

Those participating on the APPPA pro-plus listserve will recognize the name of Harvey Ussery. A new producer plus member himself, Harvey has been a champion contributor of late, offering wisdom on feed and bedding and even on how to produce a capon. Harvey is one of those people that appears to know a lot about everything--we couldn't resist finding out about Harvey's operation in Hume, Virginia.

Harvey and his wife Ellen bought their 200-year old farmhouse on two-and-a-half acres an hour-and-a-half SW of Washington D.C. in 1984. Neither had a farm background, but both yearned for a rural lifestyle. "We started out thinking we'd grow vegetables and sell them, but soon realized that we didn't have enough land to do it the right way," Harvey notes. Now Harvey and Ellen are committed homesteaders, growing most of what they eat on their rural crossroads property, which they named "Boxwood." Even though they don't sell commercially (to the disappointment of many), they have enough birds running around that Harvey has been dubbed "The Chicken Man of Hume." One of Harvey's primary goals is to help others understand food production and the specifics of raising chickens. He welcomes guests, offers workshops, participates heavily on several electronic listserves and is in the process of writing a book about homestead chicken production.

Harvey Ussery

When I ask Harvey how many chickens they actually raise up in a year he chuckles. "I was afraid you would ask that, he said. "It's not that easy to answer." Once Harvey explains his chicken system, I understand why.

Harvey's poultry system is based on two primary breeds—Old English Game, a small bird most common in cock fighting circles, and Cuckoo Marans, a dual-purpose chicken which produce dark brown eggs and are a good table fowl. There are several other heritage breeds mixed in the chicken flock, but Harvey is committed to maintaining pure lines of these two rare breeds. Harvey has in the past raised Cornish Cross broilers, but has come to the conclusion that they are "flavor challenged" and lack vigor; he now uses the old breeds exclusively.

Most birds are raised up on the farm using natural mothers. Harvey has what he calls "the Bachelor's Annex," a small building with five pen areas. When breeding season is coming, Harvey segregates five matched pairs that he has hand selected for particular qualities. The hens start to lay eggs, and Harvey collects and dates them. In the meantime, he is keeping his eyes on the Old English hens, as they tend to be particularly broody. Once a hen starts to get broody, Harvey will move her to a private brooder box,

APPPA PRODUCER PROFILE   *Ussery*

let her sit on plastic eggs for a few days till she settles, and then move in the freshest ten eggs from one of the designated breed pairs. The broody hen will sit and hatch the eggs—whether they be from chickens of a different breed, ducks or even geese.

Using this system, Harvey will produce a large number of pure-bred chicks early in the year. Once he feels he has enough chicks for all the purposes he has planned, he lets the breeding pairs re-mingle with the main flock and stops collecting eggs for hatching.

As the chicks mature, Harvey makes decisions on genetic qualities and harvests those that "don't make the grade." Size is not a factor in butchering at Boxwood. Early in June, when the birds are about nine weeks old Ellen will prepare what the French call poussin (a very young, tender fowl). Around 12-14 weeks the Ussery's will be eating "spring chicken." Once the weather turns cool, roosters will be culled for braising dishes such as coq au vin. In winter stewing hens will be turned into tasty broth, and for the holidays large, succulent capons will be served. "We used to butcher a few times a season and use the freezer a lot" Harvey says. "Now we are moving toward truly seasonal eating, where we harvest and enjoy the birds at different stages throughout the year." Harvey exclaims over the taste of his "barnyard birds." "Our dressed poultry is of an incomparably higher quality than the market alternatives."

A key to this harvesting is the process of selection. Harvey is working toward several genetic goals, and has just this year set up nest traps to assist him in helping select birds. Harvey's nest traps are basically nest boxes that have a trip that closes the door after a hen jumps in to lay. Chickens are released every hour, and their egg production noted. The four boxes Harvey has built will be put into place for the first time this winter, and so the theory has yet to be tried—but Harvey explains the concept. "I will be trapping to track four things: #1. Longevity. Birds are banded, and thus ages can be tracked. Which of those three-year hens are still laying? An old hen that is still a good producer should be kept as a breeder to pass on that trait. #2. Who is producing the most eggs? By trapping every day, high producers can be sorted from the low producers. #3. Cuckoo Morans lay dark brown eggs. Harvey is interested in selecting for the darkest eggs. The girls that are in the box with the darkest eggs will make his "keep" list. And #4. Misshapen eggs. "I really dislike misshapen eggs. Why keep a hen that consistently lays them?"

Birds that don't show performance in the nest boxes will be on the "dinner" list, those who are closest to the ideal will be chosen for the special breeder pairs; everyone else will fall in line somewhere in between and kept as utility layers. "We generally keep four to five dozen birds over the winter, and may butcher as many as 200 throughout the year," Harvey notes.

In the winter, Harvey moves his flock into a 13 x 24 ft. hen house that has electricity and water but is not lighted. (Harvey doesn't like to push the birds, and prefers to let them go through natural cycles) The house is divided into five internal areas with poultry netting, which Harvey can put together or divide depending on his needs at the time. Harvey is passionate about the deep bedding system he uses in the winter house. "Using six to ten inches of a mix of oak leaves and wood shaving bedding over a dirt floor is essential," he says. "The birds will scratch the bedding up, it will absorb the poops, there will be no smell, microbes will develop and break down the manure and litter." The house has doors and windows with chicken netting that are open to the air for circulation and sunlight. Composting bedding adds a little heat to keep the chickens comfortable.

In the spring, Harvey moves the birds from winter quarters to a series of paddocks and moveable houses protected from predators by electronetting, which he gets from Premier Fencing. He says he has "three shelter designs because they were made at three different times." The water system consists of a series

APPPA PRODUCER PROFILE *Ussery*

of food-grade hoses (from a RV trailer store) set up with flat pans and float valves. Harvey will move his primary flock to a particular area depending on the needs of the birds and the environment at the time. Last fall he brought the birds up from the far pasture in September, moving them to an area closer to the winter house. This gave him the opportunity to overseed the back pasture with the pasture mix that he likes to use every year (including grasses, clovers, etc.). In the spring, Harvey will put the primary flock out on the new growth and keep them off the area near the winter house that will be overseeded early this coming year and allowed to develop before the birds are moved back on.

Birds are moved into specific areas to fulfill certain needs. Harvey has found that his few guinea fowl are excellent at harvesting squash bugs, a challenge for organic growers in his area. He puts electric netting around the squash patch, and will keep three to four guineas there after the squash flowers fall so they can keep the place clean. Guineas won't scratch like chickens, and he has plans to also try them in the asparagus patch next spring. They will, however, eat lettuce and tomatoes, and will put small scratches on the squash (which will heal over for winter storage), so be careful when planning your live bug eradication system. Chickens are also regularly run through the orchard to clean up any insects causing problems there.

When the conversation turns to feed, Harvey admits that this is the topic he has been thinking about the most lately. He has been mixing his own organic feed for several years now. He says that he consistently hears that "you have to be really careful to get rations balanced perfectly when mixing your own feed—you can't be haphazard." Harvey points out that this implies feeding poultry is a very exact science, which then implies that you are talking about a generic chicken. Harvey notes that almost every poultry nutrition study assumes two things: 1. confinement, and 2. highly bred hybrid birds that have been bred to eat only at a trough, 100% dependent on the feed you provide. "These assumptions may be totally irrelevant for those of us running birds on pasture and working with different genetics," Harvey says. "I am not intimidated by expert opinions, recipes or formulas. I know that careful observation of my flock, backed up with a lot of reading and experimentation will give these birds the best diet, which may be very different from the standard recommendation." To this end, Harvey's latest experiments have him sprouting whole grains to more fully utilize nutrition and thus reduce quantities, making Boxwood, which must buy grains, that much more sustainable.

CHAPTER FOUR
Shelter Designs

There are as many shelter designs out there as there are pastured poultry producers. As you read the following articles, you will see that the goal in shelter design is to make a shelter light enough to move, heavy enough to stay put in a gale, and big enough to contain your particular choice of flock size. A design that is high or low, and made of plastic/ aluminum/ rebar /wood is ideal. Some have doors, others are open so birds can forage in an area enclosed by electric net fence. Get the picture? Read on....

Creative Pen Designs
By Joel Salatin   Winter 1999

All sorts of creative portable pastured poultry pens are being built around the country. These modifications accommodate variables like weather, personal taste and building materials. I have seen some that look extremely good and others that do not.

By far and away the biggest mistake I see is in making the pens too heavy. I would say that at least 75% of the people who try pastured poultry and don't stay with it are folks who build the pens too heavy. Ease of movement is the primary consideration in pen construction because that is what determines how frequently the pen gets moved. Letting the pen sit for two days instead of just one completely adulterates the pastured poultry model's amount of green materials ingestion, sanitation, chicken activity level, proper soil nutrient load and regeneration of the forage.

The weight comes primarily from two sources: wood and covering. Not only is dimension important, but so is the type of wood. A 1 x 2 in. oak board is twice as heavy as softwood the same size. Softwoods treated for rot are still my favorite medium, partly because wood is a renewable resource. Many people have trouble finding aluminum roofing, so opt for the more common corrugated steel roofing. I do not know exactly what the difference is per square foot, but steel must be more than fifty percent heavier.

Many people are using plastic covers, from polytarps to canvas to lumber wrap that is thrown away by the truckload from lumberyards. Since all of these tend to make the pen hotter because they are not as reflective as aluminum, most folks who use this are modifying pens to a peaked roof. Of course, in cooler climates or during cooler times of the year these materials can be a real asset.

Another reason for peaking the roof—one foot is plenty-- with these flexible coverings is to shed water. No matter how tight you pull the fabric, it will sag into pockets on a flat roof. Invariably that pocket will have a hole in it and then the chicks get wet underneath. Leaving the triangle open between the peak and the flat sides creates tremendous ventilation. In fact, in hot climates I highly recommend this technique even with aluminum covered pens.

Many people are trying to move the pens without a dolly, or portable axle. Some mount wheels permanently on the pens, and that works fine for only a couple of pens. But that requires leaving the pen an inch or two above the ground, which invites predators. One fellow mounted broom bristles on the back-end to create noise and help the birds move forward. He said that it made a huge difference. He also cut PVC pipe in half, bent it tight around the bottom boards and fastened it, creating a kind of runner half an inch below the wood bottom. The slick PVC allowed him to drag the pen without a dolly and he did not have predator problems from having the ends an inch up off the ground at all times.

Probably the most common choice I'm seeing is building pens out of PVC. Keith Perron and Walt Gregory out in Illinois have a very workable peaked pen out of PVC they say can be moved without a dolly because the material is slick enough to slide. The back pipe just bounces along on the grass clumps, jumping over them because it is rounded. A board will not jump like slick plastic does. Since it is so light, they jointed it so that they can pour water in the top of one corner and have it run into the bottom pieces. The water will add weight if necessary to keep the pens from blowing.

The cost of materials for a PVC pen is about the same as our wooden ones. I would like to build one of these this spring just to see if in fact it is as simple as our rectangular wooden ones. One of the advantages of wood is the ease of attaching the poultry netting and skin. With PVC, drywall screws will work, but generally a nylon or wire wrapping around the pipe is used.

John Mower in Wisconsin has built an A-frame PVC pen using lumber for the bottom skid pieces for stability. His neat idea is to cover the whole thing with poultry netting and then a polytarp, which can be rolled up and down depending on weather conditions. This is truly an innovative response to colder climates. In fact, a webbed greenhouse poly material could be used in the same way early and late in the season to extend production a couple of months.

Andy Lee of *The Chicken Tractor* fame has recently moved to Lexington, Virginia, just 30 miles away from me. What a great new neighbor to have. He is building taller and much heavier flat-roofed pens mounted on permanent wheels. He buys wheels by the case and mounts them as duals by simply drilling a hole through the bottom 2 x 4 and running a half-inch bolt through both lawnmower wheels and the board. He puts three pairs on a 16 ft. side, so each pen has 12 wheels. Although these pens probably weigh 500 pounds, including steel roofing, he can push them along the pasture quite easily.

Joel Salatin's Original (and Current!) Pen Design

One reason he went with a taller—almost four feet—and heavier pen is because he wanted multiple use. These pens work nicely for turkeys and could even handle sheep or pigs. So far he hasn't had any trouble with predators although the bottom boards stay at least an inch off the ground at all times. These pens are too heavy to push up a hill, but work fine on contour.

A couple of friends of mine have taken a design from *Countryside* and *Small Stock Journal* using cattle panels, bent at right angles and covered with poultry netting. These panels come in 16 ft. pieces and are 4 ft. high. Bending them in the middle yields two wings, each 8 ft. long. Putting two of these bent pieces together creates a square pen 8 x 8 ft. These are extremely lightweight and work well for smaller children to move. Bungie cording a polytarp over the top and down the sides provides protection.

The drawback I see to these is ease of access. Building doors into the pen large enough to accommodate an adult is difficult. Their advantage is that they can be built in a few minutes, are cheap and lightweight. Of course, they are not as big as the 10 x 12 ft. kind, but for a non-commercial setup, or where labor is not a problem like in families with a lot of children, this option is a good one.

Many folks continue to ask me about making much larger pens and moving them by tractor. Everyone I know who has tried this abandons it after running over too many birds. The noise and fumes also scare the birds and contribute to stress.

This is what I've seen lately. We're still building our pens exactly like we did a decade ago. Maybe I'm stuck in a rut. The alternative design that really fires me up right now is the PVC with a slight peak. Let's keep experimenting.

Portable Pens Compared
By Wilma Keppel Winter 1999

Too heavy. Too hard to move! Can't get the materials. Doesn't take the wind/heat/rain we get here! Has another pastured poultry producer found a solution to your problems? I spoke to growers across the Midwest to find out what's been tried and how well it works.

Chicken Tractor

The chicken tractor design by Andy Lee, a 4 x 8 x 3 ft. tarp-covered pen is the first design my partner and I tried on our steep and extremely windy farm in western Iowa. Fearing tarps wouldn't last, we covered our pen in corrugated sheet iron--much too heavy! Later pens were 8 x 8 x 3 ft. with a better frame and built-in roosts, but still too heavy. We folded the extra foot of our 4 ft. poultry mesh out from the pen bottoms. This fills gaps and discourages predators from digging. A dog leash clip holds the wire on the front side out of the way for moving. We found this design too heavy, with a weak frame that makes poor use of the materials. You do a lot of pen-moving for few chickens and the high roof lets rain blow all over the birds' sleeping area.

Salatin Pen

A lot of people use and like pens based on Joel Salatin's *Pastured Poultry Profits* design. These 10 x 12 x 2 ft. pens use 1 in. poultry mesh over a frame of 1 x 3s ripped from pressure treated 1 x 6s. The back half of the pen is covered with corrugated aluminum roofing. The front half of the roof is two doors, one poultry mesh and one metal. A special dolly goes under the back of the pen for moving. The front is then dragged by a loop of hose-covered wire. This pen costs around $200 to build.

Many women find this 150 pound design too difficult to move. However, it can take 100 m.p.h. winds with no problem.

The metal lids are another matter. Salatin's winds top out at around 50 m.p.h., enough to blow a few lids off once or twice a year. He stops this problem with a wire looped around the lid and a wooden pen crossbar. In the Midwest, blown-off lids are routine, and may travel 100 yards or more, so growers favor bungie cords or hinges. Bill Lambert turned his door into more roof, stiffening the front of the pen.

Corrugated aluminum roofing is available only by special order in much of the Midwest. Those who have used it love it, though it is twice the price of steel and can double the cost of a pen. Thin corrugated steel is a good substitute, but thicker steel is much too heavy. Aluminum panels from mobile homes are another option. Used aluminum printer's plates are inexpensive but messy and may not last.

Pen durability has been a problem for some people but not others. I suspect lumber quality plays a big part. Lumber also affects weight. Deck wood 1 x 6s build a significantly heavier pen, says Nebraskan David Bosle, who built two this way by mistake. Cheap 1 x 6s cost and weigh less, but you must sort through them to find good ones at the lumber yard. O'Neal substitutes cedar, which is even lighter and costs less in his part of western Kansas. Cedar will not hold screws as well as pine, so Keith also uses $5 of construction adhesive per pen. "Glue liberally where every joint touches," he says. Cedar is a nontoxic alternative to pressure-treated lumber.

To help his pens last, David Ossman uses a 2 x 4 in. top header down the center. This adds strength under his water buckets, keeps his roof flat and gets rid of the wire brace across the bottom of the pen that was tripping his chickens. A second 2 x 4 across the bottom front adds strength where the pen gets pulled.

Beating the Heat

Broilers are very vulnerable to heat as they approach slaughtering age. Several Midwestern growers reported heat deaths on hot, still days with high humidity, even when they jacked up the back of the pen to create an air gap.

The first line of defense is your roof. On a still 90° F day, a bare aluminum or galvanized roof will reach 160° F in full sun. A baked-on white finish or reflective paint drops that over 60° F to just 5 or 6° above ambient air temperature! White house paint works, but may not last.

In humid eastern Iowa, David Ossman lays bed

sheets weighted with rocks over his roof wire to provide more shade. Where it's drier, some growers run soaker hoses over their pens to provide evaporative cooling on hot days. More ventilation also helps. Several growers hinge the sheet metal on the back of their pens. Paul Rohrbaugh of southeast Nebraska eliminated the back panel entirely. Nebraskan Bill Henkel removes the metal sides of his pens in hot weather.

It is vital to not let hot birds run out of water. A pen of broilers seven or eight weeks old can go through 10 gallons of water between morning and evening chores. Minnesota grower Todd Lein feels management plays a big part here. "They're like Pavlov's dogs; when they see you, they think it is time to get up and eat and drink." This can stress over-heated birds enough to kill them. Instead, leave them extra water (two or three 5-gallon buckets connected together) and do chores early in the morning and just after sunset.

Salatin Variations
Growers across the nation are adapting Salatin's design to their conditions. For instance, when Brad Jepson got tired of sagging roofs dumping water into his pens, he added a peaked roof. The 12 ft. ridgeline is supported by plywood cutouts that hold a 1 x 6 in. ridgepole. To bend his metal over the roof, he cuts the ribs, folds the panel, then caulks the holes.

Keith O'Neal raises 1,000 broilers a year in windy western Kansas., where heat is a big problem. This year a dust devil blew several of the pens 25 feet without damage.

Three solid sides make Keith's pens very rigid. The birds are fed through a 2 x 10 ft. door that doubles as a wind scoop. Closed at night for predator control, it is covered in white fiberglass to let more light in. Predators can pull poultry wire off and a big dog can charge right through it, so Keith uses one in. welded wire. "Fasten it with electrician's staples or heavy netting staples, not a hand stapler," he advises.

In central Iowa, Connie Tjemeland used mostly 8 x 10 x 2 ft. pens. A tarp covers the roof and 3 x 8 ft. access door. Three bowed pieces of wood keep rain from pooling. Roll-up tarp sides have a board across the bottom to stabilize them and also get bungied up or down. Axles at one end of the pen take slip-on go-cart wheels, only one set is needed for many pens.

Nebraskan Bill Henkel used 1-1/4 in. thin-walled pipe and a MIG (wire) welder to build two Salatin-type pens. Braces are ½ in. conduit; netting is 1 in. welded wire mesh also welded to the frame. Self-tapping metal screws hold on light corrugated steel panels.

David Wallace also welded frames out of thin-wall tubing. Two sides are 10 x 2 ft. rectangles bent by a fellow who does tailpipes. These are connected by 10 ft. tubes. Two 10 ft. tubes cross in the middle of the roof, braced with diagonals David uses to hang his feeders. Aluminum panels are screwed to thin pieces of cottonwood mounted on angle iron. The one in. poultry mesh door is also framed in cottonwood.

Dennis Demmel's pens aren't blowing anywhere. Built for his 17-year-old son Paul, each is welded together from four 9 ft. x 6 in. x 36 in. gates left over from when they got out of the hog business. The roof and back, which folds out to provide extra ventilation, are flattened steel bin panels. Panels are attached with self-tapping screws and the rest of the pen is covered with 1 in. chicken wire.

The key to making a pen this heavy work is to make it easy to move. Dennis welded frames for two dolly wheels with 10 in. pneumatic tires. These clip into the mounts that originally attached the gates to posts. A shop dolly with a swing-up brace post (similar to Keith O'Neal's) holds the pen up while the wheels are added. The same dolly is then used to pull the other end of the pen. Empty pens can be pulled at 20 m.p.h. using the 3-point hitch on the tractor. Mounting the wheels takes very little time, and the pen can be moved fairly easily by one person. The wheels attach to any corner so pens can be moved in any direction without turning them.

Nebraskan Tom Larson builds and flies ultralight aircraft, and he build ultralight pens too. He wanted something durable, inexpensive and fairly easy to build. He got a $70 pen that can be built in a day. One finger will pick up an end.

Tom's pens are 10 x 10 x 2 ft. with a welded frame of ½ in. electrical conduit (outside diameter about ¾ in.). A handle at each end is made from a 5 ft. piece of conduit bent into a U shape measuring 1 x 3 x 1 ft. Tom butt-welds two 18 in. skids to the bottom corner of each end, pointing the same direction as the handles. Lifting one handle 6 in. shifts weight to the skids and lifts the back edge of the pen an inch or two so it clears the grass. This is still low enough to crowd the chickens along.

The roof of Tom's first pen was half chicken wire, but the birds crowded under the shade. Now the roof and two sides of the pen are thin galvanized roofing, pop-riveted to the frame. Four of the five roof sections are pop-riveted to conduit cross bars placed where the panels overlap. To make the door, he welds ¾ in. inside diameter washers to the last roof brace so that they stand vertically. After threading a 10 ft. piece of conduit through the washers so that it hangs out a little on each side, he rivets the last sheet of roofing to it. This door is so light the wind will sometimes get under it and pop it open, so he weights it with an old tire. The rest of the pen is covered with chicken wire.

PVC Pens
Chris Stenberg stands 5 ft. 1 in. and weighs 105 pounds. She experimented with pen designs for six years, but like many women finally gave up pastured poultry because she could not find a pen easy enough to move.

A seamstress, she developed a system of roll-up tarps with Velcro fasteners to keep the weather out of her pens and only had to throw straw under the birds twice in six years. Another innovation was using antenna standoffs to hold polywire around the bottoms of her cages 6 in. above the ground and 4 in. away from the sides. Power came from the nearest paddock wire. Predators stopped digging under her cages. The wire was high enough not to touch her boot tops, but she did occasionally have to clip the tall grass underneath.

Chris' favorite pen was a 10 x 12 x 2 ft. PVC model with a peaked roof about 18 in. higher that was covered with a tarp. It was so light that she could move it with one hand. She had to anchor it with tent stakes and bungie cords every time the wind blew. Everything was wonderful until the middle of the second summer, when a big storm pounded the pen and puddled water on the roof. Although it looked fine, the PVC had gotten brittle when the winter temperatures dipped (hers routinely hit -35° F) and it shattered into a million shards.

An engineer friend tells me that PVC's durability depends a lot on the plasticizers used by the manufacturers. These are the most expensive part of the mix, and may get skimped on. Other PVC users I talked to haven't had problems.

In Iowa, Hawkeye Steel's Brower Pasture Poultry Division has turned its hog expertise to portable pen design. The Brower pen measures 10 x 12 ft., has a flat roof, and uses fabric panels made of hoop house roof material. Pen weight is adjusted by filling the PVC frame with water until you reach the right combination of portability and wind resistance. This pen will be sold as a kit you assemble yourself, and will cost around $300.

Hog Panels and More
I talked to two producers who built pens from hog panels. Wayne Simmons of central Iowa found that a pen 10 x 12 ft. was too heavy and floppy to move, even with a dolly. His current pens are 8 x 10 ft. and 7 ½ x 10 ft. The narrower one is much easier to move. One quarter of each roof is poultry mesh, the rest covered by tarps. Tarps on the pen's sides and back can be rolled up for ventilation. "My birds panted in hot weather, but I didn't loose any," Wayne says.

John deSaanedra uses 6 x 8 ft. pens to raise broilers, banties and Rhode Island Reds in Ohio. Made from two 14 ft. x 34 in. hog panels, five 1/2 in. metal conduit roof bows and a conduit ridge pole, each fastened together with a 14-guage electric fence wire. Everything but the roof gets covered with 1 in. poultry mesh. A 10 x 14 ft. poly tarp over the top is fastened with cage rings on one side and tarp straps on the other. It gets flipped back for access to the feed and water. A door 6 in. wide by 12 in. lets the chickens on pasture if desired. To move his pen, John unhooks the anchor blocks and then scoots the ends alternately. For longer moves he recruits a helper or stands inside with the pen on his shoulders.

A similar design developed in Vermont uses a wooden lower frame and 3/4 in. conduit bows. You'll find a materials list, directions and a photo at http://www.bright.net/~fwo/sub10.html. This 8 x 12 ft. design stands about 5 ft. high, with an access door at one end. Pulled by a rope, it rides on 16 ft. wooden skids and does not require a dolly. Last year David Wallace used this pen to grow 28 turkeys in central Kansas. He mounted 8 in. wheels on the end, switching ends when he reached the edge of the field. In spite of its height, this $100 pen is very resistant to wind because the weight is all close to the ground.

An unusual design comes from Chad Anderson, a high-school student living in northern Alberta, Canada. Chad's chicken tractors are strong, wind resistant and so easy to build that he puts them together without a measuring tape. Each rides on two skids made from 2 x 4s (4 x 4s were much to heavy). A small triangle of plywood fills the end under the roof, which is two 4 x 8 ft. sheets of Oriented Strand Board (OSB). The roof angle is wider than 90° to increase the width of the pen. Chad says the roof section could be shorter, he just didn't feel like cutting the plywood.

An OSB shelf at the open end supports the 5-gallon tubs Chad uses for feed and water. This will be replaced by welded wire mesh so the birds can graze through it. Covering is stucco wire with 2 in. squares, held on one side by hooks and elsewhere by fencing staples. A good pull releases it for pen access. He has raised up to 35 broilers per pen.

And the Future…
When a new idea becomes popular, an explosion of innovation often results: witness the automobile and the computer. Virtually everything gets tried at least once. Eventually a few basic designs get settled on, and it takes major problems plus an innovator like Joel Salatin to lead the charge in a new direction. Sometimes a poor design becomes standard because of dumb luck or temporary technical problems. Over 100 years after typewriter improvements made it obsolete, we are still using the QWERTY keyboard, which was actually designed to slow typists down.

Inexpensive Mobile Pens, Storage etc.
By Bev Sandlin Summer 1997

My daughter, Montana, and I were up until 2:30 am last night brainstorming on the uses for some new pens I just dreamed up. We were looking for a more inexpensive way to build pastured pens which were less cumbersome.

I had my lumberyard bend two hog panels (each 16 ft. long and 34 in. tall) at 5 ft. at a 90° angle. They bent and delivered them at no charge. I hooked them together with plastic cable ties (these are great—inexpensive and they stand up to our winter weather here in Minnesota—ideal for connecting any wire mesh or fencing). I then attached 3 ft. chicken wire to the outside sides, made a top wire and cut in a chicken wire door on top.

I then set two old lawn mower wheels on each side in the back (attach with bolt and nut) to roll it. I purchased an 8 x 10 ft. blue plastic tarp which I attached over half of the top, sides and back with cable ties. A broken lead rope (thick and easy on the hands) was tied to the bottom of the front to pull it with. Less than $50, two hours of labor by one not-so-handy woman, and easy to handle. So far the pens have stood up to 60 m.p.h. winds without tipping over. They also pulled just fine behind the car, attached to the bumper down to the pasture.

Even though these are great, last night while brainstorming things got even more exciting. Two hog panels, bent, cabled together, wheeled and tarped could easily rotate lambs, goats, small pigs etc. Cattle panels (16 ft. x 52 in.) set up the same way would make ideal calf hutches and are a quarter of the cost of poly domes or plywood ones. More mobility? Put wheels on the front too. No mobility? Just don't add the wheels. Lots of wind? I use an 18 in. length of rebar with a 6 in. "T" on top with a 4 in. piece welded off one side of the T for staking.

You can also cut a door into the panels without compromising the integrity of the design and use that piece as the door. Heavy wire can be used to bow the tarp up and give a covered wagon effect (keeps rain from puddling on top). Buckets for water and grain can be hung off the wire or in the corners.

This basic design could be valuable on nearly any farmstead. Consider the potential for storage of the lawn tractor, rototiller, bicycles, etc. Tarp the sides and use it for a brooder pen. Wire the bottom for a rolling rabbit cage. A couple more bends (4 x 4 ft.) and you have an ideal little shelter for goats, sheep, dogs, chickens, etc. You could even use it as a playhouse for the kids!

These pens last forever, as they are galvanized; they can be stacked, hung from the ceiling or just left outside. The tarps are lightweight, inexpensive and generally last for several years.

APPPA BUSINESS MEMBER PROFILE

Brower, A Division of Hawkeye Steel
By Karen Wynne Spring 2005

The Brower company originated in the 1922 when founder William J. Brower started a mail-order farm supply business. He expanded into manufacturing in Quincy, Illinois, in the 1930s, and successfully produced a wide range of poultry equipment. Mr. Brower continued to manage the company until the mid-1970s, when his son-in-law took over. The business remained in the family until 1980 and expanded into other livestock equipment production. The business was then sold to an investment group, who soon realized that the '80s was not the best time to invest in agricultural equipment production. A few years later, Hawkeye Steel Products purchased the company and kept the Brower name for their poultry equipment division while updating and adding to the product line.

The Brower company had always sold poultry equipment through catalogs, farm stores, and cooperatives, and discovered the pastured poultry niche upon invitation to some of Joel Salatin's field days. While they have a broad line-up of equipment for sale, Brower now manufactures a number of items specifically for pastured poultry, for both production and processing needs. The company prides itself in having high quality products that are constantly being improved and fine-tuned.

Brower has a full line of poultry processing equipment for small-scale processing that they sell to small farmers in the United States and overseas and also to universities doing processing research. Brower has developed a larger 25 in. picker that can handle three to six broilers or a turkey (or two small ones), which had been a limitation in the industry. They also have developed a bleeding rack that, combined with killing cones and a tank, can greatly improve the efficiency of a small operation. The company has also created a video that demonstrates the step-by-step processing method for pastured poultry producers.

In the field, Brower manufactures everything from incubators to waterers, transport coops, mini-grain bins and roll-out nests for laying hens. They manufacture a lightweight PVC pastured poultry pen for those producers that aren't as skilled at building their own pens.

Brower continues to work to develop new equipment and maintain the company's reputation as a manufacturer of high-quality products with excellent service.

Contact Information: 800-553-1791 www.browerequip.com Houghton, IA 52631

APPPA PRODUCER PROFILE

Mike and Deb Hansen: Gifts From the Good Earth Farm
By Jody Padgham Summer 2002

Mike and Deb Hansen and their three children operate Gifts From the Good Earth Farm near Milladore in north central Wisconsin. The Hansens purchased their 80-acre farm in 1995. They have been experimenting with systems for pastured poultry for seven years, producing 1000 to 1500 birds the last several years, but expanding production this year to 4400 certified organic (MOSA) Cornish Rock Cross meat birds. They market most directly to consumers and retail outlets, with some going to restaurants. Both Mike and Deb have full-time off-farm jobs in nearby communities, Mike in Stevens Point as a Rural Planner for Portage county and Deb in Marshfield as the Director of Fiscal Affairs at the research division of the Marshfield Clinic. They have been improving systems and closely studying the economics of their production with the goal of finding a way to get Mike on the farm full-time in the next few years. Recently they have expanded their operation to include pasture raised Belted Galloway beef, which they also direct market.

Production System

The Hansen's this year will start five batches of chicks, two batches of 1000 and three batches of 800. First chicks arrived in mid April. The brooder is a 12 x 30 ft. pen inside the dairy barn, which has been retrofitted with cement work, 2 x 4s, hog panels, chicken wire and poultry netting. The Hansens have till now used red lamp brooders, but this year are converting to two gas brooders, as they feel these will be less stressful on the birds and more economical. Straight run Cornish Cross chicks are pre-ordered for the year and purchased from Welp Hatchery in Bancroft, IA (www.welphatchery.com, 800-458-4473) for $0.52 per chick. Birds stay in the brooder for exactly four weeks, living on a cement floor with 6 in. of pine shavings. New shavings are added on an as-needed basis.

The brooder watering system is a ziggity nipple drinker, purchased from Poultry and Livestock Supplies in Wilmer, MN (www.palsusa.com, 800-328-8842; cost approx $200 for 60 ft.). The drip tubes are a vast improvement over bucket systems, as the water stays clean and there is no beak dipping needed. Mike has the tubes hung on pulleys, so they can be raised as the birds grow. The brooder is carefully cleaned after each batch is moved out, with a light bleach solution which is then well rinsed. Mike notes that excluding wild birds, bird nests and droppings from the brooder area is key to maintaining chick health.

Birds are moved out to the fields using a hand-crafted transport sledge made of a metal door on wooden runners with foot-high plywood walls on three sides. The fourth (long) side is a drop down mesh door. The sledge is hauled to the brooder door, and the four-week-old birds are herded (using a piece of flexible plastic used more commonly as a manger liner) onto the sledge. After the door is closed, Mike hauls the sledge out to the field pens with his 4-wheeler. Mike claims the sledge has increased his efficiency at this stage many times over, as the birds do not need to be individually handled. In the field, the pens are propped up with a cement block and the birds are herded down the mesh door/ramp and into the back of the pen. After this batch leaves, the brooder is cleaned and ready for a new batch arriving the next day.

Field pens are Mike's design, nicknamed the "Pasture Schooner." They are 8.5 x 12 ft., and constructed with a frame made of 2 x 4s, cattle and hog panels with plastic tarps are used for weather protection (see following article.) Cost of each pen is about $95.00. Pens are moved daily with a two-wheel dolly.

The field pen watering system utilizes gravity waterers purchased used from the turkey industry for about $10 each. The water unit in each pen is connected via hosing to a centralized water system that also runs water to the cows.

APPPA PRODUCER PROFILE   *Hansen*

Pastures are rotated for three years, managing with the cattle and mowing in non-poultry years. Mike notes that the three-year rotation is also very important for organic coccidia control.

The Hansens at this point have not had any significant predator problems. They credit this mainly to the open landscape, but also to keeping surrounding habitat low and clear.

Feed

Feed is an organic starter and grower mix, delivered from Golden Grains in Sparta, WI, (608-269-5150), in an 11,500 lb. batch and stored in a bulk grain bin (purchased used for less than $100). This quantity is what their 800 birds will consume in eight weeks. The feed contains organic high lysine corn, wheat, roasted soybeans, kelp and McNess poultry mix, and costs the Hansens $300/ton, including transportation.

In the brooder, after six days the feed is served in a 12 hours on-12 hours off system, which Mike has found significantly reduces leg problems and heart attacks. Turmeric (the yellow Indian cooking spice) is sprinkled on the feed daily for its mild anti-bacterial qualities. Any evidence of coccidiosis is treated successfully by spraying a light spritz of organic cider vinegar for three to five days on the feed as it is unloaded from the storage bin. The mild change in the bird's digestive pH has successfully brought birds back to full health.

Feed in the field pens is offered in feeders custom made from plastic 15 and 55-gallon containers. 5-gallon buckets are used to fill feeders from a trailer pulled with the 4-wheeler. Feed is rationed so the birds eat the supply by 1 pm.

Processing

At exactly eight weeks, on a Monday evening, the birds are herded from the pasture pens and set into custom-made (again, from cattle and hog panels) transport pens, which sit on pallets and can each hold 80-100 birds. Pallets of transport pens are lifted with a

Mike Building a Pen at a Field Day

folk attachment on the tractor onto a custom-made trailer with removable sides. Extreme care is taken to ensure there is proper ventilation for the 800 birds before and during transport. Finding a certified organic processor was one of the biggest challenges for the Hansens. They are now very happy with Wapsy Processing, a certified organic USDA inspected processor in Decorah, IA, but the plant is a three-hour drive from the farm. Wapsy butchers, cleans, labels, bags and freezes the birds. They are put in cardboard boxes (six per box), which are then weighed. Cost of processing (excluding transportation) is $1.75 per bird. The Hansen's estimate live transportation costs add $0.15 per bird, if they take 800-bird loads. Mike drives the loaded truck to Iowa very early Tuesday morning and drops off the pallets of birds. He drives back with the truck and contracts a return hauler to bring the frozen birds back to frozen storage near the Hansens farm by Wednesday afternoon. Return frozen transportation adds an additional $0.15 per bird.

Labor

Mike says that the biggest help he has on the farm, after Deb and the kids of course, is from his 4-wheeler. He loads bulk feed into a small cart, jumps on the machine and can fill feeders and move pens

APPPA PRODUCER PROFILE   *Hansen*

in a very short time before he goes off to work. Two of the three children and Deb help to get full-sized birds into the transport pens, and Deb is chief partner in brooder cleaning and most other tasks. Deb also does the bookkeeping for the farm. The Hansens are able to keep labor output to a minimum.

Marketing

To this point, last year raising 1300 birds, the Hansens have relied on their personal and work contacts and word-of-mouth to sell their birds. The majority are sold direct to consumers, but Mike has also developed a strong market with local restaurants. Hidden in central Wisconsin, they are far from the educated food consumers of Madison, but find that their bird's taste and quality are easy to sell. They also offer grass-finished beef, and find cross marketing has been effective. The Hansens encourage sales of full boxes of six frozen birds (birds are not individually weighed at the processor) and so offer a discount for those purchasing 12 or more birds. For two or more boxes the price is $2.39/lb, fewer than 12 are $2.89/lb. Deb and Mike are gearing up for the additional sales effort this season and are planning advertising on public radio as their first big PR push.

The Bottom Line

A cost and time study from the University of Wisconsin last year included the Hansens and is continuing this season, so they have a pretty good handle on their costs. Raising 1300 birds, Deb's numbers show that variable expenses (chick, feed, bedding, marketing, trucking, processing etc.) come to $6.83 per bird. Fixed expenses (depreciation, insurance, certification, interest expense, etc.) averaged about $3.08 per bird for 1300 birds. Adding these two averages put the costs of a 4.5 pound bird at $9.84. Sold for an average of $12.00, this leaves a gross profit of $2.16 per bird. The Hansens have a goal of $4.00 profit per bird, which they feel will make their labor worthwhile. They continue to work on lowering costs (and have found that there are great economies of scale, as in reduced chick and feed price, once they moved over 1000 birds) but feel they are on the right track for production efficiency and cost effectiveness now. They see the only way to increase profit is to increase the numbers of birds produced, which will spread fixed costs out over a larger group, reducing the production cost per bird. Doing this assessment is what is leading them to larger production this year. They plan to continue expanding their operation as their time and patience allow.

Unfair Advantages

Mike is a man of huge energy and fast thinking. Combined with Deb's finesse with numbers, they are a power-packed team. Raising three children, managing the farm, holding down mind-intensive off-farm jobs and still being able to plan for change is something not everyone could do. The Hansens have capital to work with, though they tend to use invention rather than dollars to make things work. Mike's inventiveness and good grounding in physics and mechanics are a real plus in their operation. Mike readily reaches out to others and shares what he knows, and he will continue to be a leader in WI poultry.

www.goodearthfarms.com

Mike Hansen Cattle Panel Pen
by Jody Padgham Winter 2003

We profiled Mike and his system this past summer (see "Mike and Deb Hansen" Profile page 43). He has a nifty pen design that costs less than $100 (WI prices, summer 2002) and takes only one hour for one and a half people to build (that's a spouse chasing kids or a nice sized child). 8.5 x 12 ft. in size, the pens are sturdy and lightweight, and have a long life. This pen is very easy to make–I don't even think Mike measures anything.....

To make this pen: cut plywood into four 90° triangles about 6-8 in. on the short sides. Cut the short 2 x 4s to approximately 100½ in. Place the 2 x 4s on a flat surface, with the 4 in. side up, making a rectangle with same size pieces opposite each other and the long pieces on the outside of the ends of the short pieces. Screw the plywood pieces onto each corner, connecting the long and shorter 2 x 4s together. Turn the new rectangle over, so the plywoods are on the bottom.

Lay the two cattle panels side by side, on top of the rectangle, with the long sides of wood and panel parallel. Lash the panel edges together, to make a center seam, using the wire ties, which you can find in bags by the supplies for chain link fence in your local farm supply store. Use four to five ties for the length, and use a pliers to tie off the ends so no one gets scratched. Roll chicken wire crosswise across the short ends of the paired panels. Cut to length, with a little extra on each side so it can be wrapped around. Loosely attach the wire to the bottom edges of the panels with the small plastic ties.

Line the middle of the panel seam up with the middle of the short 2 x 4s. The panels will be laying flat over the wood rectangle and should rest about halfway across the longer side 2 x 4s. With large fence staples, staple the front edge of the paired panels onto the short edge of the rectangle, attaching the bottom of the panel in about the middle of the width (the 4 in.) of the 2 x 4.

Walk to the other side of the cattle panels–this end should be sticking out several feet past the wooden rectangle. With another person, push the panel up to make a hoop shape. Bring the bottom edge down tight to the 2 x 4, and staple the same way as the first end– halfway across the width of the 2 x 4.

Hansen Cattle Panel Pen

Finish attaching the chicken wire to each end of your new 'hoop house.' Hold the hog panels up to each open side of your hoop. Cut to size, creating a stepping stair shape or leaving straight. Attach the hog panel to each side of the hoop, using wire ties and stapling to the wood. File any sharp edges. You may put chicken wire on each of these sides; this is needed if you plan to put in very young birds or have a predator problem. The wire can go all the way to the top, but you will have to hinge it somehow on one side, as these high gaps are the pen access.

Now the wheels: raise up the pen and drill a hole the size of your lag bolts going into the back side corner of the long 2 x 4. Place the hole equidistant from all edges (top, bottom, back); the short sides are the front and back—the wheels go on the outside in the back. Place a washer on each side of the wheel and screw into place with the lag bolt, allowing freedom to roll.

Attach the green tarp to one of the short sides, using plastic ties. You want to put this tarp on the side that the most wind usually comes from (in relation to how you usually pull your pens.) This tarp won't fit very smoothly, but works well enough to protect the birds. The silver tarp will go from the bottom of the long side (again, the wind-ward side) and up and over about 1/3 of the way down the other side. Use large plastic ties to attach and then bungie cords to tighten the tarp down. The wheels will either be in the back or front of the tarp, depending on which direction you pull in relation to the wind.

This pen is very aerodynamic—if you put the closed side toward the wind, the pens will stay in place in even the worst winds. They are strong enough to hang waterers and feeders from the panel frame (Mike uses a reservoir and gravity feed system with red bell waterers and pulls hand-made feeders tied to the wooden frame). If you have predators you may want to use more chicken wire. If the hog panel side is too high or seems cumbersome for entry, a door may be cut, framed and hinged in the side, but it will add weight.

Note: This pen was designed as a typical pasture pen, but late last year, Mike decided to try the pens in a day-range model, surrounding six of the pens with poultry netting that was moved weekly. He is very happy with this new system, which has forced him to cut side access doors in each pen for the chickens to wander in and out.

Hansen Pen Shopping List

2@12' pine 2"x4"
2@10' pine 2"x4"
About 2' square 3/8 or 1/2" plywood
2@16' hog panel
2@16' cattle panel
50 x 3 ft. high chicken wire
Silver poly tarp, 10'x12'
Green tarp 5'x7'
10 Large (120#) plastic cable ties
20 Aluminum wire ties
20 small (45#) plastic cable ties
2 tie down straps (bungies)
2@ 3.5" plastic lawn mower wheels
2@ 6" lag bolts with washers
Wire fence staples
A few drywall screws

Pen Designs- 2004
By Jody Padgham Winter 2004

One thing pastured poultry producers are good at is pen upgrades and innovations. Throughout the year we will have profiles with different folks that have come up with good pen systems. For now, however, we'd like to share a few good ideas that have come our way in the past several months.

Cattle Panel Pen Modifications

Many of you will remember that last summer we profiled Mike and Deb Hansen and their "Cattle Panel Pen" (see pages 46 and 47). There has been a lot of interest in this pen and modifications of it over the year.

Mike himself told me last spring that he was going to try his pen inside a day-range fence. I haven't followed up with Mike, but did a modification on that theme myself this year and was very happy with the results. I followed Mike's plan pretty closely, using 2 x 4s for the base (making an approximately 8.5 x 10 ft. frame) and hooping a cattle panel from end to end over the top. Unlike Mike, however, I only put a side wall (cut hog panel) on one side. I left the other side open, with a support bar running about two feet off the ground. I then used my pen inside of electronetting and was very happy with the results. You can't see in the photo, but there are two lawn mower wheels in the back corner that allow the pen to be pretty easily dragged along.

Padgham Day-Range

Frustrated with some brooding issues, I even put one batch of chicks out at only five days old and all fared very well. The birds choose to be in or out depending on the weather, but found enough shade and shelter as they needed to be comfortable. The only problem I found was that I needed to move the pen within the electro net about every other day (though I only moved the net once a week) as the birds chose to sleep at night clustered around the pen edges and as they grew in size, manure would build up if the pen wasn't moved. This season I may need to change things, as I have an owl that started hunting here last fall that will take birds from within the net if I don't close them in at night.

I use loose feeder troughs out in the fenced yard for feed and a bell waterer hanging from the pen roof with a 15-gallon reservoir outside, sitting on an old barrel.

Plamondon Pen

Our new board members Karen Black and Robert Plamondon from Oregon have also designed a cattle panel pen. Their pen, designed about the same time as Mike's, has a wood frame door that allows easy access. Built with many of the same concepts as the Hansen pen, but with a wood frame and chicken wire rather than hog panels on the flat sides. More details about Karen's pen can be found at http://www.plamondon.com/hoophouse.html

Jordan Bentley Pen

Jordan Bentley's Hoop House

With a similar look to the pens above but using different materials, Jordan Bentley of Bucklin, Missouri, offers up the pen version he has found success with. This hoop house was designed by Armand Bechard of Conway, MO; the photo shows Jordan's modified version. 3/4 in. PVC is hooped on a 2 x 6 in. frame. Corrugated tubing wrapped around two sides of the bottom frame allows the pen to glide across the pasture with one person pulling. This 12 x 14 ft. version will house 100 meat birds up to the seventh week. Jordan also uses this same hoop to house turkeys; 50 birds up to the eighth week; 25 up to the 12th or 14th week.

Water is provided with watering cups fed by 5-gallon buckets. Feeders are used guttering and grit tubes are mounted on either end. This model also has four foot wide doors that lock into the frame to keep predators out.

Kip Glass Salatin Variation

Kip Glass of southwest Missouri uses a "Salatin style" moveable pen, but with several modifications. The height of his pens has been raised to 3 ft., which allows greater ease in getting inside to service waterers, catch birds, etc. He feels the additional height allows for better air circulation and keeps the heat radiating from the roof material off of the birds.

For roof material, Kip has tried everything from metal to silver/black tarp material, and has settled on a white vinyl material that is traditionally used for commercial tents. Kip says, "the material has to be white; at the hottest temperatures you can put your hand on the white material and not get burned- it is absolutely cooler than any other material I've tried."

Kip permanently attaches feeders to his pens. He uses a 6 in. metal rain gutter attached to the front edge of the pen under the lid, which he can put all the feed he needs at one servicing in the morning when he moves the pens. This reduces additional labor of having to lift feeders in and out and feeding more than once a day.

For water reservoirs he uses 15-gallon plastic drums lying on their sides, with the three inch filler hole on top. Tubing runs to gravity-fed bell waterers. This eliminates more labor by having to fill them only once during the day for the 75 to 80 birds in the 10 x 12 ft. pens.

Kip uses all 2 x 4 in. framing for his pens as he gets very strong winds in his area of the country. He says he has even had these pens get skidded along the ground in 50 to 60 mile per hour winds.

Kip Glass Pen

Wood Pens, PVC Pens, Hog Panel Pens, and Now Rebar Yurts!
By Tom Delehanty   Winter 1999

We hope to encourage future pen builders to continue to find the ideal pen for their application with what we have developed. We have been running PVC pens made of 2 in. pipe about 12 x 12 ft. to 10 x 14 ft. for the past four years and have found them very durable with easy repair. This past year we have been experimenting with rebar, mainly 3/8 in. We started with a smaller pen, approximately 6 ft. around and have graduated to approximately 10.5 x 11.5 ft. designs that are more oval than round. We found this approximate size to be the maximum weight that would span well without any center support.

Our center of the pen highest point is 36 in. high and slopes outward to the edge of the pen which is 2 ft. high. Our door is rebar, framed approximately 3 ft. wide and 5 ft. long, with an additional short corner door to be able to sweep the birds to the front for loading for slaughter. We have a rebar-mounted bucket holder that works well. Our door hinges are made with industrial belt strapping that works well with our barrels for feed also. We put half inch rebar on the first two of the four cross bars for the rounded roof. This seems to help the strength of the front of the pen for the doors and with moving. We have built over 10 of these pens to date and have improved the structure each time.

As we are using these pens daily, we find small tweaks that will help the final design for our application. We are going to put spindles on future pens for wheels, as we will use small 4 in. plastic wheels for moving. We remove them daily for the smaller birds, but as they get bigger we have found that we can leave the wheels on until slaughter. Birds are unable to get under the sides as long as we have an extra one inch chicken wire hanging over the bottom for field divots, etc. When we finish welding the rebar structure we wire the top with two inch wire using hog ringers (shoat rings) as they clamp tighter to the rebar. The sides are 22 in. high and we use 1 in. wire for the sides, which leaves 2 in. overhang on the bottom for uneven fields. For all the pens we have ever built we use a poly-woven plastic out of Canada that is possibly the strongest plastic anyone can find anywhere. It is highly UV-treated, and are still strong after being in the New Mexico sun for over four years now. The plastic is only attached with small ¼ in. screws that have held over 300 pounds of water on our large flat plastic pens and this plastic shows no signs of giving up yet. We use silver reflective poly woven for the roofs of all 50 plus pens we have on our farm. We overhang the sides with approximately 2 in. to shed the rain away from the pens. For six months of the year we wrap poly-woven clear around the sides of the pens for windbreak and all-round weather protection. We have found this winter that the yurts that we are running are cozy and warm, with the chickens able to keep those pens really warm on a cold morning.

We buy the wonderful plastic from a family operation called Northern Greenhouse Sales, http://www.northerngreenhouse.com (204)327-5540. They actually live in Canada, but they ship UPS from ND, which is cheaper for shipping. Marge and Bob will send you samples and ideas in their fine catalog, which inspired us to look at rebar for pen designs.

We attach the plastic to the top with hog rings only along the sides, working from one side. We believe that this pen design has a lot of advantages and feel it would benefit folks looking at running broilers later in the season. We built a jig with which we can build three of these pens a day. We will be working on building 100 pens over the next eight months so we will be able to make a better design for our year-round grazing operation here in New Mexico. New ideas will be the way to success for all pastured poultry producers and we hope you keep the faith as we all continue to learn and share.

CHAPTER FIVE
Day-Range Systems

Pastured poultry producers have in the past ten years split into two basic groups: those that raise their birds in enclosed shelters that are moved at least daily ("moveable pen systems") and those that have shelters with doors that allow access to a larger greenspace surrounded by electrified fencing ("day-range systems").

The moveable pen system was pioneered by Joel Salatin in the late 1980s and well described in his book "Pastured Poultry Profit$." In Chapter Four, Shelter Designs, we offered details on the multitude of modifications producers have made to the original Salatin design. In this chapter we explore the evolution of the day-range system.

Shelter Design Considerations
By Aaron Silverman Winter 2003

Shelter design is one of the most critical aspects of a pasture-based production system. Your shelters will have a direct impact on all aspects of your production: pasture, labor demands, overall productivity. Careful planning prior to construction, along with continual refinement, will help your system be smooth and productive.

There are a multitude of shelter designs currently employed by pastured poultry producers, and the choices can seem daunting. Shelters generally fall into three categories:

- Portable houses with no outside access for poultry
- Portable houses with outside access
- Movable houses with outside access

Portable shelters are easily moved by human power and possibly able to be transported between fields. Movable houses structures are usually more substantial, often contain floors for the birds, cannot be transported between fields, and require equipment to move (e.g., a tractor).

Each type of house has its advantages and disadvantages. As with many aspects of a farming system, there's a tradeoff between daily labor demands and capital invested. For many, one of the attractive aspects of pastured poultry production is the relatively low need for initial capital investment. As operations grow and refine, labor becomes more precious. The interaction between capital and labor will certainly play a factor in your decisions concerning your shelter design.

Seasonality also is an important consideration. Winter production requires more substantial housing than summer production. Movable houses with floors allow birds to be brooded in the field, avoiding the stress of the transfer on both chicks and people. The role of the poultry in your total farm system is also important. Movable houses that move infrequently have a greater impact on pasture than portable houses do. Their footprint may take a substantial time to recover and the impact zone around the house can be severe without careful management. Portable houses allow an easy transition from poultry to row crops and deposit manure directly on the pasture without any bedding to handle.

Over the past six seasons, Creative Growers, in western Oregon, has gone from producing about 1,000 chickens in 1997 to over 13,000 in 2002. While our house design has not changed much, how we manage the houses has undergone substantial evolution. Our process illustrates the complexities of choosing a house design.

Creative Growers is located in western Oregon, just west of Eugene. Chickens are produced seasonally, from mid-March through October. Chicken production is integrally linked with the farm's 10 acres of vegetable production, providing a major source of fertility. Early batches are raised on a cover crop that is immediately turned into high-value summer crops; mid-season birds fertilize overwintering crops (kale, garlic, leeks, etc.); late-season batches provide the foundation for the following season's field crops (potatoes, winter squash, etc.). Due to land constraints, birds are raised on three different fields over the course of a season, requiring house dimensions suitable for hauling on county roads. The houses also had to allow easy access for tall (6 ft. plus) farmers.

We established our production system with the birds remaining in the houses at all times. Stocking densities were moderate, about 80 birds per house for eight weeks. Houses were moved daily for four weeks and twice daily for the last two weeks (chicks went out at two weeks at that time). Predation was non-existent for our first two seasons under this system, and our low production volume of 1,000 birds allowed for the one and a half hours per day of servicing.

During our third season, we began to experience growing predator pressures. Our fields are lined with forested riparian zones, and we began to suffer losses when batches were close to the field edges. The main predator issue ended up being rats. After a second batch came in over 50 birds short, we began trapping and caught several huge rats cruising across the field from the creeks to grab two to three week old chicks from the houses.

The solution provided a huge jump in the evolution of our system. Surrounding batches of houses with electric poultry netting completely eliminated predation from both raccoons and rats. It wasn't until the following season that we realized that there was now no need to limit our birds to the houses. By this time our production was over 2,000 birds, and labor was becoming a limited resource. We realized that allowing the birds to range outside the houses would reduce the concentration of their impact, eliminating the need for daily moves. Providing feeders and waterers both in and out of the houses prompted the birds to spread their activity about and allowed us to increase the total stocking density without increasing their stress level.

Our current system allows us to raise 13,000 birds with the same daily labor needs as we once used for 1,000 birds. Batches of birds are raised in groups of houses, surrounded with two rolls of netting. Trough feeders and waterers are placed outside each house, with a trough feeder and hanging waterer inside each house. Chicks are transferred to the field at three weeks, at a stocking density of around 200 birds per house. During the first week in the field the impact is minimal, so the houses are moved just once. After the first week, the houses are moved three times a week, and the stocking density is gradually reduced to 100-125 birds per house. Houses are moved the length of the house, plus the length of the impact zone in front of the houses. The fencing is moved once each week, during a morning feeding. If fencing is moved immediately after feeders are filled, there are no problems with escapees.

We leave our houses open at all times, except during cold/wet periods when chicks are first transferred to the field. During these periods, older birds naturally migrate into the houses at night, and there seem to be few problems with even distribution. One benefit of leaving the houses open is ease of gathering for harvest. We found that on most nights, 2/3 or more of the birds may sleep outside the houses. When we arrive to crate them up, crates are arranged in a corral around the birds, and they are easily crated with minimal stress on the people. With two trained people, we were able to gather 400 birds in 1-1.5 hours. Leaving houses open has worked for us because of our lack of summer precipitation and low level of aerial predation.

Creative Growers Pen Under Construction

Creative Growers Portable Poultry Field House

House Base

10 ft.

Nest 10ft. pieces inside 12ft. pieces.

Secure well with 2x6 corner braces.

12 ft.

End View

- Top Purlin
- Side Purlin
- Door Jambs

Legs

1ft. 4in.
~45 degrees
2ft.

Bows

~90 degrees
4ft. 8in.
4ft. 8in.

©2000, Creative Growers

Materials List

Base
- 2 2 x 6 in x 12 ft
- 2 2 x 6 in x 10 ft
- 4 2 x 6 corner brackets

Bows
- 3 10ft. 3/4" EMT conduit
- 6 3ft. 4in. 3/4" EMT (legs) (2 10ft. lengths)
- 1 bag (10) 3/4" EMT couplers (compression fittings recommended)
- 3 bags (30) 3/4" EMT pipe straps

Side Purlins
- 2 1 x 6 in x 12 ft

Top Purlin
- 1 10ft. 3/4" EMT
- 1 3ft. 4in. 3/4" EMT
- self tapping metal screws
- short wood screws

Total 3/4 EMT: 6 1/3 lengths (10ft.)

Door:
- minimum width=24"
- suggested width=33"
- 2 1 x 6 in x 6 ft
- 1 plywood cut to width & length
- 2 hinges
- 1 door latch
- 3ft. 1" chicken wire, staples

Appropriate tarp material
plastic stripping or T-tape
plastic sheeting to cover ends
rope & polypipe or hose for handles

Building Instructions

Bows: Cut all legs. With conduit bender, bend about 90° at 4 ft. 8 in.—allowing for the 6 in. the bend will take. Check bender's instructions to ensure equal lengths on both side of bend. Bend about 45° at 2 ft. for each leg- be sure that the 2 ft. section remains straight. Connect with couplers. Secure to base using two pipe straps for each bow per side. Center middle bow.

Cut top purlin piece and connect. Attach to bows using pipe straps and metal tapping screws, with 18 in. overhanging each side. Attach side purlins using pipe strap with short screws. Height of side purlins will depend on the height of the chicken wire; side purlins are used for securing top of wire. Wire is recommended for protection in the event of the tarp coming undone during inclement weather when birds are closed in house. Door jambs are screwed into the base on the inside face of the base. Attach to bow with pipe strap and short screws. Trim to angle of bow with hand saw prior to installing.

Cut door to fit, leaving about 1/2 in. space between bottom and base. Wrap chicken wire from door jam to door jam. Staple wire to jam, purlins, and base. Apply tarp material and screw thoroughly through plastic stripping. Cover ends with plastic sheeting and staples early in season. Attach handles, and you're good to go.

APPPA PRODUCER PROFILE

Aaron and Kelly Silverman: Working with a Grower Group
By Anne Fanatico Fall 2002

Aaron Silverman and his wife Kelly raise 13,000 birds per year on pasture in the coastal mountains of Oregon. Their production is coordinated with other growers as part of their collaborative business, Greener Pastures Poultry, LLC. Aaron processes the birds in a small, federally exempt plant and sells to restaurants, farmers markets, and selected retail outlets.

Aaron's background is actually in horticulture. He began raising chickens as part of the fertility system on his 20-acre Oregon vegetable farm. Already selling fresh produce to restaurants, he found an eager market for range poultry and started expanding his operation. The poultry manure, combined with cover cropping and compost applications, builds good soil for the vegetables.

Production System
Aaron's production system is a hybrid of pasture pens and net-range. He used to use pasture pens but had a problem with rats attacking the birds through the pen. He began using net fencing to enclose the pens, surrounding five pens with two rolls of electric net fencing. The pens quickly evolved into basic shelters; they are opened up so the birds can roam the whole enclosure, returning to the shelter to sleep and lounge. During inclement weather (cold or wet) the shelters are closed for the evening. The shelters are moved forward in the enclosure every two days. Once a week, Aaron moves the entire enclosure. Eighty to 100 birds are kept in each pen.

During hot summer weather, the birds only venture out in the morning and at dusk. Therefore, Aaron provides water and feed both in the shelter and outside.

The homemade 10 x 12 ft. houses are tall in the center, so you can walk inside. The pens have a wood base, metal electric conduit frame, and are covered with chicken wire and an expensive heavy-duty tarp that is white on the outside and green inside. Trough feeders and waterers are constructed from 6 in. PVC sewer pipe. Metal gutters hung on the inside of the houses provide free choice grit and oyster shell. Aaron takes the tarp off the houses for winter storage. He closes down his broiler operation in the winter but continues making sales with frozen product.

By moving the shelters when the birds are roaming the enclosure, Aaron can move the shelters quickly. Normally you have to move pasture pens slowly since there are birds in them.

Aaron's operation averages 9% mortality, which includes the occasional disaster; approximately 1/3 of his losses are in the brooder. He has 30 field houses. To produce 13,000 birds, he employs one full-time person and one to two part-time people. Nearly all the production labor is supplied by his hired crew, as Aaron's time is devoted mainly to overseeing the processing facility and marketing for Greener Pastures.

Aaron's main production challenge is his limited space for pasturing. The farm close to his house is small and devoted primarily to his vegetable production. Aaron keeps a small flock of layers year-round in an eggmobile in his vegetable fields. While he broods his chicks at his home farm and integrates the broilers with his vegetable production, the main part of his broiler production takes place on leased fields five miles down the road.

APPPA PRODUCER PROFILE *Silverman*

Aaron uses Hi-Y, a Cornish Cross from the Hubbard breeding company. He buys chicks from Hoover's hatchery in Rudd, Iowa. Aaron uses cockerels only, since straight-run resulted in a lot of variation in grow-out weights. Due to the complexities of coordinating production with several other growers, his birds are harvested in stages, from 6.5 to 8 weeks old. Aaron averages a live weight of 5 lbs in 6.5 weeks (dress-out average of over 3.5 lbs).

A single ration is used for all of the growers participating in Greener Pastures Poultry. The ration is a diverse one, utilizing organic grains when economically available. Their ration this year was composed of corn, roasted soybeans, soybean meal, wheat, oats, barley, lime, fishmeal, and Nutribalancer from the Fertrell Company. The ration is ground in bulk and delivered in seven-ton batches from a local feed mill.

Aaron Silverman at the Farmers Market

Aaron has found that cover-cropping makes good specialty poultry pasture. Most pastured poultry producers plan forage around their ruminants-a perennial polyculture pasture with multi-species legumes and grasses is usually ideal. Aaron finds poultry actually prefer broadleaf plants over grasses. He has found New Zealand white clover good for both cover cropping and poultry forage, although expensive. It fixes nitrogen in the soil, is low growing, and does not require mowing. It develops deep roots that allow it to stay green during long dry summers in the West, but it does not form a dense mat that is hard to remove for planting crops or limits diversity in a pasture. And, he says, the chickens love it. Aaron does not have ruminants; he mows in the spring or has the fields hayed.

Brooding

Aaron broods in a stationary hoophouse. The brooder house is a 20 x 100 ft. quonset greenhouse (i.e., large half circle hoops on legs) with five foot sidewalls. The house is covered with a single layer of four-year poly, and further covered with 70% shadecloth to moderate the temperature swings throughout the day. Both sides of the house roll up. He has always brooded in a greenhouse. Until this year, when the large greenhouse devoted to brooding was constructed, half of a house was used for either plant starts (vegetables & flowers) or in-ground crops (mostly herbs that can easily be washed). Within the house, chicks are brooded in 16 x 16 ft. boxes constructed from sheets of plywood ripped down to 3 ft. widths.

Aaron practices "enriched brooding." He allows access to the outdoors at one-and-a-half weeks and feeds greens from his garden. He believes this encourages chicks to forage more when they are later placed on pasture. Aaron provides access to the outdoors with a run built onto the side of his brooder house from metal T-posts & old roofing metal. By midway through the season, the outdoor runs are just exercise areas. There's really no forage left in these areas, even with a couple weeks between uses. The greenhouse shadecloth extends over the runs, and the entire house and runs are surrounded with electric netting to exclude cats, rats, and other predators.

According to Aaron, while electrical heat is okay for small batches of birds, propane heat should be used for larger batches. If the electricity goes out, you don't want to lose a lot of birds. Also, it is easier to

APPPA PRODUCER PROFILE   *Silverman*

keep a stable temperature with propane heat. Electric thermostats save Aaron a lot of money in propane cost. Hanging Plasson waterers are used in the brooder house, attached to a low-pressure water line. Waterers are washed after every batch. Bedding is removed after three to four batches and used for composting mortalities or aged to be applied as mulch for over-wintering vegetable crops. He does not disinfect the brooder and experiences an average of three percent mortality in the brooder, including disasters (rats, cats, heater malfunction, etc.). Aaron would be interested in field brooding but it is not possible in his current set-up.

Entrepreneurship

Aaron started with 2,000 birds, processing and marketing them on-farm. He then began exploring the state licensing needed to operate under USDA exemptions. It was a challenge because Oregon authorities were not accustomed to working with a small, federally exempt plant. He considered a mobile processing unit but decided against it since it was "already going to be confusing enough for the authorities to license an exempt plant."

Aaron leased a defunct 2,000 sq. ft. locker plant nearby (built in the '50s) for just under $300 per month, after putting $20,000 into renovating the building. The group (Greener Pastures) also invested $40,000 into movable equipment and supplies. Since the plant was built to pack hogs, beef, and game, it is not an ideal layout for poultry processing, but is workable. The plant is capable of processing 500 birds a day.

The plant is currently operated two days per week. The plant is operated by a hired crew of seven people, who work anywhere from three to ten hours per operating day. Aaron's wife Kelly helps with packaging and labeling. About 20% of the plant's production is frozen and sold through the winter.

Aaron is building his operation in two phases. The second phase will involve moving to a bigger plant that is USDA-inspected. The first phase will allow him to develop production, processing, marketing and distribution systems that can be expanded in the future. It was not feasible to establish a large business at first. A successful proposal writer, he has received several grants to conduct a market survey at a local farmers market, fund legal work, to lease a refrigerated vehicle, purchase equipment and operating supplies, and perform a feasibility study for establishing an expanded processing facility.

Aaron increased his production to 13,000 birds of his own, in addition to the birds raised by his business partners at Greener Pastures Poultry, LLC. However, operating under the federal exemption limits his business to 20,000 birds per year. He plans to expand to a larger operation in the near future. He wants to operate under USDA inspection in a plant capable of processing at least 200,000 birds per year, probably working with a group of 10-15 producers. Growers net about $1.00 per bird and help with weighing the birds to ensure that the price they receive is accurate. Growers working with Aaron are required to help out with marketing—they give him either 15 cents/bird or 20 hours of time assisting with farmers markets during the growing season. Greener Pastures Poultry sells to local high-end restaurants in western Oregon, at three farmers markets, and to several butchers and natural food stores.

Aaron is leading the way in terms of collaboration. He has a producer education program that includes production guidelines and troubleshooting assistance. In addition to standardizing the feed ration and genetics, the guidelines provide beginning producers a basic handbook covering suggested methods for brooding, feeding, housing design, stocking density, brooder and field sanitation, and field management practices. He limits producers to 1,500 birds the first year so they can get production problems ironed out.

Day-Ranging
By Andy Lee Fall 1999

Like so many of the things I've learned about poultry and farming, this day-ranging system came about through accidental observation. The observation partly was seeing other people do something similar. The accidental part was to leave the door open on the chicken tractor. Day-ranging then came about as a hybrid cross between "free-range" and "pasture pens."

Typically, the free-range model we see here in the United States is a barn or hen house with a pasture surrounding it. The system is usually permanently fixed and somewhat inflexible. There are several advantages, the biggest one being that the labor requirement is low. The eggs or meat can be sold for a premium to many consumers who perceive "free-range" to mean healthy, nutritional, flavorful, and good for the environment.

Unfortunately, these advantages often don't equate to profitability because of the inherent challenges of the static building free-range model. Usually after only a few months most of the grass area near the barn or hen house has been reduced to bare ground. This is fine for dust-bathing, arguably a poultry perk, but also a mess in rainy weather when the eggs all need to be washed because of tracked in mud and barnyard manure. The bare area also quickly erodes, moving valuable, manure-laden topsoil to the nearby stream or into the groundwater.

While requiring more labor than confinement birds, the free-range system has much lower start-up costs, and I think the birds are much healthier, have a higher quality of life, and the eggs and meat taste superior to those from confinement-raised poultry.

Pasture Pens Solve Most Free-Range Challenges

The pasture pens, pioneered by Joel Salatin, and the chicken tractors popularized by Patricia Foreman and myself, seem to respond admirably to the challenges of both free-ranging and confinement models. Hawks and owls and foxes are not a problem because the chicken tractors have predator-proof roofs and sides, and the daily moves keep the predators off-balance. Moving the pens daily provides new forage for the birds and keeps the birds from over-grazing any area.

However, the disadvantages of portable pasture pens and chicken tractors often outweigh (literally) their usefulness. On level or gently sloping ground the heavy wooden pens move easily with a dolly and the PVC chicken tractors are a breeze to move. However, if the grass is weedy or clumpy, or if the slope is more than just gentle, moving the pens can be a real hassle. Young chickens take a few days to become accustomed to the pen moves, and the grower has to train them, during which more lives are lost. The labor requirement for pasture pens is fairly high, too. Each 80-bird batch of broilers requires a separate feeder, waterer and daily move.

Another drawback to chicken tractors is that the birds will graze excitedly when the pen is first moved, but within a very short time they will manure and tramp down the grass and stop grazing, especially if they are on continuous feed.

The Accidental Birth of Day-Ranging

The answers to many of these problems evaded me until last year when I let my attention wander and walked away from a chicken tractor and left the door open. Later that day I looked out and saw about 80 broilers spread out, grazing across the pasture. They were inside a wire fenced paddock, so I wasn't immediately concerned about daytime predators. They seemed content, so I left them be and went on with other chores, musing about how hard it would be later in the evening for us to catch them all and put them back.

Imagine my surprise when I went out to the pen at dusk and they were all back in it, contentedly crowding down for a good night's sleep. Since then, Patricia and I have worked out a scheme to take advantage of this phenomenon that chickens always go home to roost. At first, I only envisioned the broilers being loose in the daytime inside a larger fenced paddock. I ran a garden hose out there and installed two tubs with float valves to provide water. At that point my only daily chore was to bring them new feed.

For a while I used large tube feeders, one per 100 birds, but I saw where the bigger broilers crowded out the smaller ones. Then I took 6 in. PVC well casing and sawed it in half lengthwise. A 20 ft. length will give two troughs, each capable of feeding 150 full-size broilers or laying hens. The trough will hold up to three 5-gallon buckets of feed, and it's very easy to just walk along and pour the feed into the trough. One day I move one end of the 20 ft. troughs, the next day the other end. This moves the troughs across the paddock for uniform coverage.

The open-door chicken tractors provide shelter from the sun and rain, and a place for the chickens to bed down at night. Since they are only in the shelter part time, it isn't necessary to move the shelters daily, probably only every third day. In some cases, in a larger field, I open the chicken tractor in the daytime and let them range, then when they go back inside to roost at night I close them in to protect them from the fox. It's much better, though, if they can range freely from early morning to late evening without my interference. This they can do inside a fenced paddock, or if I use the portable electric netting.

After we grew out the 500 broilers on that half-acre paddock, we planted organic watermelons and cantaloupes, which we harvested in late summer. Then, beginning in September, we turned 400 eight-week-old turkeys into the paddock to clean up the crop residue and weeds and grass. Come spring of 2000 we'll have a very rich soil in which to plant our next market garden crop. All told, we will earn over $6,000 from that half-acre multi-cropping scheme this year.

Stationary Netting Model
By Timothy Shell   Spring 2000

Since first meeting Joel Salatin in the early '90s, I have greatly enjoyed watching and helping facilitate the growth of the pastured poultry movement. New developments like the super-strong woven poly tarps from Bob Davis and the electric poultry netting have been especially fun. For any animal husband there is nothing more rewarding, both financially and emotionally, than having contro—of the flock, the predators, of the manure, of the parasites and pathogens, of my time, of the feed, of the pasture and of the profits that come from exercising stewardship of creation in a worthy manner that benefits everyone with a clean environment, clean food, clean fun and in an honestly earned family farm income.

The ride on this journey of innovation and experiments has been both rewarding and challenging with the successes founded firmly in the failures. There does not seem to be any end in sight at this point to what there is to be learned, and the latest innovations turn out to be only the next step in the adventure of discovering better ways to raise poultry. We are grateful for the contributions of those who have gone before and paved the road to success for us with their failures and spared us the grief. We would like in this light to add our pile to the pavement to help others succeed. We want to introduce what we call the "stationary netting model," which is the special adaptation of ideas gleaned from many others to our situation, with a few of our own grafted in.

When we picked up the pastured peeper pioneering from Steve and Lisa McCumsey, we started with Joel Salatin's newly proposed netting model instead of the pens. The need for clean eggs, as clean as they were laid, led us down the trail to the stationary netting model. We first designed roll-out nests to eliminate the #1 cause of dirty eggs: hen traffic. Roll-out nests don't work well if they aren't relatively level and we don't learn that word in Virginia until we travel some out west. This year we decided to purchase a roll-out nest assembly with a hand cranked egg belt from Brower for our newest flock of breeder layers. On the 10 acres we use there are about four places to position the house level enough to have the nests function properly and I have hen houses on two of them already. I have a full time job as well and was ready to give up the hassle of dragging the sheds around. The difficulty was how to rotate around the shed with the netting. I didn't want to have a permanent fence with gates. We completely encircled the shed with one roll of netting to create a yard which we mulch to avoid mud and the "clay chicken yard" effect. We create an instant gate wherever we wish by simply pulling up a net post and propping it up with a 48 in. Premier Power Post. We make seven imaginary paddocks in pie formation and one real one with three more rolls of netting, one straight out from the yard fence like a wheel spoke, then a second spoke, then the wheel rim across the end. This yields eight paddocks on two acres, a quarter-acre each. Each week we move only two rolls of net and bed the yard down about once a month.

APPPA BUSINESS MEMBER PROFILE

Kencove Farm Fence Supplies

Locations: Blairsville, Pennsylvania, and Earl Park, Indiana
Product line: Fencing materials and tools
Years in operation: 25

Contact information:
344 Kendall Road
Blairsville, Pennsylvania 15717
800-KENCOVE
724-459-9148 Fax
www.kencove.com

Mission:
Kencove's mission is to improve the quality of life for the agricultural community, our employees, and people in need.

Hi, I'm Charles Kendall. In the late 1970s, I needed a better fence for my dairy cattle. I tried the high-tensile steel wire I had read about and was very impressed. It actually has solid advantages over any other fence I know of – low-cost, rugged strength and long life. Phyllis and I started Kencove's fence business in 1980 because we feel the great advantages of the fence materials we sell should be known and available to everyone. We have been constantly improving our services and knowledge to help you, our customers. We have really enjoyed the many friends we have met through the business. The help you have given us has allowed us to help others.

Kencove has supplied farmers and contractors with a full line of farm fence supplies for more than 25 years. Products include high-tensile wire, woven wire, electric netting, horse fence, tools, posts, gates and much more. With a comprehensive website, online ordering, and a toll-free support line, Kencove is committed to serving the agricultural community with fast, friendly service and expert advice.

Kencove's electric net fence gives great protection from predators. It can be used for poultry, goats, geese, dogs, rabbits, and gardens. Made with three stainless steel wires interwoven with twine, the lines are joined to range in heights from 28 to 48 in. Because it is made with stainless steel conductors, this netting lasts several times longer than copper conductors. The poultry netting has semi-rigid vertical stays every 3.5 in. to help keep the netting from drooping and shorting the lower strands. The plastic posts are built into the netting and have steel step-in posts to make installation fast and easy. The bent-pin post design also gives the posts more stability after it's in the ground. With 164 ft. length, Kencove's nets are 10% longer than most nets.

Typical Day-Ranging Model
By Andy Lee Spring 2000

We hatch our chicks in a GQF Sportsman incubator, then move them to an electric battery brooder for their first week. The battery brooder is in an insulated room and holds the chicks at a constant 90° F.

During the second week, we move the chicks on to a floor brooder, with electric heat lamps. They stay there until their fourth week, when they have enough feathers and body size to help them survive outdoors. Then we move them to a mini-barn that measures 8 x 16 ft., and has a plywood floor covered with about 4 to 6 in. of bedding. We use horse manure and planer shavings from a nearby horse barn for our bedding. We use this bedded model because we have 2-1/2 acres of organic market gardens where we continually need good compost.

Broilers on Pasture

The mini-barn has 3 ft. sides, and a 12 pitch metal roof. There is a door at each end, and we can walk through to service the chickens. We also use these mini-barns for our turkey and broiler breeder flocks and for our layer flocks.

We can stock up to 200 broilers in the mini-barn because we are using it primarily as a bedroom and rainy day shelter. We hold the broilers in the mini-barn for one week; this gives them a chance to harden off to the outdoor temperature swings, and it bonds them with the mini-barn so that they will always go back inside at night to bed down. This is important for two reasons. First, nearly half of their manure will be deposited on the bedding as they get up in the morning. Secondly, if we let the birds get in the habit of bedding down outside the shelter, and a cold rain comes up during the night, we will lose some of them to piling and to hypothermia.

We use one 165 ft. electric poultry netting (Kencove NP-7) to lay out a temporary paddock radiating from one of the doors. We let the broilers range in that 1,600 sq. ft. paddock for a few days, then switch the fence to the other door. Over four weeks we will move the fencing four times, and the broilers will have access to about 6,400 sq. ft. of range. This is what we call fencing the flock instead of the field.

The mini-barn stays in one position for the four weeks. When we harvest that batch of broilers, we move the mini-barn on to the next site for the next batch of broilers. This method does not spread the manure as uniformly as does the pasture pen model. However, our goal here is to capture the manure for the market garden compost.

We can move the broilers around inside the temporary paddock for better grazing and manure spreading by relocating their feed raft and water fountain daily. We use 10 ft. lengths of 6 in. lightweight PVC pipe, cut in half lengthwise for feeders. Each trough will hold a 5-gallon bucket of feed. We don't put end caps on the troughs, if it rains the water just runs out the ends, and if the feed gets wet the broilers eat it without any problem.

We fasten several troughs to a skid to make a raft that is easily moved. The water fountain is a Fortex rubber bucket with a garden hose and a float valve. One bucket will water 200 broilers.

An alternative to having a floor in the mini-barn is to leave out the plywood and simply layer bedding on the ground and position the mini-barn over the bedding. Then, at the end of the four-week cycle, just move the mini-barn and use the tractor bucket to scoop up the bedding and deliver it to the compost pile. Then go back and re-seed the area where the grass was killed. We use this method to kill fescue so we can re-seed with clovers and warm and cool season grasses.

Prairie Schooner Chickens
By David Schafer Winter 2001

When Alice and I moved into our new solar-powered home in the spring of 2000 we knew we were going to have to do something different with the chickens. Brooding with electric heat lamps was not an option. Turning electricity into heat is very inefficient and would deplete our power reserves. What to do?

I had been smitten with Timothy Shell's PVC hoophouse for his laying flock and asked Tim if he didn't think it could be converted for broilers. Ever the optimist, he gave me his thoughts and sent some detailed plans and photos. Adding side flaps, interior doors, and a propane-burning hover to Tim's layer design appeared to be an easy conversion to an effective brooder and shelter for broilers.

We figured dimensions for side flaps and canvas doors and hired a tarp shop to sew all the pieces for us. Building the PVC raft and hoophouse was quite easy and took about a day. Installing and "tuning" the tarp was pretty simple also. The structure cost around $1100. The moment of truth arrived with 500 little peepers.

In they went into the hoophouse. Since the hoophouse looks a bit like the prairie schooners the pioneers "sailed" to these parts in and these pasture-raised chicks are about as close as it gets to the native prairie chicken, we like the name Prairie Schooner Chickens. The schooner would be their home for their entire lives.

We bedded the grass floor heavily with shavings to keep the chicks off the moist spring ground. Some grasses poked through giving the brooder a very pleasant touch of green. The chicks became familiar with the plant world from the start.

Within a couple of days the chicks were sneaking out the corners and seams and poking around on the pasture surrounding the schooner. In that first batch I was diligent about shooing them back in and stuffing the cracks with burlap sacks. In later batches I only worried about it if it was very cold or near nightfall. I did make sure they learned to be inside the four "walls" of the brooder at night.

The propane ran out on day three, so I figured it was time for the chicks to heat their own space. 500 chicks can create a lot of heat. The brooder worked beautifully the first time with one minor flaw. Rain and dew ran in through the Velcro on the sides where the side flaps attached to the schooner, soaking the bedding on the edges where the chicks preferred to hang out. We put an end to that for the next batch. A tight-seamed flap sewn over the Velcro strip on the hoop house now keeps the rain and dew away from the Velcro.

As the chicks got older and hardened to the weather, we opened slits in the doors, propped up the doors, moved the doors toward the ends of the schooner, and eventually removed the doors and the eastern side flap. (Most of our severe weather comes from the west.) Since the side flaps only extend over 20 of the 40 ft. length of the schooner, even with a flap attached for weather protection the chicks can still go in and out on that side.

An electric netting designed for poultry surrounds the perimeter of the schooner giving a nice 8 to 10 ft. border. The chicks can easily run through the 2 in. squares of the netting until they're about two weeks old. No worries, mate. It's the things that eat chickens – which is almost everything – that we want to keep out and the electrified netting is awesome at that.

After several weeks we use another netting to make a three-sided paddock extending from the schooner. The perimeter fence is lifted with fence posts to allow the chicks to pass under. Our daily chores are to drag their feed and water out onto fresh pasture until we reach the end of the paddock. It usually takes a week before it's time to make another paddock.

Then we shoo the flock back into the schooner area, if they're not already there, lower the perimeter fence, take up two sides of the paddock fence and make a new paddock. It's very fast. We place their feed and water at the entrance and they're off and running again.

Speaking of running...we were astounded to see these notoriously lazy birds gallop out to their feed troughs. What a riot! It's more like a speed waddle, actually, but they also flap and try to fly as they go. Usually they charge and retreat in great waves.

Of course we never observed behavior like this in our 10 x 12 ft. pens. Another behavior we hadn't seen before is cock-fighting with the chicks actually getting a foot or two off the ground.

Besides the running/flying and the cock-fighting, the major benefits to the chicks were:
- more space to hang out in
- never running out of feed or water
- being on pasture earlier
- having a familiar place that was home for the duration of their lives
- having a large paddock to graze
- no extra handling and crating between brooder and outside pens
- having control of their daily schedule

All those add up to a more "empowered" and well-adjusted chicken. For us the advantages were even greater:
- no need to pull eight pens through a pasture
- no need to fill 5-gallon water buckets (their water now gravity flows from a tank with a valve in it to drinkers on the ground)
- never running out of pasture
- no need to shut chickens in at night or let them out in the morning
- freedom to leave the farm for 24 hours if desired
- no extra handling and crating between brooder and outside pens
- extremely easy crating and loading for processing (we park our stock trailer in their paddock, funnel them into it and crate them there or, easier yet, load them in the night by simply picking them up and putting them in crates)

In fact, the prairie schooner chicken model practically operates on its own. We found we can spend nights away from the farm if we want to. No neighbors required! Talk about feeling empowered and well-adjusted!

Okay, time for downside questions. Doesn't their feed get wet and rot? Feed gets wet, yes, and, wonder-of-wonders, the chicks prefer it that way! Well, lets face it, ground grains are bone dry. Which would you prefer, dry oats or oatmeal?

After a rainy day or night, I simply pull the feeders deeper into the paddock as always, tilting the feeder (also as always) to slide yesterday's feed closer to the schooner. New feed fills the remainder of the feed troughs and typically the chickens go for the newer feed, but not so when it's wet vs. dry. Wet feed: Not a problem.

What about overhead predators? For $40 we purchased some orchard netting to keep the hawks and owls out of the schooner area, but we never put it up. As with the 10 x 12 ft. pens, we found the predator problems were worse the closer we were to the woods. There are many ways to deal with predatory birds. My recommendation is to scare them away from the get-go. Don't let them get the first chicken. A portable radio inside the schooner is one easy idea. We found the loss to predators not significantly greater than our experiences with the movable pens. Predation: Not a problem.

What about the footprint of the schooner and all the manure build-up there? The schooner is actually a 15 ft. wide sled that is moved between batches. Because it has no floor, the manure falls between the PVC sections that make up the raft onto the ground or onto shavings. Our rule of thumb is the basic composting rule of thumb: If it smells, add more shavings.

We ended up with a 15 x 40 ft. bed of very rich material. I had good intentions of scooping that material up for the garden as we did with our previous brooder manure. Well, as these "experiments" sometimes go, I didn't do that and the experiment became: what happens to the schooner footprint if we do nothing? The recovery period seems to be about two to three years for us and you can bet those 600 sq. ft. will be fertile for a long time. Our assessment, therefore, is land damage: Not a problem.

How well does the schooner weather? Better than our portable pens did. The roof doesn't leak, there are no lids to blow away, and it is easier to weight down one hoop house than it is eight portable pens.

Our schooner has withstood 70 and 80 m.p.h. winds without any movement. The west side was stove in by the 70 mph wind and we replaced the ¾ in. PVC shoulder braces with 1 in. We put at least 500 pounds of railroad ties and rocks or cinder blocks on each end.

I knocked snow and ice off the roof the first winter, but the ice got ahead of me and sagged the roof. Once a thaw came and I got the load off, the frame popped back into shape, but I modified the tarp to be easily removed for winter storage so as to avoid that chore in the future. Weather: Not a problem.

We love multi-purpose, flexible, portable, low-cost facilities like the schooner. We found it to be an ideal sheep shade-mobile between batches of chickens. It could also double as a winter storage shed, temporary hay shed, winter home base for eggmobile layers. Your imagination is the limit.

The chicken enterprise has twice the return to investment as our beef, lamb, and pork enterprises. Plus chicken is the meat most requested by our customers. Since we added pastured poultry in 1993, the majority of our customers have been chicken customers first, then beef, lamb, and pork customers. A few years ago we nearly burned out on processing chickens, moving pens, and lugging water and feed. The schooner lifted our sagging attitudes and kept us in chickens.

In case that's not testimonial enough, two years ago Alice and I left a 500-acre (fully paid for) farm, sold 160 head of purebred cattle, and moved onto a 64-acre place of our own with only meat animals during the non-winter months. This move we did based on the strength of our natural meat business. It has been a liberation equal to learning to stockpile winter graze instead of making hay or selling our crop machinery. We are continually finding ways to make life easier and happier.

In fact, this fall our Amish butcher friend, Freeman, offered to raise some extra birds for us in his portable pens. We took it a step further and asked if he wanted our schooner—and to raise all the birds for us! It didn't take long for he and his wife to say yes.

The beauty of this arrangement is that the profit (about 60%) is large enough to divide in half and still reward us both well. Another producer who uses Freeman's butcher shop saw the arrangement and asked Freeman if he would produce birds for him also.

Now Freeman is thinking of building another schooner or two. To us it looks like new frontier just waiting for prairie chicken pioneers.

GROSS MARGIN ANALYSIS PER BIRD
2001 SCHAFER FARM

Sales	$8.75
Closing Value	0
Gross Income	$8.75
Purchases[1]	$0.63
Opening Value	0
Cost of Sales	$0.63
Gross Product	$8.12
Direct Costs	
Butcher	$1.35
Marketing[2]	$1.06
Mortality	
(Chick cost + following DCs x 10%)	$0.28
Feed	$1.70
Interest (.63 X .08% / 6 months)	$0.01
Propane ($8.00 per 500 chicks)	$0.02
Schooner (over 5 years)[3]	$0.25
Propane Hover (over 5 years)[4]	$0.05
Shavings[5]	$0.14
Total DCs	$4.86
GROSS MARGIN	$3.26

% Return = GM / (Cost of Sales + DC) = 59%

Notes

1 500 Chicks / $315.75 = $0.6315
2 Farmers Market stall, web site, KC Health Dept, KC Food Circle dues, liability, sales tax, labels, boxes, invoices, tape, phone, fuel to market = $958 / 900 chicks
3 Tarp $550, PVC $520, hardware $30 = $1100 / 5 years = $220 / yr. = $0.244 / chick for 900 chicks per year
4 Hover $243.21 / 5 yrs. = $48.64 = $.054 / chick for 900 chicks per year
5 9 bags @ $6.86 = $61.75 for 450 chicks = $0.137 per chick

CHAPTER SIX
Equipment

You don't need fancy equipment to raise poultry on pasture, but there are some things that make your days run more smoothly. We offer a few suggestions here.

Watering System Comparisons
By Jody Padgham Spring 2003

After receiving requests for a review of different watering systems, I polled several APPPA members and board members and came up with several recommendations. The overall consensus seems to be that since all situations are unique, each producer will need to try a few things before they find what is the best for their geography, labor, water and pen set up.

Robert Plamondon farms in Blodgett, Oregon. It had just snowed here in Wisconsin the April morning that I called Robert, but he already had two sets of broilers out on pasture. Robert and his wife Karen have been raising laying hens since 1996 and broilers since 1997. At the time we talked they had about 1,300 birds on the place and are gearing up for more. Robert pumps from a creek and uses automatic waterers fed from "1,000's of feet of hose." They use open pan waterers set on pallets and moved every few days for hens. This system works well, as they co-graze with goats, who love to tip over other systems. Broilers use either bell or cup waterers.

Charles and Laura Ritch farm in Alabama. They also had two batches of broilers out on pasture in early April, and have about 4,000 birds on the place overall, including about 150 layers. Charles has the advantage of having county water to tap into. He runs semi-rigid black poly pipe and garden hose to his typical Salatin type field pens. He is currently filling buckets which gravity feed to his bell-type waterers. Charles says, though, that the buckets are problematic, as they get dirty (leaves, etc.) and can stop up. He plans to directly hook his bell waterers to his poly pipe and eliminate the buckets (requiring a pressure regulator) this next season. Having tried all types of water systems, Charles uses bell waterers exclusively.

Dan Bennett and family produce about 1,5000 broilers each season, and run about 2,000 layers and 250 turkeys in Kansas. Dan pumps through semi-rigid black tubing to a float-governed pasture tank with self-feeding to the waterers in pens. Dan uses cup waterers for broilers and chicks, and a home-made open pan waterer for hens.

Tim Shell farms with his family in Mt Solon, VA. Tim concentrates on raising broiler breeding stock and producing "Pastured Peepers"—chicks bred specifically to do well on pasture. The day we talked Tim had between 1000 and 1200 broiler breeders on the farm. Tim uses exclusively nipple drinkers in all of his systems, including brooders and pasture pens. With a majority of permanent buildings, he has plumbed-in lines to direct feed the nipple system. When managed properly, Tim finds the cleanliness, convenience and lack of water overspill easily balance the time needed to learn and adjust the nipple system.

Overall tips and suggestions:
¤ Those interviewed eitheruse hose-fed systems to feed the waterers directly, or bucket-fed systems with a reservoir holding water that runs to watering units inside the pens.

¤ All agree that bird problems with water access need to be avoided at all costs. Robert recommends two waterers per pen, and, if buckets are used, separate buckets for each waterer. If one gets tipped, clogged or pulled apart, the other will work. Though ensuring constant water access, this makes pens heavier to move. (If this is burdensome, Robert recommends putting longer hoses on the waterers and setting the buckets on the ground while moving the pen.) In hot weather, Charles has family members walk past the pen line every two hours to ensure water systems are

System	Open Pan w/float valve	Bell Waterer	Nipple/Drip/Pecker	Cup Waterer
Brand names	12" Little Giant galvanized pan with float valve, or handmade 8" high pan cut from 55-gal rubber drum	Plasson Bell Brower	Ziggity Nipples Used systems	Ziggity Cups GQF Cups Little Giant Cups
Price range	$18-20 new	$10 used- $25 new	$55.00/20 bird size or much less for "build your own" or used	5 cups/$9.00 + tube, hose and hardware
Cleanliness	Can get fouled	Moderately clean	Extremely clean	Moderately clean
Water access	Easy access for birds	Consistent, easy Keep bell lip at level of bird back, with ¼" water in rim.	Can limit water access if not adjusted properly, easy for all sizes of birds when used properly.	Easy for birds to use
Problems	Frequent cleaning (2-3x / week)	Heavy to move, large size- "take up a lot of real estate"	Need to be properly adjusted. Can plug up unknowingly	Adjust to bird height. Labor intensive to clean
Overall qualities	Relatively trouble free if kept clean, hard for larger livestock to knock over, withstands freezing	Moderately easy to use and care for. Long life	Clean, but need to be used and adjusted properly for full functioning.	Good for chicks, can be a little fussy for handler
Use by interviewed producers	Plamondon hens, Bennett hens	Plamondon broilers, Ritch broilers	Shell brooders, hen houses and field pens	Plamondon broilers, Bennett chicks and broilers
Sources	Home-made, or local farm supply	www.browerequip.com; Used bells from turkey producers, www.plasson.com	Poultry and Livestock Supplies www.palsusa.com	Brower www.browerequip.com

working well. Buckets are filled as needed, as much as twice a day in the hot Alabama summer.

¤ Robert recommends keeping a complete spare water set-up so that if one goes down in a pen, you can swap it out and take the broken one to the house or shop for repair.

¤ Charles cleans his bells between each batch by running straight bleach through tubing and taking apart the moveable parts and soaking in bleach. All are rinsed well before re-assembly. Though the bells are not especially UV resistant, he is still using most after 10 years. Replacement parts can be purchased.

¤ Robert also reminds us that birds like cool water in hot weather and warm water in cold weather, and so recommends that we design systems that keep buckets shaded in the hot season.

¤ Charles has tried several systems, including a few good tries with "pecker"/nipple or drip systems. Being in the heart of commercial poultry production territory, he can get used "pecker"/nipple systems very cheap. After trying them with three batches of both chicks and broilers, Charles is convinced that the pecker/nipple system is too fussy and does not give the birds enough water. In talking to the industry people in his area, they admitted that the peckers/nipples can limit water intake, but for the cleanliness-challenged confinement industry, the cleanliness outweighed the negatives of limited water. For our pasture-based system, Charles feels the balance does not support pecker/nipple systems. "I want to be on record as being EMPHATICALLY against pecker watering systems," Charles states.

¤ Tim, on the other hand, uses exclusively nipple waterers. He notes that "each system uses different technology, needs to be managed differently, and will appeal to some and not others." Tim finds that by learning proper water pressure and height adjustment on the nipple waterers, they work extremely well in his brooders, pasture pens and layer houses, and are good for all size birds, from chicks to full-sized geese or turkeys. Tim is especially fond of nipple systems in the brooder, when dry bedding is critical. It is difficult to keep bedding dry with a bell or other open system, the advantage of the nipple system, with no overflow, is significant. "I will never use another reservoir system," Tim states. "The convenience, cleanliness and labor reduction we find with the nipple systems can't be outdone." Tim recommends one nipple for every 10 birds, and notes that it is very easy to add supplements or treatments to the water in consistent and loss-free amounts by mixing with the water intake (gravity bucket).

Turkey Bell Waterer

In summary: Dan and Robert prefer open pan systems for hens, as they are easy to maintain in a setting where the birds are moved less frequently and have more room to roam. Bell or cup waterers are preferred by Dan, Charles and Robert for broilers in a pen situation. Cups are a little fussier, with adjustments and cleaning, but are easy for the birds to use and seem to give good access to clean water. Nipple waterers are extremely clean and when properly managed are best at keeping brooder bedding dry. Good luck in making your own decisions about watering systems as you get set up for another year of production!

Scavenging for Fun and Profit
By Skip Polsun Summer 1997

Finding used equipment or materials which can be put to good use is one of the best ways to control costs and increase the profitability of your pastured poultry enterprise. It is also an excellent way to put true recycling into practice by reusing items which might otherwise end up in a landfill.

When I realized that chicken-carrying crates for moving chicks from brooder to pasture and from pasture to slaughter cost in the neighborhood of $50 each new, I became highly motivated to find used ones at a cheaper price. I didn't know where to find them, so I tried to think of how to get in contact with someone who would know.

I had recently had a conversation with the Arkansas Extension Poultry specialist, so he came to mind right away. I called him and, sure enough, he did know about a company that hauls chickens for Tyson on a contract basis. He referred me to someone else to get the specific name of the company, and two calls later I was talking to exactly the right person. The hauling company had used, slightly damaged plastic crates which they were just going to throw away. They said I could have as many as I wanted free of charge. I did have to drive one hour each way to pick them up, and I didn't know for sure what I was going to find when I got there.

Their discard piles had many crates that were beyond repair or additional use, but also quite a few that were quite satisfactory for my needs. I thanked them very much and drove home with a pickup load of crates which have made my life easier ever since. Total cost to me: three phone calls and a half day trip to go get the crates. The crates I brought home would have cost me $1,000 if I had purchased them new.

I hope my simple story is a motivation to you to be creative in finding reusable materials or equipment!

APPPA BUSINESS MEMBER PROFILE

Gillis Agricultural Systems, Inc.

Willmar Office 800-992-8986
Storm Lake Office 800-792-6828
www.gillisag.com
sales@gillisag.com

Mission statement:
"To be the premier agricultural equipment supplier in the Midwest by providing quality products and services to the poultry and livestock industries."

Bob Gillis, one of four brothers that own and manage Gillis Agricultural Systems, Inc., was kind enough to accept my last-minute request for an interview. The company, an APPPA member since 2003, operates two offices in Minnesota and Iowa that provide a wide variety of pastured poultry equipment including feeders, waterers, netting, incubators, brooders, processing equipment, and many other supplies. Gillis has been serving Minnesota, Iowa and adjoining states for three and a half decades, and now also offers products through its catalogs and website.

Francis Gillis, father of present owners, established the business in 1971 in Willmar, Minnesota, to serve the poultry industry in the Upper Midwest. Soon the business expanded into servicing dairy and hog producers. In 1975, a second office was opened to provide more local service to poultry and hog customers in Iowa. By 1977, four Gillis brothers had joined their father in business, each occupying different positions in sales and management. After 45 years of involvement in poultry and livestock equipment, Francis Gillis retired in 1990 and the four brothers became co-owners.

Later that decade, the company developed its website and increased its emphasis in catalog sales to a national market of both hog and poultry producers. In 2002, the first catalog for free-range and gamebird products was produced for national markets.

Gillis strives to offer all the latest quality innovations in equipment for birds on pasture. The company offers a free catalog, its website, www.gillisag.com, offers many web pages full of equipment for every stage of poultry production. Gillis' 22 employees work hard to back the products with outstanding service. And after 35 years in the poultry equipment business and the combined experiences of family and staff, they have plenty of expertise to share!

Reefer Madness
By Scott Jondle Winter 2003

The Scott Jondle family just completed our third season raising pastured poultry in western Oregon on our 210-acre Abundant Life Farm, just west of Salem. This year, we raised 2200 broilers in 11 batches and 150 turkeys. Until this year, one of the challenges we faced was how to cool our processed broilers and turkeys until they are either picked up by a customer or are put into one of our eight upright or chest freezers for storage until they are sold. At first, we tried processing in the morning and selling in the afternoon. But when we got up to 200 broilers, we simply couldn't finish soon enough to get cleaned up before customers started arriving. So we decided to process the day before customer pickup, which greatly lowered the stress level, but added the complication of cold storage for these 200 broilers. We tried keeping them in the initial ice water cool down vats overnight, but soon concluded that was fraught with risk. In addition, early on we discovered, through a near disaster, that stuffing a bunch of broilers in our three-door refrigerated cooler (normally used to store our eggs) wouldn't work either, especially on a hot summer day.

Refrigerated truck could be used to store chickens

By this time, you're probably asking why not get a walk-in cooler. That certainly is an option, but I think we ended up with a better solution. A seed of an idea was planted when I talked to a custom meat slaughter owner who sold meat next to me at the Salem Farmers Market. She told me about the refrigerated trailer (often called 'reefers'--the kind pulled behind Peterbuilts passing you at 75 mph on the freeway) they purchased to store meat when their walk-in was full. What intrigued me the most was that this trailer could be set for any temperature between –10 to 70° F, instantly going from heating to cooling to freezing with only a twist of the temperature dial. Talk about a multi-use unit. In addition to this remarkable flexibility, it can be placed anywhere on your farm, and can be moved if you later decide the initial placement wasn't ideal. Because it's not a permanent structure, no permits, inspections, or other bureaucratic nonsense is required. And because it runs off a diesel engine, you don't need an electrical hookup (more about that later). That sounds great, you're thinking, but what about price? A new reefer costs $70 thousand, and that doesn't fit into many farmers' budget.

I was pleasantly surprised to learn that a good condition used reefer is very affordable.

To start my search process, I looked in the Portland Yellow Pages under Trailer-Truck and found there were many dealers selling used reefers. After making several calls, I found that reefers range in length from about 25 to 53 ft. At first, I was asking about the smaller reefers (25 by 8 ft. will store a lot of broilers), but soon discovered that the smaller sizes are in more demand and thus command a higher price and are harder to come by. So, being price sensitive, I shifted my search to the larger sizes. I soon had a list of three or four dealers that had trailers that sounded worth looking at. A few days later, I spent a day looking at reefers in Portland and found a 48-footer with a Thermo King refrigerating unit at the largest trailer dealer in the area. This trailer was a 1992 model, probably a trade-in, looked brand new, had low hours (7,700) and was being heavily discounted because it had been sitting on the lot for too many months. It was too long for local deliveries, and not quite long enough and too heavy (with two side doors) for long haul operators. The cost—$3,900 plus $125 to transport it 75 miles to our farm. I don't think this is a unique bargain. I suspect checking out similar dealers in any large city may yield similar results. It was delivered before the start of this year's broiler season and was set up about 25 ft. from our processing area. Since the floor of the reefer sits about four feet off the ground, I had a carpenter friend build a 6 ft. wide deck and stairway along the side of the trailer for access. This deck is wide enough that we can sell our broilers from it. I also had it made so the railing across from the side door is hinged so it can be opened to allow my tractor with a front-end loader to unload broilers, which are placed 25 to a container in food grade, plastic, 55-gal drums that I cut in half and put wood/rope handles on. This saves the agony of having to wrestle these up the stairs.

Here are some features that make this reefer ideal.

1) It is a multi-temp model, which means it has two separate temperature controls and two separate cooling units, one in the front and one near the back, running off the same compressor. This is a fairly common option, I believe. By using an insulated bulk head divider (which I picked up free from another dealer that wanted to get rid of it) that can be placed anywhere inside the trailer to divide it into two compartments, you can set the forward compartment to freeze, and the rear compartment to cool–at the same time. At least one side door is almost mandatory, since it is much easier to open then the large rear door and lets much less cold air escape when opened. Get two side doors if you want to freeze and cool at the same time.

2) Probably the most important feature is the auto on/off option. This option turns off the diesel engine when the set temperature is reached and automatically turns it back on when needed. This feature will save you many gallons of diesel fuel during the course of a season. When running in the continuous mode (where the diesel engine is always running), I've been told it will burn about 0.4 gals/hour. In the auto on/off mode, the 50-gal diesel tank will probably last most of the season.

3) The refrigerating units have a life expectancy of around 26,000 hours. If you can find one with at least half its life left, you are in good shape.

4) When turned on, a reefer puts out a huge blast of cold air that quickly gets down to cooling or freezing temperatures. Even in the dog days of summer, it didn't take more than 15 minutes to get down to 33° F, so you don't need to turn on the unit until you are ready to fill it with meat. And several hundred broilers can't even begin to tax its cooling capability. I should add that I am cooling only the first 15 ft. of the reefer because of where I placed the insulted bulkhead divider, so that helps get down to temperature fast.

5) Having freezing capacity completely independent of the electric grid is a comforting thought. If we should ever experience a prolonged power outage, we could transfer all our cold storage meat into the reefer, and have enough room left over to store all our neighbors frozen stuff as well.

6) Another option that some people will find handy is electrical standby. This allows you to plug the cooling unit into a 220 V power source and run it without using the diesel engine. There are two drawbacks. These reefers cost more and they require three-phase power. Since I don't have three-phase power close to where I parked it, I decided not to get this option. (If you don't have three-phase power, converters are available, but rather expensive). The advantage is cheaper operating costs, especially at the current cost of diesel fuel. If you have a need to run a reefer for prolonged periods of time, this option would definitely be worth looking into.

Here is how we use our reefer. After we have cleaned up from processing, we remove the broilers from the ice water cool down vats and place them in the plastic half drums mentioned earlier. We place a wet cloth over the top of them to keep them moist. Using the miracle of hydraulics, my front-end loader lifts three drums at a time onto the reefer deck. The drums are placed on pallets inside the reefer to insure adequate cooling air underneath. I set the temperature control dial to 33° F, place the mode switch in auto on/off, and flip the master switch to on. After about a 15 second preheat, the engines roars to life. That's all there is to it. The next morning, we set up a table on the deck, take out each drum one at a time, weigh and bag each broiler, and place them back in the reefer. Our customers come for pickup that afternoon. At the end of the day, the temperature control is set to 0 degrees and the unsold broilers now begin their freeze down. After a couple days, the rock hard broilers are then transferred to one of our electrical freezers and the reefer is turned off until our next batch is processed three weeks later. We used this same procedure for our turkeys. Try finding cool storage for 150 turkeys!

Using the reefer only three days every three weeks minimizes fuel usage and keeps operating costs to a minimum. I put a little over 200 hours on the engine this year; at this rate it should still be running when our great grandkids are running the farm.

We also raise beef, pork, and lamb. When these are ready to pick up from the custom slaughterhouse, we bring the frozen meat back and put it in our reefer (set to 0 degrees) to store until our customers come to pick it up. We like to maximize the opportunities our customers have to come to our farm, because they can usually find other meat to buy as well. And, this procedure eliminates the customers having to deal with sometimes surly slaughterhouse personnel. And we get paid without having to chase after it.

By now, you may be thinking this sounds to good to be true. So, what are the drawbacks? Well, like any piece of machinery with a motor, sooner or later you will need some service work done. I've gotten to know rather well the independent mobile reefer mechanic, who was recommended by the dealer. He's been out four times, once to do a routine service check after I first bought it, once to replace a leaking water pump and worn fan belt, and twice to chase down an intermittent electrical short. Having watched him work, I now feel confident enough to do the yearly service stuff myself (change oil and filters). On the positive side, he turned out to be a great guy and has become one of our best customers.

In summary, we have found our reefer to be a very cost effective way to give us the periodic, large cooling/freezing capacity we require to successfully raise and market pastured poultry.

CHAPTER SEVEN
Eggs on Pasture

You haven't had a good egg till you've had one from a bird that has lived on pasture. Bright, firm yolks and clear, runny whites indicate fresh, nutritious food that chefs and consumers will beg for. Some parts of raising hens are similar to that of broilers (life in the brooder, for example). But once they mature and are ready to lay eggs (at around 20 weeks), hen management is very different from a broiler system.

What About Eggs?
By Joel Salatin, Fall Winter 1997

When the book *Pastured Poultry Profit$* came out we were only producing eggs in the "Eggmobile" and the "Raken House." The "Eggmobile" is a portable henhouse that we move a couple of days behind the cattle to sanitize the paddocks. The chickens range free from the trailer, eating grass, bugs and especially fly maggots in cowpies, spreading out the pasture droppings in the process.

The "Raken House" is a combination chicken house/rabbitry, which was originally built to house our son Daniel's rabbit operation. A loose-housed laying hen facility, it has two tiers of production: laying hens on the floor and rabbits in cages at eye level. The chickens scratch through the rabbit droppings and aerate the bedding, producing wonderful compost and a delightful woodsy odor.

About three years ago, we added a commercial-sized pasture egg operation to the farm when we started the apprenticeship program. After all, we needed something for the apprentices to do! Actually, we needed an enterprise that would cash flow the added expense of teaching, housing and feeding a couple of young men.

We built a fleet—29 to be exact—of field pens for the pastured eggs that were almost identical to the broiler pens. We put an 18 in. wide hinged door on the top of the pen, all the way across the 10 ft. dimension, that would provide access to 10 nest boxes tucked in the top corner of the capped end. Although the door creates a crack in the roof, any rainwater falls in front of the 12 in. deep nestboxes and the layers do not mind.

The nest boxes have a 6 in. front board to keep the birds from scratching out nest material and a 6 in. partition on 12 in. centers to individualize the nests. Without partitions, the birds all lay in one spot and tend to break the eggs. A perch board in front drops down in the morning and comes up in the evening to prevent the birds from roosting in the boxes and soiling them at night.

We had to build a dolly with shorter prongs to move these pens because the nest bottoms are only 12 in. off the ground. The bottoms are simply pieces of corrugated aluminum or sheet metal—plywood is too heavy.

It take as long to build in the boxes as it does to frame the entire pen. The boxes add significantly to the weight of the pen. Anything to lighten them—using plastic boxes or whatever—can help.

We put 50 birds in a pen and shoot for at least 36 eggs per day. That is fairly easy during the first year of production, but is impossible the second year. We like to keep the birds two years because one of the biggest hurdles of the egg enterprise is dealing with the spent hens. We are actively cultivating a market for this new product—stewing hens hardly exist now because the industry so debilitates the birds that in 10 months producers must pay someone to come and haul the spent hens away. Ninety percent of second-year eggs are jumbos.

Anyone over 50 remembers "chicken and dumplings" and eagerly tries one of these birds. But younger folks have never heard of this critter. As a result, we're in the "giving samples" stage of our marketing on this one, but the results have been extremely good. Our goal is for outgoing flocks to pay for the incoming ones.

More Pastured Egg Information
by Skip Polson Summer 1998

When I make presentations about pastured poultry there are usually questions from the audience about producing eggs on pasture. The level of potential interest in egg enterprises seems pretty high. Therefore, I want to share what I have learned with other APPPA members. I hope you will find this useful.

My family has kept laying hens on pasture for the last two years. The facts summarized here are derived from the pastured layer enterprise on our 25-acre farm in central Arkansas during calendar year 1997.

¤ Number of mature (laying) hens: 80 for the whole year (these began laying in the fall of 1996); plus 110 for October through December; this is equivalent to having 107.5 hens for the whole year.

¤ I raise my layer pullets just like I do the broilers: two to three weeks in the brooder, then out on pasture in a pen moved once every day for the rest of their lives.

¤ We don't normally have snow cover here, therefore I am able to keep them out in the pasture pens through the winter. I do not have any way to provide artificial light to influence their laying rate during the winter. They go to sleep when it gets dark, wake up when it gets light, and most of the time I wish I did the same!

¤ For the past two years I have used Production Reds (this is what the hatchery calls them). I do not know their precise genetic makeup. These birds look like Rhode Island Reds but they are not nearly as heavy. They have a calm disposition and they are much smarter than the Cornish Cross broilers. They learn quickly to move with the pen; and they aggressively scratch and peck and hunt for bugs.

¤ They consumed 31 pounds of feed per bird to reach laying age: 18% protein, non-medicated grower crumbles commercially available at my local feed store. The price was 13 cents per pound during 1997.

¤ When they begin laying I switch their feed to 16% protein, non-medicated lay pellets (also from the feed store). The price was 12 cents per pound during 1997.

¤ The mature birds consumed 0.53 pounds of lay pellets per bird per day averaged over the whole year (194 pounds per hen per year).

¤ They laid 232 eggs per hen per year. Not too bad really, but probably a little less than the published averages for the breed. I think it's wise to be conservative in estimating how many eggs you think you'll get, because I'm sure the published figures are based on commercial operations where all the conditions are highly controlled and artificial light is provided. I think even an outstanding manager is unlikely to achieve the published figures when keeping their birds on pasture. The birds are unquestionably healthier and happier, but they won't lay quite as many eggs.

¤ All my current pasture pens are wood framed, essentially the same design as Salatin's except that I make mine smaller (8 x 8 ft.) so they will be easier to move and handle. My layer pens are just like broiler pens with these exceptions:

    a. They have nest boxes inside (six used [broken] plastic milk cases scrounged from a local dairy for $1 each).

    b. Their construction requires three additional pieces of lumber (1 x 6 in. by 8 ft.) ripped in thirds. This lumber is used to make roosts and frame the lid for access to the nest boxes.

    c. One pair of 2 in. hinges is used for the lid to access the nest boxes.

¤ I spend one hour of labor per day on these hens. This includes the daily pasture chores, gathering eggs, washing them, boxing them, and delivering them to the stores in Little Rock, AR, (a quick side trip once a week on my way to work).

If you use these facts to calculate my costs, and a selling price of $1.50 per dozen for the eggs to calculate the gross income, you will see that I essentially recouped my out-of-pocket costs for the year and didn't get paid at all for my labor.

Not a good business arrangement!

EXPENSES DURING 1997:
Pullet chicks	$ 110
Feed for growing pullets	443
Feed for laying hens	2,503
Total	$3,056

INCOME DURING 1997:
2,080 dozen eggs @ $1.50 $3,120

NET: $ 64

What can be done about that? Several things: raise the selling price of the eggs (I've done that, they're now going for $2 per dozen), find a way to cut feed costs, study your chore routine to make sure you're being as efficient as possible with your labor, consider ways to provide some artificial light in the winter to increase egg production, cut other costs (like egg cartons) to the bone, and consider keeping the birds in an eggmobile (or other extensive pasturing arrangement) instead of the pasture pens. If they're allowed to forage more, they will consume less purchased feed, thus reducing the largest cost item of this enterprise. Keeping more birds all together in an eggmobile (or some other system) also allows you to care for more birds with less labor each day. But I don't have enough space for an eggmobile, and lots of other people don't either.

¤ A key consideration for any extensive grazing system for layers is how to keep them rotating and grazing where you want them, so manure will be spread and so that overgrazing and denuding the ground will be avoided. This is the primary failing of most extensive poultry systems.

¤ The manure the hens put on the field is definitely a plus. My pastures are lusher and have been improved because the birds are there. My sheep benefit from the improved grazing. I haven't tried to put a value on that, but I know it's real. My sheep are better off, and feed costs for my sheep enterprise are less than they would be if I didn't have the laying hens and broilers out on the pastures.

¤ I totally agree with Joel Salatin that the broilers definitely pay better than the layers. Exactly how much depends on all the cost, revenue and labor factors involved in each specific enterprise. To make any money at all on pastured layers you must constantly keep costs and labor as low as possible, and egg production and revenue as high as possible. That is obvious economic and business sense. Each person in this business has to look for the ways they can make their costs exceptionally lower and their revenues exceptionally higher than everyone else in the business.

Wisconsin Eggmobile

¤ The quality of eggs produced by our hens on pasture is high. Our customers like them a lot. As with all unconventional products, however, you have to be prepared to tell people how these eggs are different than the industry's eggs. You have to set your price high enough to give you the return you desire, and you have to do that without apologizing. And we all have to keep learning how to produce eggs efficiently on pasture so that our selling prices do not need to be unnecessarily high. There is still a lot to learn, and there is a lot of room for additional innovation.

Hoophouses for Hens
by Joel Salatin   Fall 1998

Several years ago when we added nearly 2,000 laying hens to our pastured poultry enterprise we faced the challenge of winter housing. The six-month on, six-month off seasonal broiler production model suddenly became woefully deficient in a year-round egg production model.

I had never set foot in what I considered a really acceptable stationary confined poultry house. But leaving the birds out in pens all winter was equally unacceptable. Although we knew the birds would survive, the logistics of wading through three-foot snow drifts with egg baskets and water buckets did not seem like an exhilarating way to farm.

Hoophouses came to our rescue. With each passing year, we are more confident that these are, if not the best, at least the best solution we know about.

Chickens only lay eggs after all other energy needs are met. If they use all their caloric intake to maintain body temperature, they do not have enough energy to produce eggs. If a house is tight enough to accomplish this with the chicken's body heat, it lacks ventilation. Supplemental heat sources like wood, petroleum or electric heaters are too expensive.

Hoophouses use solar energy, and even though they cool down at night, the chickens snuggle together to stay warm and the cold does not affect them too much. You and I can handle a cold bedroom as long as we can snuggle down under plenty of blankets. But in the day, when we are moving around working, eating and relaxing, if we can't get warm we are soon miserable. Surrounding, or ambient temperature, is more critical during our production time.

Hoophouse kits are available from a number of suppliers. We purchased one and the pieces came in on a commercial truckline. We opted for the cheapest per-square-foot model that was still high enough to back a tractor or truck under. This model is 20 ft. wide and nearly 12 ft. tall at the apex. The hoops slip into 3 ft. columns anchored in the ground.

We knew the house would need deep bedding for proper manure management. That required hefty sideboards because otherwise the bedding would just push out underneath the plastic. We opted to split locust posts and nail them on upright stakes so we could have 18 in. of bedding in the facility. Then we went along just outside this containment wall and pounded in the 3 ft. pipes, called columns.

In the instructions, these columns are supposed to be in concrete, but we just pounded them into the ground with a sledgehammer. In order not to pean over the top, we made a sock to pound on out of bigger pipe and a heavy piece of metal. A friend who duplicated our efforts used a pipe ductaped to a piece of wood. A minute of dressing with a round file readies the column to receive the hoop pieces.

We use webbed nine-mil. poly for the cover because it is tough. Although it is more expensive than regular hoophouse plastic, it can last up to 15 years. It is the same material used along the open edge of confinement poultry houses. It rolls up or down depending on ventilation requirements. Using self-drilling screws, we attach a rot-resistant wooden firring strip at ground level. Then we pull the plastic to another piece, roll it up tight, and drywall screw the wrapped piece to the stationary piece. It's poor-boy, but works quite well.

Because the chickens can easily scratch the plastic and rip holes in it, we fasten a firring strip about 6 ft. off the floor along the inside of the hoops. We attach 6 ft. x 1 in. poultry netting to the top of the bedding containment wall and this wooden strip, keeping the chickens from being able to access the plastic.

Because chickens move and plants do not, these hoophouses need both ventilation and a physical barrier. As a result, we put in four locust poles at the ends of the hoophouse: two are 12 ft. apart for the door and two are 6 ft. beyond those in either direction to hold a sliding door rail. The sliding door allows us to regulate the ventilation from fully open to cracked, one end of the house or both ends.

Swinging poultry netting doors, mounted with conventional strap hinges and pins on the same inside locust poles, keep the chickens in when the sliding doors are open. To keep the bedding in we just place a couple of boards across the 12 ft. opening. When we clean out with the front-end loader we just pull off those retainer boards and drive in. Ditto for putting bedding in.

Buying everything brand new except the locust poles, these structures cost us $1 per sq. ft. to build and cover. We orient them so that the prevailing winds hit the end, thus stimulating ventilation on warmer days. We use regular hanging waterers, nest boxes and trough feeders. Even though the waterers freeze on exceptionally cold nights, as soon as the sun comes up they thaw and begin to work.

If we are in a period of single digit temperatures or below zero° F, we keep a couple of those indestructible black rubber pans to give the birds water until the sun warms things up enough. A frost-free hydrant in each house, connected to the farm well, gives plenty of water.

Chickens do not like things damp, so I suggest making sure that any runoff from the hoophouses can make a quick exit away from the house and not go into it. We have two hoophouses 20 x 120 ft. The houses are about 3 ft. apart. We've had snow piled up level to the top of the houses in this crack, and that is a lot of water when it melts.

In houses birds drop just as much manure as they do when out on pasture. Ideally the birds will stir their carbonaceous bedding faster than they cover it with manure: this stirring encourages slow decomposition and ties down ammonia. We have found that at 3 sq. ft. per bird the manure load is heavier than the birds can incorporate, causing a capping of manure on top of the bedding. Once the bedding is capped, the chickens do not scratch through it.

At 5 sq. ft. per bird the manure load is light enough that the birds have no problem keeping the bedding clean with stirring. At 4 sq. ft. some areas may cap and others will not. Capping can be eliminated with something as simple as a mattock or garden hoe or as elaborate as a garden tiller (do it at night when the birds are sleeping).

Another technique we've used is pigs under 100 pounds in 6 x 10 x 3 ft. portable pens that we just walk through the house. The pigs aggressively root up the bedding too tough for the chickens to till. Use only small pigs; large pigs enjoy chicken dinners too much. Two pigs per pen seem to work well to stir up the bedding without recompacting it. The chickens enjoy roosting on the pen and eating earthworms the pigs dig up.

The raw material for the bedding can be anything carbonaceous, but I would stay away from anything with long fibers like old hay or straw. The best is sawdust, although woodchips work well as do shredded leaves. Anything fluffy and friable works fine.

To stimulate the chickens to stir the bedding, we fling whole wheat on the floor each morning. The chickens enjoy the treat (about 20 pounds per 300 birds) and invariably they miss some that ends up in the bedding. Some of these kernels actually sprout in the bedding, offering sprouted grain to the birds when the pigs expose it.

Clean bedding is the key to clean eggs. Keeping the nestboxes well away from waterers gives the birds plenty of time to clean off their feet before entering nests. Our birds seem perfectly content to sleep on the floor since that's the way they sleep in the field. Certainly roosts could be provided, although that would tend to concentrate manure. We keep sawdust or bags of peanut hulls close by to spot treat manure concentration areas.

We find that it takes about 50 days for the carotene from the pasture to wear off in the birds' bodies. This makes yolk color suffer the last 50 days of the housing period. We fortify the ration with about two percent alfalfa meal to push more greens into the birds. Of course, production is never as high during the winter as it is at other times during the year so we need not worry about sales pitches to new prospects during this lower quality time.

We did try some marigold petal dust from Mexico one winter. Even though it did darken the yolks, the color appeared brownish rather than the more natural orange. Our customers recognize that the color will change and simply accept it. One of our gourmet chefs says that when he was in chef school in Switzerland, they had seasonal specialties to capitalize on the color and handling quality differences of the eggs as they changed throughout the year. He tells me: "They talked about April eggs and May eggs and June eggs, and adjusted the recipes accordingly."

Certainly large scale sprouting is something we need to be looking at. But since we haven't done it I won't act like my current ideas will work. We'll have to wait and see.

Fortunately in the spring it only takes about seven days for the pasture yolk color to come back up in the eggs once the birds hit tender spring grass. At the first sight of green, the birds can go out. The early grass seems almost like a tonic to them and is by far the most palatable forage they encounter for the whole season.

After the birds go out, we plant summer vegetables like tomatoes, squash and corn in the hoophouses. This gives additional cash from the hoophouses and allows us to get a jump on outside-planted varieties, insuring a premium price.

We never clean the houses down to the dirt floor, but skim off only what is necessary to keep the bedding below the retaining wall. A midpoint division in each house gives us four sections. Not only does this allow us to keep flocks separate, but it also reduces the number of birds in any one group. This reduces stress and increases production.

The hoophouses keep the birds warm and productive. Even when the temperature is below 0° F we have no problem maintaining 70° F or better. The plastic allows maximum light, further stimulating production. The direct sunlight keeps the bedding dry and reduces pathogens. Plants double hoophouse usefulness and the entire structure, at $1 per sq. ft., is cheaper than any other weather-proof building, even if all you want to do is store machinery.

Hoophouses properly modified for chickens offer a cost efficient, and enjoyable winter housing model. A retired neighbor came over one day after we got birds in the houses and stood one Sunday afternoon with his wife for at least an hour, just watching the birds and enjoying the sunlit ambiance of the facility. All he could do was shake his head and recount the smelly, dusty, dank houses of his youth. He was visibly entranced by the notion of chickens in a hoophouse and made me appreciate the discovery even more.

The hoophouse model is so appealingly comfortable on cold winter days I think I'll move my office out there. I wonder what typing method chickens use on a computer keyboard: hunt and peck?

Making Eggs

Marketing Pastured Eggs
By Joel Salatin Summer 1998

We sell about 30,000 dozen eggs per year for $1.60 per dozen from Polyface Farm. About 75% go to 20 restaurants and the balance go to our farm patrons. In our area, brown supermarket eggs sell for $1.29 in the winter and 99 cents in the summer.

Standard commercial white eggs sell for about 60 cents a dozen. People always ask; "How do you get restaurants to pay $2.00 per dozen (includes delivery) or more for your eggs?"

When we first went marketing we had no brochure and no relationship with any chefs, but we had samples. We called five white tablecloth restaurants in Charlottesville, VA, about 50 miles away, and set up appointments one hour apart.

Here is how the phone conversation went:
"Hello, XYZ Restaurant."
"Hello, this is Joel Salatin from Polyface, the farm of many faces. May I please talk to the chef?"
Pause.
"Hello, this is Soufflé Escargot speaking." This is usually in a BIG hurry- chefs are notoriously busy.
"Hello, this is Joel Salatin from Polyface, the farm of many faces. I produce the world's best egg, and I'd be honored to show it to you. I'll be in your area tomorrow; would it suit you if I stopped by for just a minute about 3 p.m.?"
"Yes, that would be fine. Bye."

Notice that I didn't give any sales pitch. I showed respect for his time by not yakking unnecessarily. I also did not let him give me a "no" on a presentation. After we lined up these appointments, we took a sample dozen for each restaurant and a couple cases to be prepared for any immediate sales. Every restaurant we went to was buying commercial white eggs for 60 cents a dozen from a volume vendor. From a traditional marketing standpoint, what we were doing was suicide: asking the customer to pay triple-and-a-half the price, plus change vendors. At the end of the day, we had three out of five as new customers. That is an astounding batting average. The other two drooled over the eggs, but backed off on the price.

The following are things the chefs have taught us. You may want to incorporate these into your marketing techniques.

At the first restaurant the chef-owner said almost nothing. He just took the sample dozen out of my hand, grabbed an egg and cracked it directly into a saucepan of boiling water on the stove. Immediately his expression changed from ho-hum to wonderment. The egg cooked nicely for about 30 seconds and he then scooped it out with a slotted spoon. By this time he was downright animated.

He placed it gently on a saucer and began stroking it like a child would stroke a chicken. Daniel, my son, and I were taking this all in with appropriate awe. Finally I couldn't stand it any longer and asked: "What's the big deal?"

The chef spun around, grabbed a regular white industrial egg and broke it into a saucepan. It immediately exploded into a thousand pieces. In a few seconds, the white resembled pieces of Styrofoam packing and the pale yoke floated by itself. Imagine my surprise when he said to me: "One of our centerpiece specialties is a water-poached egg for Sunday brunch. We are having a staff meeting tomorrow morning to discuss discontinuing it because we can't get the eggs to hold together. Do you have any more with you?"

He promptly bought a case for $2.00 a dozen and has been one of our most loyal customers ever since. He religiously takes 60 dozen a week. It is fun to market GOOD stuff.

Do you want to get the attention of a pastry chef? Here's how. Go in and ask for a saucer. Break an egg into it and then reach down and pull the yoke right up out of the albumen. The chef's eyes will fall out onto the floor. Then go ahead and start gently throwing the yolk from hand to hand, perhaps a dozen times, as you talk. By this time the chef thinks you are practicing magic—honestly!

As it turns out, pastry chefs have begun shunning recipes that call for yolk and white separation because many of the yolks break during the separation. To think that an egg can take the kind of treatment you just put it through is nothing short of miraculous. The whites whip up 30% more than conventional eggs, making spectacular meringues.

One of our pastry chefs in Washington D.C. makes a six-inch cake, each batch yielding four. When she switched to the pastured egg, the same recipe and same number of eggs made five instead of four cakes. She pays $2.40 per dozen for our eggs and says she can make more money with these than with 60 cent eggs. How about that?

She also says that her marketability window doubles from 36 hours to 72 hours because the pastries stay moist and hold up longer. Imagine what this means in terms of lost product and salvage prices.

One chef explained to us why our eggs were more economical than conventional ones when used on a griddle for breakfast. The conventional eggs are several weeks old and each flat has at least one cracked one. This cracked one leaks onto its neighbors and over a few days, this sticky white makes the other eggs adhere to the cardboard flat. Trying to wiggle these affixed eggs loose generally causes a couple more cracked catastrophes and additional mess.

Of the ones that the chef can extract intact, roughly 50% rupture their yolks when broken onto the griddle. They must be scraped off and discarded. Then the chef starts over. In the kitchen, lost time, frustration and lost product can all be eliminated with pastured eggs. Over the years, we have routinely given five-dozen or more eggs to a breakfast chef to try so they can see these differences themselves.

One chef broke out in rashes every time she touched an egg. She said it was due to the toxins in the eggs. Our subcontractor delivery fellow kept encouraging her to try going gloveless sometime when she handled our eggs. Finally one week he went in and she met him at the door, hands upraised in triumph, and gloveless: "Look! No Gloves!" she announced.

Remember that chefs are food sculptors, artists, craftsmen. I am convinced that the reason we've been so successful with chefs is because they handle more numbers of eggs and they do the same recipe over and over, thereby noticing differences more quickly than the average homemaker. When we first began selling at the local farmers market, about five vendors had eggs at 80 cents a dozen and ours were $1.50. Here was competition just a few feet away. We certainly did not want to alienate our farmer friends, but we did want to differentiate ourselves from everyone else.

We went to the supermarket and purchased both generic eggs and the high-priced "healthy" ones. At the farmers market--or any place you want to impress homemakers with the difference, like garden clubs or food fairs--we broke out three eggs in saucers and labeled them with white index cards. People would come by and ask what we were trying to show.

At first, I began explaining omega-3 and omega-6 fatty acids and colorful carotenes, but a little science goes a long way--too long. The customer's eyes would glaze over and that was the extent of the conversation. We've now perfected the inquiry to a Socratic method by asking the customer questions:
Customer: "What are you showing?"
Me: "What do you see?"
Customer: "Well, this egg is real tight and has a dark colored yolk. Those two look a little spread out and pale."
Me: "What is the first sign that your child isn't feeling well?"
Customer: "Pale color."
Me: "Well?"
Customer: "Oh. I'll take two dozen."

This little interchange takes about 10 seconds and the customer self-discovers the answers, makes the right deductions and understands. It's simple and unthreatening. Once these folks try the eggs, they come back and tell us that their children never liked eggs until they started serving ours. I've heard that oodles of times, but I never get tired of it. As for the "spread out" part, you can do with that whatever the decorum of the situation allows.

The bottom line is this: pastured eggs have superior taste, structure, handling qualities and beauty. Let the eggs sell themselves by making simple comparisons even a child can see and understand. As the industrial model deteriorates, the pastured egg alternative becomes easier and easier to market because the distinctiveness is more obvious. Pastured eggs are a great enterprise for children. But whether we're big folks or small folks, pastured eggs offer great dining and marketing experiences.

Eggs on Pasture? A Kansas Story
By Jody Padgham Summer 2003

There are not a lot of people in the country doing large-scale pastured egg layers, but Dan Bennett and family have found layers to be a satisfying way to augment their pastured broiler and beef business. Dan spoke December 2002 at the ACRES USA conference in Indianapolis, IN, about their egg laying operation.

During the 2002 season, the Bennetts ran 2,000 laying hens and produced 14,000 broilers and 24 beef steers on 50 of their 210 acres in rural Kansas. They direct market eggs, beef, pork and chicken to neighbors and stores and food co-ops in nearby cities (Lawrence, Kansas City). (For more on the overall Bennett operation, see the Bennett "Producer Profile" on page 206.)

The 50-acre pasture, gently rolling as only Kansas can be, is divided into five-300 ft. wide paddocks with a permanent single wire electric fence. 400-500 lb yearling beef calves are bought in the spring, and rotationally grazed through the paddocks to enjoy and condition the grass ahead of the chickens. At times Dan has to mow the early season grass, as the cows can't keep up with the spring flush in all the paddocks. Broilers in pasture pens and hens in hoop houses with a day-range system follow the beef.

600-day old pullets are brought into brooders twice a year. Pullets (young laying hens) must grow out six months before they lay; farmers must adjust for this lag time in their production plans. Dan broods all his birds on the farm and insists that bringing in older birds spells disaster, as you run the risk of bringing disease in with mature birds, generally not a problem with day-old chicks. The chicks stay in the brooder, with clean dry bedding, fresh food and water, for three to four weeks. Dan feeds chicks supplements of grass, sand and beef liver to ensure health and stamina in the birds.

Twelve by twenty four foot temporary garages ("Cover-it" brand) are used as pasture hoophouses for the laying flock. With one end closed and the other left open, 500 birds use the hoop and the pasture around the house created by 450 ft. of 'feathernet' electric poultry netting. In the summer the houses and netting are moved every week so that the birds get regular access to new grass.

Even in the rough Kansas winter (windy and chilly) the birds stay out—but the pens are pegged down in one place for four months. For winter pasture, a ring

of electrified net is run around each house, creating a 'staging area' that must be covered with bedding as the grass is soon worn away by bird traffic. From this staging area, a portion of the fence is lifted so the birds can access a "petal" of netting, which radiates out at an angle from the house. Planning on eight "petals," Dan moves the fence to access new pasture every week, making two rounds around each house during the four-month winter season. In winter the hoophouse is bedded with wood chips, with new chips added each week. By the end of the winter the pack is about a foot thick. Frost-free hydrants are installed next to each of the four winter house locations to give birds access to water. In addition to trenching in the water line, electricity was also installed at each house location which allows electric heaters in each waterer to keep the ice away.

Nest boxes in the hoop houses are a 1940s vintage hanging roll-out nest. 56 nests are placed for 500 hens, with four units of 14 hung from the top of the hoophouse. Family members go into the houses every afternoon to collect eggs from the roll-out. The Bennetts collect directly into cartons, and candle the eggs right through the cartons. They have found that the roll-out nests keep the eggs clean, and reduce egg-pecking and other problems found with bedded nests. They do end up washing a few eggs each day and do so by layering eggs in a wire egg basket and soaking in hot water with a little soap. Eggs are then hand-dried. Bulk eggs are packed by flats into 15 dozen boxes. New cartons are purchased in bulk quantities from www.eggcartons.com.

Dan feeds his hens a layer mix utilizing the Fertrell Poultry Nutribalancer (allowed for certified organic systems). He also offers free-choice oystershell and kelp. Dan takes feed out to 175 lb. capacity range feeders he got from Brower, Inc., using an old auger box on the back of his truck to haul from the bulk storage to the field feeders every day. During the summer months, an above ground water system of permanent black pipe with hookups every 200 ft. fills rubber tubs with clattel float valves.

Unlike many others, Dan does not use supplemental lighting in the winter. The white hoop house walls let in a lot of natural winter lighting, and Dan also likes to give the hens a rest in the winter. Production peaks in the summer at about 110 dozen a day, but can drop to a low of 40 dozen a day in the winter for the 2,000 hens. They average about a 50% lay rate over the year.

On any pasture operation, disease problems almost always originate in the brooder. By maintaining extremely careful and clean brooder management, the Bennetts have very few losses. Raw cow's milk is used to strengthen the chicks if any disease problems erupt. Predators are well-controlled by the electrified poultry netting, which also must be carefully monitored and maintained. Dan brings in 1,200 new chicks each year, (which costs $1,300-1,400) and culls birds after two years. Spent hens can be sold for stewing hens in some ethnic communities for $2-3.00 each or to hobbiests as pet layers for about the same. Dan alternates colored breeds, so he can easily sort by age as the birds mature.

The Bennetts have found a strong market for "natural pastured eggs" at the food co-op in Lawrence, about 30 miles away. (Though raising birds using organic practices, they are not certified.) They deliver eggs once a week in their truck with a refrigerated plug-in compressor cooling the box and charge $1.89 per dozen, wholesale. They started selling eggs at farmers markets, which is a great way to get initial contacts and customers that then will come out to the farm. The Bennett's have found that eggs are a great way to develop customers who will then come to the farm to also buy beef, pork and chickens.

Bennett Nest Boxes Inside Hoop Shelter

Dan has kept a very close watch on the numbers of the Bennett Ranch egg operation. By the first year, raising 500 hens, the revenue of $11,700 was outpaced by expenses of $14,200. The loss of $2,500 was seen as the initial "investment in infrastructure," which included pens, watering system, electrical work and feeding equipment. In the Bennetts' second year, moving up to 2,000 layers, gross income jumped to $36,600, expenses were at $21,700, allowing for a $14,900 margin (which goes to cover labor time). Dan feels this is a sustainable figure, which can improve over time. Now starting their fourth year, the Bennetts are happy with their egg system and the income it brings into their overall operation. Dan and family continue trying new refinements and modifications to create time and money savings in their farm system.

APPPA BUSINESS MEMBER PROFILE

EggCartons.com
Division of Kings Supply Company

EggCartons.com, division of Kings Supply Company
24 Holt Road – PO Box 302
Manchaug MA 01526
Toll Free 1-888-852-5340 info@EggCartons.com

Paul Boutiette, the owner of Kings Supply Company, a maintenance supply company, searched for over two years and could not find egg cartons on the Internet for his eggs from his petting zoo hens (he also owns Kings Campground in Sutton MA). He hatched an idea! He found a factory that made egg cartons and then began selling egg cartons on the Internet to others in the same predicament. Thus, EggCartons.com! Since its inception in February of 2000, over 8 million egg cartons have been sold. We are breaking sales records annually, currently over one million dollars per year! Any visitor to www.EggCartons.com can order egg cartons and poultry supplies on-line and have goods shipped out quickly and efficiently. Customers can pay by credit card, PayPal, check and other usual payment methods.

Sales now include not only paper (pulp), foam and clear plastic egg cartons; our product line also includes many additional poultry supplies including incubators, egg washers, egg baskets, waterers, feeders, books & gifts.

Slogan: "Your Eggs Deserve The Very Best!"
Philosophy: Take care of the customers and the sales will take care of themselves.
We firmly believe that products can be sold at a discount and customers can still be afforded quality service. Excellent customer service is our strong point and is evident by how our qualified staff treats customers' needs first. Getting the product out the door quickly and getting it right is demanded by today's buying public, and we honor that demand. We ship same day on most orders and provide on-line tracking of the order. Customers appreciate that immensely. Customers don't order egg cartons and goods from us because they "want" the goods, they order these items because they "need" them, and we realize that.

Quality egg cartons from major manufacturers including Pactiv Corporation, HartmannUSA, CKF, Inc. of Nova Scotia, Interplast of Canada, SODESA of Spain, and others. We gladly ship as few as 25 egg cartons per order to orders of 50,000 cartons and more. Egg Cartons are pre-printed with our friendly "Farm Fresh Eggs" printing or totally blank. We sell customizing kits so customers can stamp their own cartons to personalize them.

We also wish to highlight our "The Incredible Egg Washer" which is our own patent pending egg washer that we have developed and we manufacture here in the USA. It washes as few as one dozen eggs and up to eight dozen eggs at a time quickly and cleanly. It sells for under one hundred dollars and is the perfect size for the home or small farm egg producer. It is manufactured under the name Fall Harvest Products.

We immensely enjoy this business and our jobs here, and we have fun everyday at work. This is because we are working for customers who are the salt of the earth. The good people who are our customers make our work place a great place to be.

What got us started?: "Don't Wait For Your Ship To Come In – Swim Out To It!"

Lightweight Nest Boxes
By Skip Polson, Summer 1997

Adding nest boxes to pasture pens to accommodate hens for egg production can easily make the pens heavier than desired. As always in pen construction, the goal should be to make them as light as possible so that the task of moving them can be as easy as possible. Despite all my efforts, wooden nest boxes (even very thin plywood) simply added too much weight to the pens.

The best solution I have found is to use plastic milk cases for the nest boxes. These are the plastic boxes which the dairies use to deliver milk to the grocery stores. Because the delivery people stack them on top of each other six or seven high, any of the boxes which are bent or out of shape will not work because the stack will fall over. Therefore, the local dairy here in Little Rock recycles any of these crates which are bent out of shape or broken in any way. The defective crates are collected in a trailer for later hauling to the plastic recycler. They are happy to sell these warped cases to me for $1 each. Installing these in pasture pens has proven to be a very acceptable solution (Ed note: Many "big box" stores now carry similar square plastic crates in the storage section. Be sure to get sturdy ones.)

These boxes are hung inside the closed end of the pen, arranged so that a flap can be lifted to access for gathering the eggs. I hang the boxes with their open side up, after cutting out an access hole for the hens in one side of the box. The hens do not need a very large opening to enter the nest box. It is very important to keep the opening high in the side of the box so that the bedding in the nest can be as deep as possible. The deeper the nest box the longer the bedding will last and the less likely the hens will be able to scratch all the bedding out of the box.

Milk Carton Nest Boxes

We also install a step for the hens to climb up on to get into the box. This step is pulled up after we gather eggs in the afternoon. In the up position, the step forms a partition in front of the nest boxes. This prevents the hens from roosting in the nests at night and filling them with their manure. Clean nests help keep the eggs clean and greatly reduce egg washing time and effort!

Egg Processing by Hand
By Anne Fanatico Summer 2003

Cleaning

A slightly dirty egg can be brushed off or rubbed with sandpaper. Loofahs and sanding sponges are also used. Eggs can be cleaned by hand by immersion in water. Dirty eggs can be rinsed in water that is slightly warmer than the egg (cooler water may force bacteria through the shell into the egg). The USDA's Egg-Grading Manual (USDA. 1990. Egg-Grading Manual. Agriculture Handbook No. 75. USDA Agricultural Marketing Service, Washington, DC. p. 36) recommends that wash water be at least 20° F warmer than the eggs and the water should be at least 90° F. Some producers use commercial dishwashing powder and bleach (a traditional food service sanitizer). Eggs are generally put on dish racks to dry. Producers usually wash and process eggs by hand in a kitchen setting. Clean eggs stored at 45° F and 70% humidity will keep for three months. In a standard refrigerator, where the humidity is lower, a washed egg keeps for five weeks. Producers usually use cracked eggs for baking.

It is important to note that washing eggs washes off the cuticle or "bloom"—a waxy layer that seals the pores and keeps out bacteria. Older egg production books do not recommend washing eggs at all. It was important in the past not to remove the cuticle since refrigeration was not always possible. Washed eggs always had to be consumed immediately. Many small-scale producers and the conventional egg industry now remove the cuticle. Cleaned eggs can be dipped in a solution of water and chlorine bleach to sanitize them. It may depend on your state authorities whether washing requires a follow-up chlorine rinse. Some producers use a plastic watering can and bleach and water mixture. The conventional egg industry often sprays on a mineral oil to re-seal the pores and keep out bacteria.

Candling

Eggs may be required to be candled when sold commercially in the U.S. to insure interior quality of the eggs (i.e., no blood spots, cracks, etc.); however, small operations may be exempt from USDA regulations. Even if you are exempt, candling may still be important to ensure your customers do not unknowingly receive fertile eggs with developing embryos, eggs with blood spots, or cracked eggs. If you gather frequently and use cold storage, embryos will not have the chance to develop into fertile eggs. According to producer Robert Plamondon of Oregon:
"Candling is essentially a joke, except as a way to test for cracks. If you follow the rule of 'never sell an egg that you collected from somewhere you didn't look yesterday,' there's not much point to candling. I never see anything interesting when candling, except for cracks."

If you plan to hand candle, there are many different setups that might be useful. Murray McMurray (Murray McMurray Hatchery P.O. Box 458 Webster City, IA 50595-0458 (800) 456-3280.) sells a hand candler in their catalog. Kuhl Corporation (Kuhl Corporation 39 Kuhl Road PO Box 26 Flemington, NJ 08822-0026 (908) 782-5696 www.kuhlcorp.com) has a hand candler. Brown eggs are more difficult to candle and may require a more expensive model.

According to Colorado State Extension, "a suitable light can be handmade by cutting a 1 ¼ in. diameter hole in the end of a coffee can. Insert a light bulb fixture through the lid, using a 40-watt bulb. View the interior of the egg by holding the large end up to the hole cut in the bottom of the can. As the light passes through the egg, twirl the egg several times. If blood spots are present, you will see them." (Geiger, G., W. Russell, and H. Enos. 1995. Management: The Family Egg Supply. No. 2.510. Colorado State Extension. p. 3.) Another low-tech way to candle: tape a 3 in. length of empty toilet paper tube to a flashlight.

Candling is an important technique in hatching eggs to determine embryo development.

The USDA grades eggs for quality. Quality is based on shell quality, the air cell, the white and the yolk. For example, the highest quality AA has a clean, unbroken, unstained shell; the air cell is 1/8 inch or less in depth, the white is clear and firm, the outline of the yolk is only slightly defined and free from defects such as blood spots.

The term "grading" is also used when putting eggs into various classifications depending on their sizes (peewee, small, medium, large, extra large, jumbo). Refer to source of weights. Many producers do not grade but mark them as mixed or unclassified.

Egg Processing by Machine

When your operation gets larger, you may decide to use machinery to help automate egg processing. While there is a lot of information on large-scale egg processing equipment for the conventional egg industry, there is little information on intermediate-scale processing.

Immersion Washing

According to Robert Plamondon "washing is a complete pain when done by hand."

There are small-scale immersion washers on the market. Plamondon describes the KlenEgg washer from NASCO (NASCO 901 Janesville Ave. PO Box 901 Fort Atkinson, WI 53538-0901 (800) 558-9595 www.nascofa/prod/Home) and the Kuhl washer. Pictures of the products are available on the Internet sites. The KlenEgg is essentially a galvanized bucket big enough for an egg basket with a heating element to keep the water warm and a tube with holes in it at the bottom. You hook the tube up to an air compressor and "you've got yourself an egg Jacuzzi." The air compressor blows bubbles into the water, agitating it and cleaning the eggs. "With a suitably small compressor, this would work fine in the kitchen." It costs about $250, is "ugly and jury-rigged," but works pretty well. The heating element is a 115V, 1500-watt element. You can fill it with hot water from the kitchen sink, and it is small enough to pick up and dump used water. In fact, for kitchen use, Plamondon does not recommend plugging in the heating element— "by the time the water's cold, it's probably also dirty." Another popular immersion washer is from Kuhl. It sits on casters and uses an electric motor and a plastic propeller as the agitator. It costs about $900. Since it is heavy and drains out the bottom, it is not suitable for use in the kitchen, but rather for places with floor drains. According to Plamondon, it doesn't get the eggs any cleaner than the KlenEgg but costs over three times as much. It has a heating element that you can get in 115V or 240V. Plamondon has used his Kuhl washer as a scalder as well.

Immersion washers have a bad reputation because there's always the temptation to run one more basket through when you should really stop and change the water. The rule of thumb is that you should wash no more than three dozen eggs per gallon of wash water.

There used to be a lot of makes of farm-flock egg washers, mostly immersion washers. They'll date back to the 1940s and 1950s, but my experience is that machines from this era tend to run forever. Look for oiling points and put some 3-in-1 oil there, and replace the power cord, since the original is sure to have cracked insulation, and it'll probably run for a lifetime. There must be tens of thousands of these in barns and sheds on old farms.

New egg washing equipment is available from Murray McMurray for about $1,000. Legality of immersion washers is questionable in some states; however, as mentioned before, a sanitizing rinse with bleach may be used.

Non-immersion Washing

Other types of washers use brushes instead of immersion to clean. Plamondon has distributed information about the AquaMagic washer/sanitizer/drier, which also has attachments to candle and grade. "Because it uses brushes and water to clean the eggs instead of water alone, it cleans them a lot more effectively." Plamondon has a picture of an AquaMagic on his website at www.plamondon.com. It comes in two models: the model 10 (formerly called the model 60) and the smaller model 5. These are large machines—the Model 10 is 16 ft. long and 2 ft. wide. The Model 5 is 7 ft. long. They are available from the National Poultry Equipment Company and have been made for at least 40 years. It takes 240V at 20 amps plus a water and drain hookup. It has its own water heater.

The AquaMagic candles, washes (with a water spray and brushes) and dries (with fans and more brushes). The washer section works MUCH better than immersion washers, and the drier section means you don't have to leave eggs sitting around to dry. The washer comes with a little pump that pumps detergent/sanitizer solution out of a bucket and mixes it with the warm wash water. It comes with a chute loader, which is a ramp that you fill up with a row of eggs. They roll slowly down the ramp as the washer picks the eggs up one at a time. The washed/sanitized/dried eggs come out the far end onto a table, where you pick them up and put them into flats or cartons. For a little extra, you can have a candling light added to the chute loader, where a bright light shines up through a slot in the chute, allowing you to candle the eggs as they pass by.

Plamondon suggests buying the small unit and requesting the water heater to be omitted. He says the water heater is a very fancy unit, capable of an unnecessary amount of temperature regulation (down to one degree). Just hook the unit up to an existing source of hot water. With the savings, you can have the candler light added to the chute loader (about $700) and still have some for shipping. The total would be slightly over $6,000.

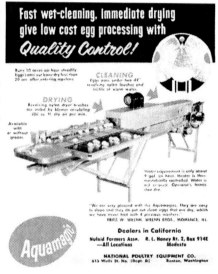

Plamondon's Aqua-Magic

Plamondon points out that although this machinery is expensive, it can last a long time since there are few parts to break down. In fact, used units that would still operate well may be found stored in old barns. Parts are still available for these old machines, since many core parts have remained interchangeable over the decades. However, "you'll want to think twice about buying an old unit in dubious operating condition if you aren't mechanically inclined." Plamondon bought his large used unit for $50!

APPPA PRODUCER PROFILE

Hens in the Snow? Winter at Springfield Farm
By Jody Padgham Fall 2004

Was anyone else at the PASA meeting last February as struck as I was by David Smith saying he used a SNOW BLOWER to cut a track for the hens so that they could get out to their food in the winter? When a member wrote to request we do a story on over-wintering laying hens, I knew I needed to talk to David about the details of that snow blower and the other systems he had developed to keep hens producing throughout the winter.

David Smith, his wife Lily, their daughter and her husband manage the 65 acres of Springfield Farm 30 miles north of Baltimore in Sparks, MD. The four of them perform all the labor to raise 400 turkeys, about 5,000 broilers and 2,200 laying hens each year. They have an ever—expanding market, being so close to metropolitan Baltimore, and expect to continue growing their production over the next several years. Although their operation is larger than most, the many systems David has put into place since they started in 2000 have relevance to anyone interested in raising pastured eggs year-round.

Having started out with an "Eggmobile" like Joel Salatin's, David soon recognized that he needed a less labor-intensive system. He moved to a variation of Hermann Beck-Chenowith's "day-range" model and has ended up with a system based on permanent 24 x 24 ft. hen houses scattered out on pasture. Each house, 7 ft. high, has eight ground-level hen doors and utilizes two acres of pasture. Electric feather poultry netting is used to protect the birds from predators. Three net fences are hooked together and run from a given door in a pie-shaped wedge out to the pasture. Every two weeks the fences are set up outside a new door and the birds come out into a new quarter-acre paddock, eventually making a daisy-like rotation entirely around the house. David runs two sets of fencing, with one idle, so he doesn't have to lay out fence at night.

Each hen house, which costs about $6,000 each to build, holds 750 hens. The east and west ends of the peaked roof buildings are completely closed, but the north and south sides have a 2 ft. high wood wall that is completed to the roofline with poultry netting. The north side is covered with a clear heavy plastic curtain that can be raised or lowered; the south side is open. The floor is dirt, with wood shaving bedding brought in twice a year. The floor is cleaned to bare dirt in March or April and refreshed with 4 in. of shavings. Six months later 2-3 in. of shavings are added on top. The composted litter is spread out in the fields when it is removed in the spring. Buildings are wired for electricity and each has its own frost-free pump.

Nest boxes are clustered in the middle of the house, with approximately one box per 10 hens. There are no roosts in the house, but the hens are given ladders so that they can climb or fly up to the rafters where they tend to sit. (Wings are not clipped!) Some birds roost atop the nest boxes, and boxes are not closed up at night. Shavings are refreshed in the nest boxes every few weeks "or when the egg cleaning crew complains," David says. Eggs are collected by hand from the nests and taken to the egg washing room where they are hand-washed using a stainless steel egg washing tub and dried in baskets. They are then placed into cartons or flats and either put in the sales fridge for on-farm sales or delivered to stores or restaurants in boxes of flats.

Back to the hens: water and food are outside year-round. Hoses are run from each house halfway out the paddock to rubber water pans, 20 in. in diameter and 8 in. tall. Three pans are hooked together for 750 hens, and each has an automatic float valve. As long as the temperature is above 28° the hoses

APPPA PRODUCER PROFILE — Smith

are kept hooked up. Springfield Farm is located in Zone 6, where the average January temperature is 40° F, but on occasion the temps dip below zero. If the ice in the pan freezes, they break the ice and the sun during the day will thaw things enough for the water to flow. Once the temps regularly fall below 28° F, the hoses are brought in and water pans are filled with buckets. Pans are checked several times a day.

Feed troughs are far out in the paddock year-round. Around mid-November when the grass stops growing the regular rotation is halted and one double sized paddock (half-acre total) is chosen for the wintering area. Extra feather net is set up to enclose the winter area, and all unused fence is taken into the storage shed and hung up for the winter. The two paddocks used to over-winter will be skipped for the first two rounds when rotations begin again in mid-March to allow the pasture to rest. David notes that the winter paddock is "as good as new" by mid season the next year.

Springdale Farm Hen House With Fence

They do get snow in Maryland, though rarely more than 4-6 in. and it usually melts after a few days on the ground. David says that they don't have problems with the electric feathernet in the snow—they set it very tight and adjust it as needed throughout the season and it will maintain a charge even in light snow. When the snow comes down heavy, the Smiths will move feed and water inside until they can get the tractor out to snowblow a path out to the feed troughs. (They have found a lucrative business in plowing for neighbors and so may not get to the chickens till late in the day.) The chickens don't like to walk in snow deeper than 2 in., which is when the snow blower comes out. Once a path is cleared, the cold temperatures and moisture doesn't keep the birds in.

David really feels it does the birds good to be outside year round. They seem not to mind the cold and enjoy scratching even in frozen ground. They certainly enjoy sunbathing in a wind-free corner on a sunny day.

At Springfield Farm they gather around 900 eggs each day. David figures that the feeding, egg collecting, washing and sorting can be done by one person in five hours. Fence moving, building cleaning and maintenance and delivery will take another person two to three hours each day. David states that the entire egg operation takes about seven and a half hours per day, split between two people. He figures the return to their labor on eggs is about $30 per hour, which is very sustainable.

One strong advantage that the Smiths have is their proximity to a large, educated customer base 30 miles away in Baltimore. Even before they had a hen on the place David had an agreement with a restaurant to deliver "farm fresh, natural" eggs at a good price. Springfield farm has had significant media coverage, including a spot on TV, and the Smiths find the demand for their eggs, broilers and turkeys far exceeds their supply right now. They deliver about 500 dozen eggs each week to restaurants for $2.25-2.50/dozen. That will soon go up to 700 dozen as one of their customers opens a new location. Customers coming to the farm (who also buy chicken and other products) buy an average of 50-60 dozen per

APPPA PRODUCER PROFILE   *Smith*

week at $3.00-3.50/dozen. It is easy to see that this unique market makes a venture of this type viable. Most of you reading this article will find a much different market situation in your area, but the demand for "farm-raised, natural" eggs in the winter will be strong in any developed market.

Costs at Springfield Farm are figured to be $1.10/dozen, not including labor, marketing or transportation. The $6,000 per building costs are being depreciated over a 10-year period, and poultry netting is estimated to last five to seven years. David estimates that his egg operation set-up costs will be recouped over a five year period.

Managing the life cycle of all these birds is an impressive, and understated, responsibility David takes on. He has found that it doesn't pay to raise replacement hens on-farm. Instead he currently buys ready-to-lay hens from APPPA Business member Moyer's Chicks. David estimates it costs him about $5.00-6.00 to raise a chick from day-old to its 20-week laying start, and he can save $1.00-2.00 by buying those birds in. He would like to use more traditional breeds, but has been happy with Moyer's Red Sex Links. David has to closely manage the flock. His farm is unusual for an operation its size in that David keeps the hens through their first molt (12 months after full production starts) which will last for up to three months. David keeps his birds an average of 30 months, after which he sells them as stewing hens to restaurants, which at this point buy several hundred per year.

The egg market will double during the Thanksgiving and Christmas baking season, and the Smiths work to build a two week reserve in order to keep pace with orders during this time. This is where David's planning expertise shines--he manages a very careful replacement schedule to ensure that he will have maximum production when he needs it. David will bring in a new flock in March so that prime laying time will coincide with peak demand.

Inside the Hen House

David has found there are several keys to maintaining good production in the winter with his hens. The first is light. He begins to light the houses in September and maintains 12-14 hours of light per day all winter using timers. David doesn't heat the houses, but with 750 birds he is definitely getting the advantage of compiled body heat. David also feels that it is very important that the birds get outside as much as possible, which is why he goes through the energy of snowblowing trails and keeping water open in mid winter.

David cautions that over-wintering hens is "not a no-brainer" but that with proper care and consideration, production should not drop significantly while market demand is definitely strong.

Keeping Eggs Clean
By Skip Polson Spring 1999

Washing eggs certainly does take a lot of time and is a tedious process when you have a lot of them to do by hand. This labor requirement is one of the down sides to egg production which is sometimes overlooked by prospective producers who think this enterprise would be more to their liking (and easier for them) than butchering broilers. It is a significant reason that egg production does not pay as well per hour as broilers (unless one gets very well organized and labor-efficient).

In my experience, the primary determinant of how clean the eggs are when collected is the cleanliness of the nest boxes. Keeping the bedding in the nests fresh and clean is the best thing one can do to minimize the number of eggs which have to be washed. Thus, some system of closing the nests at night to prevent the hens from roosting in them and filling them with manure is really helpful. I have used hinged perches for this purpose. During the day they are in the down position, serving as a step for the hens to enter the nests. At night they are in the up position (folded up and back against the front of the nests), blocking the hens from entering the nests. Consistent use of this system requires lifting the perches up every afternoon and putting them down again in the morning.

Dirty nests would be a problem in any system of production (whether total confinement, pasture pens or eggmobile). On rainy days, when the hens have wet feet, almost all the eggs will require washing. On dry days, most will not have to be washed, as long as the nest boxes are clean.

Nests can also be designed so the eggs will roll out of reach of the hens as soon as they are laid. This reduces egg breakage and the incidence of hens eating their eggs and leaving a mess in the nest which makes other eggs dirty.

Once an egg enterprise reaches a certain size so that it comes under the licensing requirements of the state, I think that most states probably require that all the eggs be washed and washed with a cleaner that meets certain chemical composition requirements at that! When one reaches that scale, there are egg washing machines and related equipment commercially available which make this task easier.

Environmental Stresses on Laying Hens
By Jeff Mattocks Summer 2003

Here we are again, the middle of the summer, cruising along with our pastured poultry. Just when things look great, while we're not looking, our broiler weights decline and our layer production slides. Before you know it, this becomes a significant issue.

Hot and Humid
Unfortunately there are very few options to alleviate the negative effects of hot humid weather on our poultry. We have known for a long time that poultry of all kinds eat for their energy needs and not for proteins. When it gets hot, humans and other forms of life all tend to eat less or change diets to match the environment. Chickens are no different!

What we see in our poultry is a reduced feed intake resulting in reduced production. The effects of heat stress on broilers are seen immediately in heart attacks (flipovers).

The affects on laying hens are not seen until the week or so following heat (post stress period). Studies have been done where birds were taken from 64° F to 95° F for three days. During this period an immediate difference observed in the birds was a significant reduction in feed intake. After the stress was removed and temperatures returned to 64° F the feed intake returned to near normal or previous intake. However, egg production dropped an average of 30% in all of the groups tested.

The following information has been paraphrased from *Commercial Poultry Nutrition*, by S. Leeson and J.D. Summers.

To combat heat stress they recommend:
1. Increase energy level of the feed.
2. Reduce crude protein.
3. Increase vitamin mineral premix.
4. Use some Sodium Bicarbonate if shell quality is a problem.
5. Use supplemental Vitamin C when heat stress occurs.
6. Increase the number of feedings per day to try to feed at cooler times of the day.
7. Keep drinking water as cool as possible.
8. Use large particle mash feed if available.

These recommendations are for large commercial flocks, and they do not all apply to pastured poultry layers. My recommendation is to reduce the energy inputs (generally corn, vegetable oil, wheat or roasted soy). You can replace any of these with alfalfa meal, oats, barley and other low energy inputs. For instance 200# of corn replaced by 200 lbs. of oats in a ton of feed will reduce the energy by 42,000 calories per pound of feed. This change will make a big difference to the chickens. A change of 100 lbs. Alfalfa meal for 100 lbs. Corn will yield the same 42,000 calories per pound of feed.

You can see that making feed adjustments is not "rocket science" but I wouldn't recommend making changes that you haven't done before or consulting someone who can do the math for you.

Cold
The winter of 2002-2003 I had to learn new lessons about laying hens in cold weather. In Lancaster County, Pennsylvania, we had the coldest winter of the last 50 years, with over 30 days below freezing. I didn't give any extra thought regarding my egg-producing customers until they started calling. Several things occurred that winter that did not happen during the past six years.

The hens started to:
- Get irritable (overcrowding)
- Pecking and eating feathers (lack of protein or Methionine)
- Losing feathers (beginning of a molt)
- Eating nearly twice as much feed (cold or poor feed quality)
- Egg sizes greatly increased (too much energy in the feed)
- Egg production decrease (could be one of many problems)

I started looking for the reason why my producers were seeing all of these problems. Each of these symptoms is normally caused by different deficiencies.

This was more than I was familiar with so I called some experts and gathered some opinions. They all led me back to the same conclusion: HEAT! I had never given the temperature a consideration. I thought that with all those hens the houses should be warm enough even when the outside temperature was below freezing. I was wrong. While looking for the answers to these problems I found a piece of information that really caught my attention. One of our local egg producers started to supplement heat during the week of Christmas. When the heat was added the feed consumption dropped 7 lbs. per 100 hens per day. Hens typically eat between 22 and 24 lbs. per 100 hens per day. Before the heat was supplied the hens were eating 30 lbs. per 100 per day. It may not sound like much but when you multiply this for 5,000 chickens, you have 350 lbs. of feed a day. This adds up to 2,450 lbs. per week and five tons of feed per month. At an average of $280 per ton times five is a savings of $1,400 per month. I can buy a lot of propane for $1,400 a month plus the chicken house will be a much nicer place to work vs. being outside. Along with all this the hens will be happier with supplemental heat than by eating extra feed to stay warm. The other benefit to you is less manure, less ammonia smell and dryer litter.

I have recommended to all of my local customers to find a way to supplement heat. The heat should be stabilized between 55-65° F. If you can stabilize the temperature in this range feed consumption, egg production and egg size will stay normal. I took my largest local customer for a ride to another chicken house that was doing better. When we arrived at the other chicken house the ONLY difference between his house and the visited house was the temperatures. His home house was below 32° F at night and 45° F during the day.

The visited house was not lower than 55° F at night and up to 68° F during the day. Feed consumption was lower, and egg size was normal (80% large and extra large, 10% Jumbos, and 10% mediums).

Winterizing Laying Hens
By Jeff Mattocks Fall 2004

Earlier this year a member asked what they need to do to prepare their laying hens for winter. This is a great question, but it may be a surprise to some to hear that our response is a little late--if you are counting on the hens for winter income the winterizing process actually should begin in the third week of August. All is not lost, though, as there are still important things to consider.

There are several steps involved in properly managing layers for winter profitability.
1-Supplemental lighting
2-Diet change
3-Temperature corrections

Lighting
Supplementing light is needed to maintain a consistent length of daylight, which will maintain a consistent rate of lay. Laying hens require 12-16 hours of day length to maintain production, with the optimum length being 14.5 hours. Light supplementation is the first winter change and should begin in the third week of August. The third week of August is when day length is starting to get noticeably shorter, for us in the Northeast it is very near 14.5 hours of natural daylight. From here on into the winter, light should be added at 15 min. increases per week to maintain the 14.5 hours total light. This doesn't mean that the light needs to be on for 14.5 continuous hours. I use a timer that allows two "on times" per 24-hour period. I set the timer to come on at 5 am and go off at 8 am, and also set it for 5 pm till 9 pm. This provides our flock with 16 hours of total light and seven hours of supplemental light. Meanwhile this same incandescent light provides some temperature correction.

The minimum light required is a 40-watt bulb for an area 10 x 10 ft. I personally use a 60 or 75-watt bulb for an area 4 x 8 ft., which is on the extreme end of the spectrum—I wouldn't suggest anything higher. If electricity is not an option, I recommend the use of a Coleman gas lantern. The length of burn or light can be controlled by measuring the fuel that is supplied. The typical lantern will burn for six hours on approximately 8 oz. of fuel. Therefore, the lantern would require 1.33 oz. of fuel per hour of burn. You should verify this when you purchase a lantern, as each lantern is slightly different. The lantern can be used at both ends of the day but appears to work best (for labor reasons) only in the evenings. Using the lantern in the evenings will allow it to slowly burn down and out, somewhat like a normal sunset. This allows the layers to get to the roost before the switch goes out and is a very mellowing environment. A gas lantern will supply approximately the equivalence of a 60 watt incandescent light. The other benefit of the lantern is that it provides really good supplemental heat.

Diet Changes
Diets should change seasonally as a response to temperature and instinct. As the 24-hour ambient temperature declines, the hens will start eating more to compensate. They are eating for their energy needs, which is what all forms of poultry will do. During the summer months typical layers will eat (not including spillage and waste) .25 lb per bird per day. This will be the average for a 75° F, 24-hour average temperature. During the winter months when the temperature average is 40° F we can expect feed consumption to increase to .33 lb. to .40 lb. per bird per day. This is solely due to energy requirements to maintain warmth and add fat for insulation.

At this point the instinct portion of the reasoning comes into play. Nearly all forms of life that experience temperature variances to any extreme will tend to increase body fat for winter self-preservation. Laying hens are no different; with what little instinct they may have remaining after intense breeding, they will store up fat for winter. The side affect of this is an increase in egg size. Most producers will manage for a correct egg size in the summer months and if the size increases it may take you out of your established market. Sales trends show that marketing Jumbo

Hens Enjoy the Snow on a Sunny Day

eggs or larger is difficult and/or not cost effective. There is not enough additional mark-up to compensate for the increase in management and feed consumption. So, this all adds up to the fact that during the winter month the layer ration should be changed to increase the energy and slightly decrease the protein content to adjust for temperature and instinct.

Some Sample Ration Examples

Winter Laying Ration   **15.5% protein**

Ingredients:	LBS
Alfalfa Meal	100
Limestone	100
Oyster Shell	75
Poultry Nutri-Balancer	60
Shell Corn Grain	1115
Soybeans, Roasted	550
Total	2000

Nutrient Name:	Amount	Units
Crude Protein	15.5%	%
Crude Fat	7.1%	%
Crude Fiber	4.2%	%
Calcium	3.87%	%
Phosphorus	0.73%	%
Energy	1,284	Kcal/LB

For those of you who are forced to use soybean meal (48%) rather than roasted soybeans, the following ration is an example of a winter ration. I feel that soybean meal is a cheap by-product that should only be used due to availability, not cost. Roasted soy will enhance the layer flock's health, production and egg flavor. Roasted soy also provides protein along with energy values equal to corn.

Winter Laying Ration   **15.5% protein**

Ingredients:	LBS
Alfalfa Meal	100
Limestone	100
Oil, Soybean	75
Oyster Shell	75
Poultry Nutri-Balancer	60
Shell Corn Grain	1165
Soybean Meal, 48%	425
Total	2000

Nutrient Name:	Amount	Units
Crude Protein	15.5%	%
Crude Fat	3.1%	%
Crude Fiber	4.4%	%
Calcium	3.84%	%
Phosphorus	0.70%	%
Energy	1,274	Kcal/LB

In these rations you will notice alfalfa meal. The alfalfa meal is important to help keep the color in the yolk from paling over the winter. You will also notice that the protein has been lowered to 15.5% for winter. This is to maintain the egg size by reducing the grams of protein ingested. These changes also increase the energy values to slightly above recommended rates to ensure body warmth.

Temperature Corrections

In the article "Environmental Stress on Laying Hens" (page 88), I talked about temperature effects on hens. I can't stress enough the importance of moderating the temperature swing between night and day during the winter months. Your hens will be much happier, more productive and more sociable with a moderated climate. Most producers provide some form of solid or enclosed structure for wintering hens.

In a small shed or stick-built structure, heat may be supplemented in several different ways. Deeper bedding provides more insulation. I would recommend a minimum of 4 in. to start the winter, increasing to 8 in. by mid-winter. You will want to start reducing the bedding as spring breaks.

Higher stocking density will increase shared body warmth. During the fall months 1 sq. ft. of space per hen is adequate. By December, this space should be reduced to .75 sq. ft. and by January or February to .50 sq. ft. These recommendations are for floor space square footage requirements. I like to keep my ceiling height down to 6 ft. for winter housing to minimize total square footage. During the coldest months (January or February) I will utilize a heat lamp on a timer to come on just before the incandescent light goes off- 8:30 pm till 5 am. This is only to reduce the 24-hour temperature fluctuations. If I can keep the hen house at 40-45° F my hens are much happier and will continue to lay at an acceptable rate.

We should always remember that our animals are not DUMB animals. They only require understanding and compensations. As we learn and understand more as to why our animals do the things they do, we will get smarter and our appreciation for nature will grow. Try to be accommodating and meet the characteristics of nature halfway.

CHAPTER EIGHT
Turkeys, Ducks and other Poultry

Pastured turkeys are very popular with consumers - especially for holiday feasts. As Joel Salatin says below: "Turkey is a special bird for a special day, and folks are more than happy to add a little extra from their wallet to pad the festive occasion.... I think for anyone already doing pastured chickens, turkeys are a perfect additional enterprise, another profit center from your current resources." Many pastured poultry producers are going one step further and raising turkeys with heritage, or heirloom, genetics. These birds have a slower growout and produce a smaller finished product, but the taste makes the extra effort entirely worthwhile, say many. Ducks, pheasants and geese also thrive on pasture and offer good market diversification.

Turkeys
By Joel Salatin   Spring 1998

In the last two years we added turkeys to our poultry portfolio and from what we've already seen, we could have another enterprise fully as profitable as the pastured broiler. In 1997, we produced and processed 350 white turkeys.

First, let me give credit to one of our apprentices, Tai Lopez, for pioneering the first year and developing this addition. We encourage our apprentices to develop new enterprises under the shelter of our umbrella. This builds their confidence for a repeat performance on their own and it stimulates us to diversify by simply maintaining an ongoing endeavor instead of doing all the research and development to get set up.

Turkeys are both similar to and different from chickens:

1. Brooding. Chickens need 90° F for the first couple of days and by three weeks can handle freezing temperatures if they are gradually hardened off. Turkeys need 95° F for the first few days and cannot take freezing until at least five weeks old. They can take 50° F nights as early as four weeks, but not at three weeks. Hold them inside an extra week or two compared to chickens.

2. Ration. We use one ration at about 19-20% protein for our broilers throughout their eight weeks of life. Turkeys, on the other hand, need a 28% protein ration for their first five to six weeks. In the wild, poults only thrive on open ground where they can eat plenty of bugs. From weeks 6-10, they can do fine on the broiler ration. After week 10, you can drop down to a 15% ration by upping the corn percentage. One way to jump up the protein early on is to feed eggs for the first two weeks. We just hard-boil them and mash them onto the top of the feed, shells and all. The poults quickly learn to devour this supplement and it acts as a tonic. Feed no more than 25% egg as percent of the ration.

3. Personality. Turkeys are much more people-friendly. Whereas chickens tend to flit away as you come close, turkeys actually come to you. In fact, folks who have tried complete free-range on turkeys complain that it doesn't work because the turkeys keep following them back to the house. They just love people and that makes them enjoyable for children when the poults are small.

4. Foraging. Turkeys will eat up to 50% of their ration in forage, as opposed to chickens where 20% forage is considered exceptional. This means that the more ground the turkeys can cover, especially if it is succulent forage, the less feed they will eat. Their long legs allow them naturally to cover more ground. They simply eat grass more aggressively. Even 10-day-old poults will eat 10 in. blades of grass. Poults readily consume lawn clippings or handfuls of grass. To stimulate foraging and reduce feed costs, we drop the bird numbers to only 12-20 per pen and move them twice a day the last month. This also means that the difference between pastured chickens and their supermarket counterparts is not as noticeable as the difference between pastured turkeys and their supermarket counterparts. The differences are simply incredible.

5. Grit. Part of the tradeoff for being aggressive foragers is the requirement for more grit. Turkeys consume prodigious quantities of rocks. You can buy this poultry grit from a feed store, but we just scoop gravelly material from sandbars along the creek. As the birds get bigger, they will eat rocks the size of marbles. We just go along once per week and drop a half-gallon or so in a pile on the ground before we move the pen. This procedure keeps the turkeys from being frightened like they are if you drop something strange into the pen. They find all the little rocks and eat them readily.

6. Emotions. Turkeys enjoy a routine more than chickens. Although they have a bad reputation for stupidity, I think turkeys are much smarter than chickens. When I go out to move the pens they see me a hundred yards off and begin to pace the front of the pen, talking wildly about what is about to happen. Anything out of the ordinary sends them into a panic. Even putting some whole corn in the feeder will make them flutter to the far side of the pen and refuse to come to the feeder for half a day. Their ability to respond positively to routine is the same trait that makes them panic at any routine change.

7. Light. Turkeys do not like shade or shadows. When we first began raising turkeys we noticed that they would not graze or rest in the shaded cap end of the pen. When we moved the pen, we could not even find any manure there. They only used half of the pen effectively. So this past summer we made half a dozen pens without any roof or sides-just poultry netting. Only the door over the feeder was solid. Everything else was wide open. The turkeys immediately spread out, grazed evenly, and performed superbly. This distributed manure evenly as well. Obviously, these pens were extremely cheap to build and lightweight. The Thanksgiving bird went into our regular broiler pens because they needed the shelter, but by that time the sunlight was more slanted so rays penetrated farther into the pen.

8. Heat. I know what you might be thinking: "If you took off all the shade, what in the world did the birds do when it got hot?" Good news, folks. Turkeys in small groups seem to handle any heat just fine. Oh yes, they will pant just like a dog or any animal, but in blazing 90% humidity and 100 °F days they would happily lounge right out in the full sun rather than get back in under the shade. Clearly turkeys can handle heat much better than chickens.

9. Cold. Again, you would think that a bird that is able to handle heat would collapse in the cold. Wrong. We take these birds right up to Thanksgiving, through low teens, and they do fine. They do not gain quite as fast as they do when the temperature is warmer, simply because they convert calories to maintaining body temperature rather than add muscle, but they certainly stay healthy and content down to the lower teens. Of course, the waterers freeze at that point so we use those indestructible black rubber 12 in. tubs for water. The 4 in. tubs are okay, but the birds step in them. The taller ones stay clean.

10. Housing. We use the same 2 ft. chicken pens for pasturing turkeys and although the birds can be hard on some of the small bracing, the pens work fine. Although they can't completely stretch their necks as high as they can go, they do not rub the top under normal movement conditions. A turkey neck is quite crooked, or swooped, and really does not come much higher than the top of the bird's back. They move much easier than chickens, cramming up against the lead edge to get the freshest morsels.

The first year with Tait we did try putting the birds in an electrified netting yard, but they would not stay in. In fact, we found out that the Premier company recommends its products for everything but turkeys. The birds would stick their necks through and graze contentedly while they received a violent jolt. So much for brains. We also tried a portable house in the field, free range, hoping the birds would come back to it at night. Some came to the house, others went elsewhere and some followed us home. None went inside-they roosted up on the roof.

One fellow in Ontario told me he gets along alright with a roost somewhat like a glorified drying rack. He keeps water and grit there and lets the turkeys follow the chicken pens, scavenging dropped feed. The portable roost keeps the birds protected from predators. Our experience is that the birds end up roosting on the chicken pens and caving in the poultry netting.

Oh well, you never get finished refining.

I did meet a fascinating fellow in Alberta a couple of years ago who had built a 50 x 100 ft. portable turkey corral out of sucker rod. In oil country, this sucker rod, which is a malleable 1-inch square tubing and rolled up on huge reels, is free for the taking. He made 12 ft. gates and fastened them together with a couple of links of chain so the whole thing would conform to the ground like a big inch worm. I do not know if he skidded it or if it had wheels, but the breakthrough was the segmented long dimensions, allowing it to conform to the ground. He said he didn't have any trouble with coyotes and it worked great. We hope to do some designing on such an idea. I think it has merit, especially since turkeys seem to enjoy walking so much. Giving them a huge area would be helpful and being able to move more at a time would be efficient. The problem is that you have to move it with something other than human labor.

But think about this: what if you built a 10 x 10 ft. stable for an ox or two, or maybe a couple of milk cows, with three sides tight and one end slatted? The animal(s) grazes through the slats on one end and the whole stable is mounted on wheels. As the animal grazes, it just pushes the stable forward. Then you hitch it to this turkey corral and you have a portable facility moved multiple times each day without any human intervention except planning. Like I said, we never get done refining.

For now, we're quite pleased with just getting extended use out of our existing pens-we didn't have to build anything new. It just added another $700 gross volume per pen when the pens sat idle anyway (October-November). Not a bad use of resources.

11. Cashflow. The single biggest drawback to turkeys is the longer time between investment and payback. We dress turkeys at 16 weeks, as hatched. Some folks want 20 pounders and others want 12 pounders. By raising both toms and hens we can let one batch meet both desires since the different sexes spreads the weight differences. These 16 week birds average 16 pounds or better. The poults are more expensive as well, running around $1.50 apiece. You have a lot more money tied up longer, but the net return to labor is the same.

Young Turkey Poults on Pasture

12. Mortality. Except for the higher brooding temperature requirements, we have not found any difference between turkeys and chickens early on. The big differences come about week five. Once turkeys get outside and hit six to seven weeks, they are almost indestructible. Rather than becoming more and more lethargic, they become more and more aggressive. They seem to blossom and just get healthier as they get older. Adult mortality simply does not exist. From that standpoint, they are pure joy to raise.

13. Processing. Turkeys are much more efficient to process than chickens because, for roughly the same procedure, you have three to four times the volume of meat and the price is better, making the dollar return per hour way higher than broilers. We use four turkey cones for killing, running batches of two. This fall we did 50 birds per hour—at an average of 20 pounds (because of a hatchery glitch, we had 18-week-old Thanksgiving birds) that amounted to 1,000 pounds of poultry per hour. Our new 1959-model state-of-the-art rotary scalder handled two at a time just fine, but the picker stalled a couple of times. We had to keep pulling out the biggest feathers, which did not want to drop down out of the picker drum. I explain the main processing difference as one between playing with Leggos and Duplos. The processing advantage more than offsets the production disadvantages.

14. Marketing. Here a huge difference between turkeys and chickens exists. Turkeys are very much a seasonal thing—unless you get into turkey ham, turkey salami, turkey sausage, turkey hot dogs, etc. In fact, roughly 75% of all the turkey produced in the U.S. now goes through further processing into these other products. Eating a whole turkey is just not something the average person does except at Thanksgiving and Christmas. We sell a few in the summer, but most go during the week of Thanksgiving. Many people get another one to freeze for Christmas. This is fine with us because it is when we have extra, unused housing anyway. I can tell you that fresh pastured turkey at Thanksgiving is a hot item. All we did was add it to our spring customer letter and order blank. Immediately we had another $10,000 enterprise.

15. Pricing. Turkey is much more price forgiving. The reason is that because it is a seasonal thing, people are not so much buying a staple, like chicken, as they are buying a nostalgic, entertainment, celebration vehicle. The special holiday turkey mystique is real, and nobody faults you for charging three or four times market price-after all, that only amounts to $10-$20 for a celebration meal. The perception is that that is nothing. Chicken, however, enjoys no mystique. It is a staple, and therefore hits price prejudice much sooner. Turkey is a special bird for a special day and folks are more than happy to add a little extra from their wallet to pad the festive occasion.

16. Diversity. This is not so much a comparison as a point to think about. Since we sell beef, pork, rabbit, eggs, firewood and chicken, we find that each time a customer visits the farm they buy more than the single item that drew them here. Before turkeys, we finished the year the year with broilers in late September and then beef and pork in late October. That was the last big farm visit until the following May. Now we get more of our customers back one more time during a festive (read that "what else can I buy?") time of year and this stimulates sales of other things, including crafts or woodworking items we may have. Rachel's handmade potholders sell great at this time of year. This additional farm visit can add significantly more than the face value of the turkeys to the farm's annual income.

17. Stacking. Again, this is not a comparison but a point. Since we can run the turkeys over the same ground as the broilers, it gives us another high dollar-per-acre item without buying another acre. At 20 turkeys per pen netting $15 per bird, that's $300 net per pen. Moving them twice per day, that's covering 240 sq. ft. per day over an eight-week period (early weeks are only daily moves) for a total of 56 x 240 sq. ft. or 13,440 sq. ft. per pen. That's roughly a third of an acre, which means we're netting $900 per acre on top of the $1,200 per acre with broilers. Not bad for not requiring any more housing, machinery or land. If you don't run them on the same acreage, it means you cover more ground with fertilizer. Either way, you win.

All in all, we are extremely excited about the future for turkeys and plan to expand it as much as is feasible. I think for anyone already doing pastured chickens, turkeys are a perfect additional enterprise, another profit center from your current resources.

Turkeys are Strong Foragers

Turkeys Have Feelings Too!
By Dan Bennett Fall 2003

No, I'm not claiming that turkeys possess any higher form of intelligence than other animals nor am I overly concerned for the well being of this specific species. I'm simply responding to the unusually critical attitude of many farmers when the topic of turkeys comes up! I picked up an old farm book a few years ago at our local library excited to find a chapter on Turkeys in the contents. As I turned to glean some good old fashioned advice, I read the single word under the title, "Don't"! Luckily, I rejected this author's advice and over the past five years have had a very positive experience raising turkeys on pasture.

In this article, I hope to cover the highlights of my experience with turkeys. I will discuss production, marketing and the overall role turkeys can play as a part of your farm or ranch. I will use this article as a basis for my talk at the PASA conference in Feb. of '03 during which time the APPPA annual meeting will take place.

We've all heard the stories about turkeys being so dumb that they will drown out in a rain storm because they obviously "don't have the sense to come in out of the rain"! I believe by having the right ration, the proper time in the brooder, and the proper shelter in the pasture--pasture raised turkeys can be a very rewarding experience. I would even go as far as to say that the white broad-breasted hybrid turkeys are hardier and have twice the personality as the hybrid Cornish Rock Cross broiler. In my book, that means they are more fun to have around.

If you've read any of my articles before, you know I'm a Joel Salatin disciple and my turkey production is no different. In fact, I was just looking in my turkey file and ran across Joel's article entitled "Turkeys" written in the Grit a number of years ago. (See page 92.) Most of my ideas are not new—in fact, this may simply be a re-write of what you've read before!

In terms of production models, I believe the pasture pen pioneered by Joel is not only the best for broiler production; it's the best model for turkeys as well. Now I know the argument, "but those poor turkeys don't have the room to run, spread their wings and fly in those pens." This was recently preached to me by my good friend Jim Protiva (who by the way is very successfully raising 1,000 turkeys this fall using a day-ranging model in southern Missouri). More power to you Jim, my Kansas turkeys prefer pens!

As I mentioned earlier, the time in the brooder is critical for turkeys. You need to have good brooder skills to excel at turkey production—you need to watch them very carefully the first few weeks. I use the same brooders and brooding techniques for turkeys as I use for broilers with the exception that we use crushed boiled eggs on top of the feed to get them going. You may have seen my recent article in the Grit! on brooder management. (See page 15.) Use those techniques effectively and you will have no problems with turkeys. We do give them plenty of greens in the brooder as they seem to enjoy it so! Turkeys do need to stay in the brooder longer than broilers. I use the rule of thumb given by Joel Salatin –five weeks in the brooder is best. I agree with him and if the weather is not just right and I have room in the brooder, I'll leave them in longer.

We start with all hen turkeys to help with uniformity. I get them very reasonably from Jim Phillips at Stover Hatcher in Stover, MO, www.stoverhatchery.com.

At five or six weeks, I put them out on the pasture and just watch them grow. Our pens are the Brower PVC 10 x 12 ft. poultry pen which gives turkeys 3 ft. height in the covered end of the pen. We start with up to 40 turkeys per pen and then after the broilers begin to free up pens, we reduce density to about 20 per pen. We move the pens daily because, for one, we were told to, but also because those turkeys eat grass more aggressively than anything I've ever seen. If we didn't move them daily, we wouldn't have any pasture left! These turkeys are so eager to move each day that they practically push the pen along themselves as they see that fresh grass approaching.

At 16-18 weeks, we end up with a very uniform group of dressed turkeys in the 15–20 lb. range. I'm convinced that the pasture pen approach adds to our uniformity and reduces our death loss on the pasture to very near zero!

We've made two big mistakes in our five years of raising turkeys. Number one is putting them out on the pasture too soon and not keeping a close enough eye on the weather. Remember the game "pile on" we used to play as kids–when all the kids would pile on you? Well, turkeys seem to like that game more than we do!

Number two is packing too many turkeys in the trailer for the trip to the processing plant. They need their space and if you don't give it to them, they will steal it from their weaker neighbor–if you get the picture! I'm sure we've not learned it all; but be on the lookout for these two "gotcha's"!

For feed, I must give credit to our friend Jeff Mattocks from The Fertrell Company. Jeff recommends using his standard Fertrell Broiler Ration as a base feed. For the first four weeks, he recommends adding 4 lbs. of fishmeal per 25 lbs. of feed for the added protein that turkeys need. For the second four weeks, he recommends adding 2 lbs. of fishmeal per 25 lbs. of feed. After that, just use your standard broiler mix. We try to keep this feed program in front of them free choice at all times. Water is a given; they drink a lot of it!

Let me now talk a little about marketing. I believe our turkey availability in the fall is the major reason November is always our best sales month of the year. Many of our new customers come to us first to get one of our fabulous turkeys. It's just natural for them to buy some chicken, beef and pork while they're at it. Many up and coming health-conscious consumers think about an "All Natural" Turkey for their Thanksgiving table first. Once they've found you, many will take the next step of buying your other products regularly. Turkeys can be a great entree to the rest of your product line!

To sell out of our turkeys each year, we have done little more than to highlight their availability in our fall newsletter. We also have the luxury of a state inspected processing facility in our area which enables us to sell through the health food stores as well. I've heard of some producers who advertise through health food stores; the stores take the orders, earn a commission and the producer delivers directly to the customer. When the product is this good, creative producers will figure out how to get it in the market place and creative consumers will figure out how to get it on their tables.

In terms of how a turkey enterprise can fit into the overall operation of your farm or ranch, I believe it's as natural as it can be. If you're raising broilers right now, turkeys can use underutilized pens and equipment late in the year. The potential to increase revenue through additional sales is real for not only the new product but also for additional sales of your existing product lines as well. I can't understand why any pastured poultry producer wouldn't do it. Give it a try; it's worth it for the fun factor alone. The increased profits will be gravy!

We will raise 200 turkeys this year but after having such a great time writing this article, I may raise twice as many next year–it just reminds me how much fun they are. Turkeys truly do have feelings too and they reward you for taking good care of them. When was the last time your broilers made you feel this good?

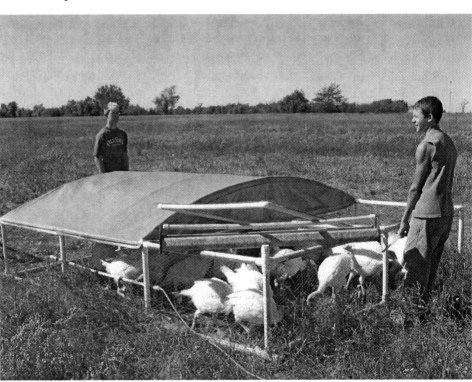

Bennett Boys Moving Turkeys

APPPA PRODUCER PROFILE

Alexander Family Farm
By Dan Bennett Winter 2004

Alexander Family Farm is truly a family farm in every sense of the word and is built on a solid biblical foundation. Kim Alexander holds firmly to the belief that at Alexander Family farm they will grow food in a way that is as close to nature as possible. In addition, the family desires to produce as much of their own food as possible. Kim sees this as one of the basic tenets of agrarianism and believes that this approach is as important today as it's ever been. We applaud Kim's efforts to achieve this lofty goal!

Kim Alexander has been working to build the family farm for many years and reached a milestone in 2003 by becoming a full time farmer. Alexander Family Farm is located in Dell Valle, Texas, just outside Austin.

The Alexanders currently have around 1,000 layers producing at the most recent count 50 dozen eggs per day. They have two milk cows from which they sell milk and have a herd of 45 beef cattle. Last season they produced close to 2,000 broilers and 265 turkeys. They slaughtered six beef cattle and 10 grass-fattened lambs. Diversification is obviously a key to the operation of Alexander Family Farm.

In addition to diversification, however, Kim is a master at recognizing an opportunity and seizing it. When he saw a market need for lamb, he recognized that 4H lamb projects in his area afforded an opportunity and quickly bought the show lambs, cleaned them up on pasture for two months and then had them butchered, realizing a nice profit. Kim also does a great business during his slow months of January, February and March by offering custom poultry processing to the county fair producers around Austin. Kim earned over $10,000 last year for his efforts and sees the potential for over $15,000 this year.

The Alexanders turkey production includes raising 30 Bourbon Reds in cooperation with "Slow Foods." They charge $3.50 / lb. for the heritage breed at an average weight of 10 lbs. and $2.25 / lb. for their whites at an average weight of 25 lbs. Kim charges a $15 deposit for each turkey, which he feels is critical for the following reasons: it assists cash flow at a time when output is high; it makes sure people are serious; it takes some of the pain out of the customers' final payment; and it prevents over-selling or under-selling by ensuring an accurate count of turkeys. Their turkey business received an incredible boost by an article about Alexander Family Farm in the Austin paper the week before Thanksgiving. This not only drove the demand for turkeys higher than last year's supply but also paved the way for an expanded customer list for all products in future years.

Like all other facets of the Alexander Family Farm, Kim's broiler production is unique. In fact, be on the look out for a manual Kim plans to write in the near future describing his broiler production practices which falls somewhere between pasture pens and day-ranging. Kim believes his approach is especially effective at handling weather fluctuations and the extremes of heat. Kim raises his broilers in batches of 300-400 throughout the season, which begins in February and ends in November.

The Alexanders use small (11 x 20 ft.) steel pipe framed shelters on skids. They place 300-400 birds per shelter, and move it within a poultry net yard daily. They use 5 ft. wooden home-made feeders and large PVC waterers fed by a gravity system. Their pasture is composed of cactus, mesquite, crab grass, burr clover, Bermuda and native grasses. They bush hog yearly to keep the mesquite back and drag to keep the cactus back. No other pasture treatment is used. They currently have no irrigation; however, they are looking into tapping into a spring on the property. They are currently experiencing severe drought conditions. Kim says "the average yearly rainfall in south central Texas is 35 inches. You ought to be here on the day it comes!"

APPPA PRODUCER PROFILE — *Alexander*

The Alexanders raise their broilers to eight to nine weeks of age, the heritage turkeys to seven months, and the white turkeys to six months. Their overall losses are holding at about 10% and most of that loss is brooder-related. Kim notes that sons Joshua and Jeremiah are the brooder house managers and do an excellent job at 13 and 9 years old respectively.

In terms of cost management, Kim may have no equal! An old saying seems to fit the Alexander family – They have learned to do so much with so little for so long that they are now able to do almost anything for nothing. Kim jokingly refers to himself as a "junk yard" farmer. Be on the look out for a book entitled "Junk Yard Farming" by Kim Alexander on the New York Times best-seller list in years to come! Kim's unfair advantage is his ability to keep his costs down by making and adapting things that work for him.

Kim broods his poultry in two 12 x 12 ft. A-frame brooders. He uses a home-made hover with electric heat and broods for three weeks. He and the boys make heavy use of wood shavings to ensure cleanliness in addition to using nipple waterers exclusively. They build up a compost bedding in their brooders and only clean them out every few years or so. Kim has invested in one piece of new equipment in his brooders – $15 small tin feeders from the farm store to start the chicks.

The Alexander labor pool is, as you might imagine, all family. Kim is full-time and his sons make up another person. On processing days, it's all hands on deck and, as mentioned earlier, they do a fair amount of custom processing as well.

The Alexander family built their own processing facility—a 20 x 30 ft. open shed patterned after the Salatin design. Kim built most of his own equipment (junk yard farming once again)—although he recently stumbled upon an old Ashley scalder and they are now doing 120 birds per hour. Customers come out on the day of slaughter to pick up birds—the unsold birds go in the freezer for sales between processing days.

Kim's marketing approach has evolved over the years from primarily word-of-mouth to now using multiple approaches. Twice a year the Alexander family send their newsletter to over 200 customers. They joined the Austin farmers market in 2003 and sell eggs, take orders for meat and sell a number of family made crafts. These include mesquite wood spoons, soap, candles and baskets. He also does slide shows to community groups which has been very effective for him. Kim is doing a fair amount of direct marketing at this point to sell his growing volume of 300+ dozen eggs per week. The Alexanders' see their egg volume peaking at 500 dozen per week.

Kim recognizes his record-keeping is slim and a weakness at this point in time. He will solve this challenge by assigning this "real life" bookkeeping project to his older home-schooled children.

Kim sees his fair advantage as his personal beliefs combined with his "can-do" attitude, a willingness to work hard and his good location. His greatest challenge is the central Texas climate.

Kim is a strong believer that the low capital / sustainable agricultural model has given us all a way to make a living in agriculture. The Alexander Family Farm makes it all work with a high degree of integrity and no compromising of their basic beliefs. Kim believes that the people in this movement have been a huge factor in his success. Their openness and availability to him have been a huge help and encouragement to him. Kim's friends and colleagues have openly shared in their successes and failures and Kim has been a willing listener.

It's Turkey Time Again
By Don Brubaker Summer 2005

I trust that you all have had a great winter, and as we all have anxiously awaited spring, we anticipate another turkey growing season. Whether you have grown turkeys for many seasons or you are just beginning a new enterprise on the farm, it is always good to revisit your management practices before you get started.

Poult Arrival
It is important to always start your poults in a clean and disinfected brood pen. You should start the heat 24 hours before the poults arrive so that the temperature is 90-95° F at poult level. You need to decide whether you are going to heat just the floor or the whole room. To heat the floor your heat source needs to be knee-high; to heat the room you need to have the heat waist-high. You should observe the poults several times throughout the day to see if they are comfortable. You can tell whether or not they are ok by the way the birds are positioned around the heat. If they are under the heat huddled together, most likely they are cold. If they are in a nice donut around the heat they are most likely comfortable. If they are really spread out along the walls and to the corners, they are most likely too warm and you will need to adjust the temperature.

The area that most producers neglect is air quality in the brood pen. In trying to save fuel cost the producer keeps the brood area too closed up and in the process of trying to save a nickel they spend a quarter. What I am saying is that most of the health problems that occur during the grow-out started in the brooding pen. Always have an adequate supply of fresh air at the right temperature and low humidity and you will be good to go. In order to do this you will need a means to move air, either a fan (a timer fan is good) or a curtain to help exchange air.

There are many types of bedding that you can use but I will just mention three types in this article. One of the best beddings to use is kiln-dried pine shavings, but sometimes we find that the poults cannot tell the difference between the shavings and the feed. To help with this you can use peat moss, which works very well but does have a tendency to be dusty. You can also use indoor-outdoor carpet for the first week or so and then just remove the carpet and wash and disinfect it for the next time.

Clean water is the single most important thing that goes into producing a healthy turkey. Make sure the water supply is clean and stays clean throughout the grow-out cycle. You should have your water checked on a yearly basis. This will head off a lot of problems. You should also clean the drinkers on a daily basis. You can also add a cleaner (citric acid) to the water lines on a weekly basis. Citric acid is available at your local agricultural store—directions are on the bag. Along with clean water, having enough space for the poults to drink is just as important. As a rule of thumb you should have one inch for every poult at the start and increase by a quarter inch every week until they are ready for market. Don't forget that clean water and enough drinking space will go a long way to having a healthy flock of turkeys.

Starting with a good feed will get the poults growing from the moment the poults are put in the brood area. You should start with a 26-28% protein ration and feed this for the first four weeks. Make sure you have enough feeder space. Poults have the best feed conversion in the first four weeks of their life. Sometimes poults have difficulty distinguishing between feed and bedding, especially if you are using shavings. You can help them by putting marbles or coins in the feed to make it shiny. This will attract the poults to the feed.

Grow-out
By the fourth to eighth week grow-out period, the turkeys should be well on their way and growing well. At this time the temperature can be 70-75° F and at this time the birds should be on a 21% protein ration. Continue to feed and water as discussed previously.

Finishing
By eight weeks of age your turkeys are ready to be moved outside on pasture. I find it best to day-range turkeys. This can be done by providing a simple moveable shelter. The shelter should provide 1.5 sq. ft. when you first move them out and by the time they are 20 weeks of age they will need 5 sq. ft. per bird. You can keep them in a given area by using a portable

fencing system made by Premier or Kencove—either one will work great. On pasture, start with the same space outside as inside; as the turkeys get older, provide enough space to keep the pasture fresh. The size of the paddocks all depends on the time of year and quality of the pasture that you have for them. Some people claim that the turkeys will work their way through the fence so some will double up on a fence and stagger it to make the holes smaller. You will only need to do this until the birds are big enough to not slip through.

To feed the birds on pasture you may use tube feeders. Trough feeders can be made from 4-6 in. PVC pipe by cutting out one-third of the pipe so they can't waste as much feed and put end caps on it. Some even attach several troughs on a skid frame so they can easily move it around the field.

To water your turkeys on pasture, you may use bell waterers supplied by a 5-gallon bucket suspended overhead. Or a 4-6 in. PVC pipe as described above but with a float on the end to control water level.

Pasture Nutrition
Raising turkeys on pasture can be a lot of fun and enjoyment for you and your family so there's no need to make feeding them too complicated. Many of you are raising broilers on the farm so I recommend just using the broiler diet and adding extra fishmeal to meet the protein needs (see below for recommendations).

19% Broiler Starter/Grower:
1015	lbs.	Shelled Corn
625	lbs.	Roasted Soybeans
100	lbs.	Oats
100	lbs.	Alfalfa Meal
75	lbs.	Fishmeal, 60%
25	lbs.	Aragonite (calcium)
60	lbs.	Nutri-Balancer
2000	lbs.	

You may add 4 lb. of fishmeal to 20 lb. (5-gal. pail) of Broiler Grower 19%. This addition will provide a 26% protein mix for turkey and game bird starter. To be fed from day one thru day 28.

You may add 2 lb. of fishmeal to 20 lb. (5-gal. pail) of Broiler Grower 19%. This addition will provide a 21% protein mix for Turkey Grower #1, to be fed from day 29 thru day 56.

Once your turkeys are out on pasture they should receive regular Broiler Grower until slaughter.

If you are one of those that feed more than 200 turkeys a year it will make more sense to have a more specialized diet for them. Here are some recommendations that will work great.

Turkey Feeds for large-scale producers (200 or more) and turkey-only producers:

TURKEY STARTER:
    1 day to 42 days: 24% Protein

300	lbs.	Shelled Corn, Ground
100	lbs.	Fishmeal, 60%
1000	lbs.	Roasted Soybeans, Grnd
500	lbs.	Spelt
7.5	lbs.	Dynafos
15	lbs.	Aragonite
80	lbs.	Fertrell Poultry Min.
2000	lbs.	

TURKEY GROWER 1:
    43 days to 70 days: 20% Protein

870	lbs.	Shelled Corn, Ground
50	lbs.	Fishmeal, 60%
980	lbs.	Roasted Soybeans, Grnd
12.5	lbs.	Dynafos
20	lbs.	Aragonite
70	lbs.	Fertrell Poultry Min.
2000	lbs.	

TURKEY GROWER 2:
    71 days to 98 days: 18% Protein

1037	lbs.	Shelled Corn, Ground
860	lbs.	Roasted Soybeans, Grnd
12.5	lbs.	Dynafos
20	lbs.	Aragonite
70	lbs.	Fertrell Poultry Min.
2000	lbs.	

TURKEY FINISHER:
    99 days to Market: 15% Protein

1340	lbs.	Shelled Corn, Ground
580	lbs.	Roasted Soybeans, Grnd
20	lbs.	Aragonite
60	lbs.	Fertrell Poultry Min.
2000	lbs.	

Once the turkeys are eight weeks old you can start feeding them free-choice whole wheat, which they will love. Start with just one feeder and add feeders as they consume more wheat.

Processing and Marketing

Many of you that have been raising turkeys already process your own birds which is a good way to add to your bottom line, but still many of you may have someone else do the processing for you. Either way, let's take a look at some more ways that you may be able to add to the bottom line.

Most producers just sell whole birds at Thanksgiving and Christmas, but consider more value-added ways to sell your turkeys. One of the neatest ways is to smoke the turkey-they taste great that way. You will need to check out the regulations in your area so it can be done legally. You may even be able to do it yourself, which would be great.

Another way would be to make ground turkey—many prefer the ground turkey breast but you can also grind the legs, wings, and thighs for the dark meat or mix both white and dark together. You can also sell the breast whole, bone-in, or boneless and then grind the dark meat. Many households like just a whole turkey breast for Thanksgiving or Christmas. So offering to grind the rest would give both you and the customer a valuable product. You won't be stuck with dark meat and they can use the ground turkey in place of hamburger. A new fad is deep-frying the whole turkey. Keep that in mind in your advertising. Good luck!

Research Adventure
By Ryan Rich Spring/ Summer 1998

Have you ever considered raising Thanksgiving Turkeys? If so, did questions like these pop into your head? How many turkeys should I raise, how much grain and water will they require, who will do the slaughtering, who will I market to, will the birds be delivered or picked up, is this production method good for the environment, is there enough cash flow, and most important of all, do I have enough patience and drive to undertake this project?

These are the considerations Emanuel Farrow and I thought of when applying for a Sustainable Agriculture Research and Education (SARE) grant last winter. This grant program helps small farms and organizations conduct research to enhance production methods, protect the environment, and educate the community. Partial funding for the work reported was provided by a grant from the USDA-SARE program.

Emanuel and I raised 2,000 chickens in 1996 in "chicken tractors" on pastures at Future Farm Organics, a NOFA (Northeast Organic Farming Association) certified farm in Vermont. This year we decided to raise turkeys because their weight and foraging capabilities proved to be more efficient than chickens. The turkeys required less time over the season because we didn't need to start raising them until early August for the Thanksgiving market. This allowed us to devote time to a developing vegetable CSA and a pastured beef operation.

We used the "chicken tractors" in 1997 as "turkey tractors." The grant proposal to raise 200 turkeys for Thanksgiving was accepted. Even though our farm was certified organic, a lack of USDA organic meat standards made it costly to not be able to market our poultry products as organic, on labels and in stores. So we decided to buy five-week-old poults for $5.50 a piece from a free-range poultry farm an hour away. Half were purchased in August weights. This eliminated the cost of brooder pens, lights, electricity, sawdust, and starter grain. Calculations revealed it was a cheaper way to start in grain costs alone. Not to mention that out of our 200 turkeys, we only lost five to pneumonia, and that was within the first week of transporting them from the free-range barn of chips, sawdust, and manure to our nice, lush pastures.

Our production methods included moving our 10 x 12 ft. hoop house "turkey tractors" every morning with a dolly. We started with about 32 turkeys per pen in six pens. We were lucky and had a gravity-fed spring line run to two 15-gallon tanks in the field which made watering each pen with a 5-gallon gravity feed easy.

Twenty pound feeders were used in each pen. Organic grain was our biggest cost at $440 a ton (about $140 a ton more than conventional grain). The grain came in reusable bags instead of bulk for sheer ease of transport to the field in a garden cart. It also gave us a precise knowledge of the amount of grain used per day. Seven tons were used in total.

Turkeys were watered and fed twice a day (morning and night.) About the second week in October a seventh pen was added to thin the birds down to 25 per pen. There was no problem mixing the flock in that seventh pen.

On November tenth, the temperature dropped considerably, the water line froze, and three feet of snow fell in the fields. The turkeys held up in that weather better than we did. They were moved a few more days until it rained a little and then the pens froze to the ground. We carried water 100 yards in 5-gallon buckets from the dairy barn on a snowy path. One-gallon buckets were used in the pens because the turkeys would drink the water faster than it would freeze. We went from spending four hours a day to seven feeding and watering.

Slaughtering was our second biggest cost. Since this was our second year in production, our "on the farm" market was not big enough to do all of the slaughtering ourselves. We sold quite a bit to cooperatives and health food stores throughout Vermont. The difficulty was getting the birds to the facility. No trailer was available, and we didn't want to use poultry crates which could have injured or bruised the turkeys. So we put the cap on the pickup truck, loaded and transported forty at a time. The slaughter plant was an hour and a half away. It took three days to get all of the turkeys processed. On top of feeding and watering the rest of the birds, marketing and selling, there was just enough time to eat lunch. We were now working eighteen hours a day and still hadn't sold all of our turkeys.

Marketing Thanksgiving turkeys is hard for farmers who don't have a well-established, devoted clientele. I was advised by other farmers to start with less and build up slowly. But my gung-ho attitude and the small margins in agriculture always made me think bigger, especially when my largest production vision didn't compare to the immensity of conventional poultry farms. We sold about 30 turkeys for $2.75 per pound right off the bat in August. That made us confident that all the birds would be sold in no time. The problem was that as turkey producers we were thinking Thanksgiving at the end of the summer, while consumers don't think turkey until the week before Thanksgiving. By the first week in November, 100 had been sold. That meant there were two and a half weeks to sell as many birds as we had in the last three months. I was on edge to sell, sell, sell, while Emanuel started plotting a Christmas turkey market. On the Thursday before Thanksgiving the phone started to ring every 20 minutes. Sign up sheets given to stores were filling up; people who had heard of our organically raised pastured turkeys were insisting on getting one for their families, and there were even questions about Christmas turkeys. The next day (Friday) we were completely sold out and had to turn away 30 potential buyers. What a relief to have some of the stress finished.

Then it was time to get the USDA-inspected turkeys in vacuum-sealed bags to the people. Fifty birds were picked up at the farm by customers. The distribution of seventy-five turkeys to six stores was easy. Each stop the truck was eight to 24 turkeys lighter. About 75 were delivered to individuals at a charge of $5.00 a stop. The delivery fee wasn't worth the time spent driving all over New England. It's a good thing that Vermont Novembers are so cold. It reduced refrigeration costs to zero. Had it been warmer, fresh turkeys would have been much more of a problem, and frozen would have been a safer, but less profitable way of distribution.

As part of the grant, soil tests were taken before and after pasturing the turkeys to see their effects on the soil and grass quality. We hypothesized that integrating turkeys on hayfields would increase soil and hay quality, and would decrease weed and insect problems.

One of the most important figures in the table is the Effective CEC (Cation Exchange Capacity). The increased CEC indicates the ability for plants to have positively charged ions (cations) more available for uptake. Micro-nutrient test results were average for VT soils, which are usually not deficient. Next spring, once the manure has had longer to assimilate into the soil, more informed soil and forage tests will be taken.

There was an environmental concern with raising pastured turkeys. Even though organic grain was used, it came from as far away as Washington and Brazil. There weren't enough local sources to fulfill the demand. We consider grain a high input, because it has to come from off the farm and be transported far distances. It compromises our local production philosophy. On the other hand, we felt diversifying farm operations enhances farm health as well as economic stability. Farms mimic natural systems more and more, and the farmers become less dependent on one crop.

These are some of the issues to consider before taking on a Thanksgiving turkey venture. We hope that our experience can benefit any future poultry plans you may have. The most important thing to remember is attitude. Throughout the entire operation one must remain steady enough to get through the rush, the erratic hours, and the many unexpected obstacles and lessons. If that can be done, as Emanuel and I found out, the chances for success are greatly increased, physically, emotionally, and economically.

Soil Test Results	Before Turkey Pasturing	After Turkey Pasturing
PH	6.0	6.3
Available Phosph. (ppm P)	9.0	12.6
Reserve Phosph. (ppm P)	172	230
Potassium (ppm K)	60	98
Magnesium (ppm MG)	53	92
Aluminum (ppm Al)	71	54
Calcium (ppm Ca)	691	102
Effective CEC (meq/100g)	4.1	6.1

Raising Historical Turkeys
By Jody Padgham Fall 2002

Royal Palm, Blue Slate, Bourbon Red, Narragansett, White Holland, Black, Broad Breasted, Bronze. What is this, a list of beverages available in a smoky bar? No, it is the breeds of historical (or heirloom or heritage) turkeys being raised by Mike Walter at the Walters Hatchery in Oklahoma.

Mike has been committed to the preservation of historical turkey genetics for ten years, starting when he was only 20. When asked why heirlooms are important, he explains that these are the birds that were historically raised throughout the country. The average grocery store bird, typically a white turkey, comes from genetic stock developed in the 1950s. "In the 2000 turkey census there were only 108 registered Blue Slate hens left in the U.S.," Mike tells me. "It is important to keep the diverse genetics of these birds around, and intentional preservation is critical."

Commercial birds are now of only one genetic type, and in fact don't even reproduce naturally any more. They are now bred using artificial insemination. Commercial birds grow fast, are efficient eaters and have been bred to be disease resistant. The historical genetics Mike is working to preserve represent genetic lines of birds that take much longer to grow out (six to eight months vs 16-18 weeks for commercial birds), and as a result, have a much fuller flavor. These are the kind of birds your grandparents ate and still rave about.

Walters Hatchery
The main focus of the work Mike is doing is genetic preservation of these turkey varieties. Mike will be keeping 300 breeding pairs, a mix of the above mentioned varieties, to produce poults and breeding pairs to sell for next year. Eggs are laid and incubated from April through mid-June. Walters Hatchery is a state licensed facility, and Mike ships day-old poults and nine-month-old breeding pairs around the country. (www.historicalturkeys.com)

Mike dispels the rumor that turkeys are hard to raise by noting that the key to a successful historical turkey production operation is in starting with healthy birds. Mike is very proud of the quality of Walters Hatchery birds and chicks. Each breeding pair is blood-tested for several strains of avian disease every nine months. "Problems are almost always imported through introduction on new birds," Mike says. If you eliminate this potential by buying clean birds, you will have an excellent chance of success.

Walters Hatchery also produces historical birds for the slaughter market. Having the extra 400 birds around for slaughter not only brings in significant income to support the breeding work, but also allows Mike to pick only the best birds to retain or sell as breeding stock.

Connected to the popular "Slow Foods" movement, Mike has received national recognition for his birds and is able to claim a hefty $3.50 per pound for his dressed historical birds, which usually range from 15-18 lbs. for hens and 18-22 lbs. for toms at slaughter. Those not picking birds up at the hatchery also pay shipping, which can be as much as $52.00 per bird (fresh and iced, shipped next-day air). "Once people try these birds, they always come back for more," Mike comments. "Even people in the neighborhood feel it is worthwhile to pay the extra price for the longer grow-out and fuller flavor."

Raising Turkeys
Mike was willing to share a few tips on raising turkeys.

¤ Most people make the mistake of raising turkeys like chickens which is like raising horses like cows. They are different than chickens, and need to be raised in deference to their natural tendencies.

¤ Keep new poults on clean wood shavings or wire mesh for the first eight weeks. It is critical to keep the area clean and fresh to discourage salmonella. Reduce the temperature 5° F per week in the brooder until you reach the outside ambient temperature. (This will have to be approximate in some climates.)

¤ Move birds out to range after eight weeks. Mike keeps his birds in 7,500-ft. octagonal-shaped open pens. He has eight pens on his place and rotates each batch of birds between two pens. Bird density is something he is still working on, and he is not comfortable at this time making recommendations. Mike suggests working with your state Extension personnel and Department of Agriculture when designing your pen system--especially if you are close to water or drainage of any kind.

¤ Grown-out birds range on seven to eight-acre fields, with less than 1,000 birds per range. Mike sets up shade areas and cover for the birds, but they rarely use it. The birds are personable and easily herded, and, when produced from disease-free genetics, are very strong.

¤ Mike feeds his turkeys a ration that was developed in 1936 at Ohio State. (He would like to keep the particulars of that secret.) He recommends finding old turkey ration formulas, but emphasizes that turkeys need 28% protein all the way until slaughter. His mix uses less soybeans, and is heavier on wheat and milo than most, and is as natural as he can find but not organic.

¤ Go to a USDA-inspected facility for slaughter, so that you can ship birds across state lines.

¤ If you are buying breeding pairs, visit the birds or ask to see pictures before you commit to buying them. The hens will contribute shape and confirmation to their offspring, toms contribute color. Genetics will dictate the amount of meat a bird will carry--some people breeding birds will produce for quantity of stock and not quality. You will want to keep an eye out for good confirmation in your breeding pairs. Hens will lay eggs every 36 hours from April through August and can lay for an average of nine years. Toms will be fertile for about six years.

¤ Disease-test any breeding pairs you will be keeping at nine months. Blood can be drawn by a vet and sent to labs at most poultry science departments at Universities. (Mike is a medic and does his own blood drawing, which makes this economically feasible at his scale.) Keep any new birds isolated from the existing flock unless you are sure they are disease-free or they have been tested.

Mike markets his birds fresh, as historical turkeys, but with as much emphasis on "the greatest tasting turkey you will ever have." He requires his customers to put down a deposit in April for birds he will have butchered this year on November 23rd and shipped overnight on Nov 25th. Pre-ordered bird orders are invaluable for planning, but Mike will also grow out 200-300 extra birds for later sales. He butchers birds only for the Thanksgiving and Christmas market, but sells chicks and breeding pairs throughout the summer months to balance out his income.

We complete our conversation with Mike confirming that he can and will talk about turkeys all day. I recommend you check out his website (www.historicalturkeys.com) and consider buying a few breeding pair of your own to try your luck with these beautiful birds. Walters Hatchery is managed by Mike Walter and owned by the Walter family.

Additional Note:
Joel Salatin did a workshop and stopped by the APPPA table at the Ohio Small Farm Field Day in September this year. He has been doing a lot of work on turkeys and had a few interesting things to say:

Royal Palm Turkey

¤ Get the young birds out on grass early--even at only one to two weeks. Salatin raises his young birds in insulated houses that are out on fresh grass (not used for grazing in the last year).

¤ Turkeys have a hard time with fences. If you want to get them used to moveable fence, you have to start early, otherwise they will just run through it. Salatin puts fence around the insulated houses (with doors) the two-week-old poults are using. He uses poultry netting that has been DOUBLED so that the holes are small enough and the turkey poults can't squeeze through. He has found that they learn well enough that the fence is a barrier at this early age that they will then respect a fence.

¤ Salatin confirms Mike's perspective that turkeys are indeed different to raise up than chickens, and that you need to try as much as possible to duplicate their natural habitat. He notes a major difference being that turkeys like a lot of space to roam and do best in open-topped LARGE ranges (such as Mike's seven-acre runs). The density of the birds can be high, but they want to all be able to range over a lot of territory every day.

¤ Turkeys are very personable and friendly and will sit on your back porch if you don't contain them, so keep this in mind!

Raising Heritage Turkeys
By Kip Glass Fall 2003

After two years of raising different breeds of heritage turkeys, I'd like to share my experiences in raising and selling these wonderful birds.

Why are heritage turkeys gaining in popularity? First, there are several groups currently promoting old breeds of animals and plants. If we don't keep the older plant or animal genetics around, we loose the source from which our current hybrids have been selected. One group recognizing and promoting this is "Slow Food USA" (slowfoodusa.org). Heritage turkeys are the breeds that were raised in the past. These turkeys can breed naturally, unlike current commercial hybrids which must be bred artificially since their large size interferes with the natural mating process.

Before hybrid birds were developed, turkeys took over 22 weeks to mature. Modern commercial operations using the Broad Breasted White (BBW) turkey raise them for an average of 16-18 weeks. When a turkey is allowed to mature past 22 or 24 weeks of age, it develops a fat layer under the skin that adds to the flavor and self-basting of the meat. If we let the BBW turkey mature to that age it would weigh in excess of 30 lbs., larger than most consumers would be interested in. I can tell you, once you have tasted a mature, pasture raised, heritage breed turkey, you will not be interested in another dry, hybrid, confinement raised turkey.

So far I have experimented with the following heritage breeds: Bourbon Red, Royal Palm, Black Spanish, and the Broad Breasted Bronze. In the near future I would like to also try the standard Bronze. My preference so far is toward the Bourbon Red. It is an aggressive forager and dresses out into a very visually appealing bird. You will need to do your own research and experimentation to discover which breeds you personally like and what breeds you feel give the kind of product you want to sell, or what breeds your customer base prefers.

Now, let's get down to my experiences of costs and trials of raising these birds. First, the cost difference in poults is very significant. My BBW turkey poults are $1.75 each, the heritage breed poults average $7.00 plus. As far as brooding, I can't see much difference in how it needs to be done, but I feel the heritage breeds are a little hardier and probably have fewer genetic frailties than the hybrids have. You will want to start them off right, because if you loose any of the heritage poults you have instantly lost over $7.

As for pasture, you can either raise them in pens or let them range. I feel you are depressing these turkeys if you don't let them range. They love covering the ground and just aren't happy in a small pen. Currently, I am encircling an area with three—165 ft. electrified poultry nettings four ft. tall. They can fly over the fence, but seem to do it by accident when exercising their wings and playing around. Usually they can get back in on their own or will climb right back in if I lay the fence over. I'm using an open sided and open-ended 20 x 16 ft. shelter with roosts underneath the covering for 80 birds. Now the large 15-week-old maturing birds are starting to roost on top of the 10 ft. high structure. I'm still trying to figure out how to stop or capitalize on this. I may just not have one side covered next year or do a separate roost skid so they have the option of being under cover or not. I move this skid and netting approximately every 10 days to allow them fresh range.

Another extra expense to consider is labor. Since you are raising the Heritage turkeys to approximately 24 weeks you have roughly eight more weeks of labor to factor in compared to the BBW turkeys. Of course, in an open range situation the labor isn't that difficult.

I haven't been that precise in the feed requirements of these birds, but it is obvious that because of their more aggressive grazing, they don't require as much feed.

Currently the price we are getting per pound is $2.30 for the BBW turkeys and $3.40 for the heritage turkeys. We have outsold two to one the heritage turkeys. I don't know if it is our incredible sales description or the fact that people want to experience something different. People that purchased the heritage turkeys last year said it was the best turkey they have ever experienced.

As for income comparisons, I'm not too precise, as of yet, because of the unknown feed consumptions. But, as stated, I know I'm feeding less feed per heritage bird compared to the BBW. Also, something you must factor in is that the heritage breeds won't average out as heavy a dressed weight as the BBW.

Let's compare a 16 lb. BBW turkey to a 14 lb. heritage turkey at my prices.

The BBW would be $36.80 (16 lbs. x $2.30) gross income, compared to $47.60 (14 lbs. x $3.40) for the heritage breed, a $10.80 difference. Let's deduct out of that gross income difference $6.00 for the difference in poult cost and you have $4.80 more profit per bird for the heritage breed. Of course, the heritage breed requires less feed, so more than likely at my prices per pound, I will make roughly $6 per bird more profit with the heritage turkey.

More experimentation and more accurate record keeping needs to be done, but, initially, there seems do be a potential for extra income in raising the heritage breeds.

Don't forget the intangible benefits of helping to preserve a vanishing breed of turkey and the added enjoyment of raising birds with character.

In closing, give them a try; you might enjoy them as much as we do, and so will your customers.

Blue Slate

Naragansett

Ducks and Pheasants: Taking the Salatin Model One Step Further
By Chuck Benhoff Spring 2004

This article was first published in the April 2000 edition of the Stockman Grassfarmer. It is reprinted here by permission of the author. When I spoke to Chuck around the first of April 2004 he told me that, although the article is a few years old, not much has changed on their farm.

FARMVILLE, Virginia:

In 1995 my wife Tammy and I jumped into pastured poultry with 1,500 laying hens and 24 mobile Salatin-style cages. Our goal was to produce 100 dozen eggs per day which would be marketed in the Richmond area of central Virginia. We quickly had a volume of over 100 dozen per day and, to my surprise and relief, the eggs were easily sold. Most went to a large chain grocery as well as restaurants, health food stores, etc. The eggs were of such high quality no buyer questioned the $1.50 per dozen price tag.

Our cost per day was roughly 50 cents per dozen. Simple math shows a $1.00 per day profit, seven days a week, part-time farm enterprise. The entire project was run day-to-day by Tammy. In 1997, our second son, Carl, was born and the labor and time involved in the pastured egg business was impossible for Tammy to maintain. My off-farm job did not allow time to take over the project, so reluctantly we decided to sell the chickens.

This left us with 24 empty pasture cages. We kept about 100 laying hens for personal and family use and raised about 100 broilers each summer. Several neighbors bought or traded cages and started home flocks of pasture layers. We still had about ten cages sitting empty on the edge of a field, overgrown in weeds.

For several years I had been intrigued with the idea of raising other types of poultry and waterfowl in pasture cages. Joel Salatin raises broilers, layers and turkeys. I wanted to try some high-value exotic fowl for possible sale to upscale restaurants. We decided to raise two breeds of duck and one meat-type jumbo pheasant.

Due to the cost of the pheasant chicks, we ordered eggs and hatched them in our incubator. Ducklings were shipped live and both the ducks and pheasants were started on pasture during the first week of July.

If weather permits, we usually start chicks in our yard on grass, which has been recently mowed. Chicks can navigate on short grass at about four days of age. We use 3 x 4 ft. mini-cages to get chicks started on the tender short grass. Being close to the house, we can run an extension cord to the mini-cages and keep the chicks warm with a heat lamp at night. This also discourages any would be predators which may take advantage of the baby chicks. Rats, opossums, raccoons and skunks have all wreaked havoc on our chicks at some point in time.

Ducks in Pasture Pens

The ducks and pheasants were started the same way and at about ten days went to pasture in full-size cages alongside 100 broilers, 50 laying hens and 85 replacement pullets. Ducks were started on 24% protein waterfowl pellets and the pheasants on 26% protein turkey starter. After four weeks, both were switched over to market bird finisher, which is the same feed we use for our broilers.

From the start, the ducks were easiest to handle. Not once did a duck ever escape from a cage. We lost only one duckling, who was a runt and only lived a few days beyond delivery. The growth rate of the ducks was amazing, comparable to Cornish Cross broiler.

We raised two breeds of duck: White Pekins and Rouens. The Pekins showed much better feed conversion and higher meat yield. The Rouens are beautiful birds colored exactly like a mallard. The breeds were mixed in a pasture cage.

We had absolutely no problem raising the ducks. At about six weeks of age we gave them a tub of water to play in and keep cool. Without the tub, they quickly emptied the 5-gallon waterer. The cages were moved every other day and the ducks cleaned up nearly all of the grass and clover that they could reach.

The Pekins finished in 12 weeks at eight to ten pounds liveweight, and the Rouens were about 25-30% smaller.

Pheasants, on the other hand, are a bird of a different feather. I had been told that a pheasant could never be confined in a 10 x 12 ft. cage, although I could not find anyone who had tried. We chose white meat-type pheasants in the hope that the domestication process had settled the birds down. Pheasants are notorious for picking on each other, especially the roosters.

We quickly learned how fast and far these young pheasants could fly upon escape. Their aerial acrobatics were only outdone by their foot speed. The young pheasants were so small, they regularly escaped the pen when moved. Any small opening under the cage was quickly taken advantage of by numerous sparrow-sized pheasant chicks. Up to about six weeks, the young pheasants could be caught by the children, but once they were older a net was generally necessary to corral the escapees.

We clipped the pheasants' wings, hoping to ground their flying tendencies, but the wing feathers quickly grew back. During the first eight weeks we lost only one pheasant due to a huge black snake.

The pheasants' growth rate was slow compared to the ducks or broilers. At eight weeks, we considered just turning them loose, they were so small. During the next month the little birds turned it on and at 12 weeks weighed three to four pounds each.

At this age the pheasants were getting very aggressive with each other. Many on the lower end of the pecking order had bare spots on their heads. This probably could have been alleviated some by splitting the group up. We had 40 in the cage and that was probably too many. These pheasants appeared to be much smaller than their wild cousins; however, we would find out later that looks can be deceiving.

All the ducks and pheasants were killed at 12 weeks. We processed them exactly as we would a broiler chicken. The ducks needed about twice the scald time due to their thick, oily feathers. Once properly scalded, both ducks and pheasants plucked easily.

The Pekin ducks dressed out as plump, maybe even too fat, but beautiful five to seven pound roasters. The Rouens were more narrow and less fat, approximately three and one half to five pounds. Pheasants were two to two and a half of nearly all meat.

We saved the duck livers for a Swiss chef who had moved to the area. When he came by to pick up the livers, he admired the dressed ducks and said the only duck he had been able to find was in Charlottesville at $4.00 per pound ($4 x 5 lbs. = $20 per bird. Is anyone calculating profit margin yet?) Including cost of chicks, we had a total of about $5 in each duck and $4 in each pheasant. I am sure the cost of chicks and ducklings could be much lower if we had purchased in larger quantities. We started with 45 ducklings and hatched 41 of the 100 pheasant eggs.

The duck and pheasant added a much needed variety to our farm-raised menu of beef, chicken and pork. The end product is a culinary delight. The duck is tender, moist and mild in flavor. The pheasant has the distinctive flavor of the wild birds with abundant meat.

As you may have guessed, next year more of the old laying cages will be filled with ducks and pheasants. We will stick with the Pekin duck and probably experiment with another variety of meat pheasant.

Each evening we sit down to a meal of pastured pork, beef, chicken, duck or pheasant, food of such high quality and taste that it cannot be purchased in any store or restaurant. I must admit, I feel sorry for the average American whose idea of a home-cooked meal includes a cardboard wrapper, plastic plate and microwave.

APPPA PRODUCER PROFILE

Geese in the Pasture
By Jody Padgham Spring 2004

If you say the word "Goosemobile" anywhere in South Dakota, it is a good bet that many of the 700,000 people that live in the state will know that you are talking about Tom and Ruth Neuberger's mobile meat market, which travels the state featuring their pasture-raised goose, duck and other meats.

Tom and Ruth came up with the Goosemobile idea about 20 years ago when they realized they needed to go out to their customers, rather than expect the dispersed rural population of South Dakota to come to them. The Goosemobile started its route offering only goose. Now customers in the 175+ towns they visit around the state during several sales days in November and December can purchase chicken, duck, goose, pheasant, turkey, beef, pork, lamb, buffalo, goat and ostrich. Customers are invited to pre-order, but can also come with only a desire and a checkbook to purchase the "100% all natural, grassfed, low-fat and locally grown" meat. The Neubergers raise all but the buffalo, goat and ostrich themselves, and process all the poultry in their inspected on-farm processing plant.

The Neubergers have been raising geese and ducks since 1979 outside of Canistota, South Dakota. At the high point of their production there were 3,000 pastured geese and several hundred ducks on their 160 acres. Since the 1980s their production has slowly dropped, and this year they plan to raise 150 geese, 300 ducks, 3,600 chickens, 150 turkeys, several head of beef and a few dozen pigs and lambs.

Neuberger's Pasture Pens

It would be easy to talk for days about all of the innovations, from marketing to processing, that the Neubergers have developed in their decades of grassfed meat production and marketing, but the subject I picked their brains about most heavily in a recent phone conversation was their success with goose production.

Why Geese?
Ruth notes that her fascination with geese began when she read that geese produce the "purest" meat, because they are very selective eaters, and especially won't tolerate medicated or tainted feed. Geese are also relatively disease-free, and she felt they would be an easy bird to learn to raise. In the 20 years since that first batch of 100 goslings arrived, it appears that Ruth and Tom's fondness for the honkers has not diminished.

Tom outlines two principle reasons they have kept with geese for so long. First, they provide a good profit. Since they eat mostly grass, they are not expensive to feed, and produce a high-quality, high-value meat. Second, they provide a quality byproduct in their feathers and down, which Ruth turns into pillows and comforters and sells under her trade name "Dakota Down."

APPPA PRODUCER PROFILE *Neuberger*

Geese are easy to raise, and do very well on pasture with little supplementary feed. Goslings are available at many hatcheries—the Neubergers raise white "meat type" geese (Embden) but some people also raise Toluse (grey). Goslings will run $2.00-4.00 each, depending on the quantity ordered. Tom gets his goslings in late April or early May and slaughters in November for holiday and freezer sales.

Brooding

Tom notes that brooding geese is much the same as for chickens, with the same concerns of cleanliness, warmth, feed and water, with the one caveat that he's noticed: geese will "pile up" more easily than chicks. The geese will actually climb on top of one another in an effort to escape, and so mass suffocation is very possible. Tom cautions that for this reason heat lamps do not make good goose brooders. He recommends a canopy brooder; the one he uses can accommodate up to 200 birds.

One thing you will notice right away with geese is that they LOVE water. When they are small this means that you need to isolate the water into a corner—and get used to the fact that they will do everything in their power to keep that area wet. It helps to keep the waterer up on bricks, with the lip of the trough at just the height they can reach. Because the young birds will do what they can to stay wet, it is imperative you provide a good heat source so that they can get dry.

Never feed goslings or geese medicated feed. Commercial goose and duck starter will be non medicated, and should work well.

Geese Travel Through the Snow

In the Pasture

Goslings should stay in the brooder until they are feathered out, which will take about a month. They can then be moved out to pasture. A goose's preferred food is grass—they don't like to eat leaves and will move to alfalfa or weeds as a last resort. Bluegrass, brome or wheat all work well for goose forage. Each goose should be given about a pound of grain per day as a supplement—Tom prefers whole oats to corn, which he has found will bring on more fat in the finished bird.

The Neubergers find that a single strand of electric fence will hold geese in a pasture, but they have now moved to permanent paddocks lined with trees that have woven wire perimeters. The domesticated breeds are too heavy to fly far and so will stay where you put them. Tom moves the geese into a five-foot high wire enclosure at night to protect from predation, but in general has found they don't need shelter from the elements. In fact, Tom tells a story in which all his efforts to protect his over-wintered breeding flock one year went unnoticed, as the birds instead chose to huddle in the most wind-swept spot in the middle of a fierce winter gale. Geese seem to have a unique way of coping through all kinds of weather. The geese should be moved when the pasture needs to be renewed. Timing will depend on stocking rates and pasture condition and composition. Tom has found that sheep and geese have very complementary eating habits and pasture well together.

APPPA PRODUCER PROFILE   *Neuberger*

Due to the love affair geese have with water, setting up a pasture watering system is especially important. The Neubergers have not found it necessary to set up intentional ponds or watering baths for their geese. It is important to set up the watering system so that the birds can get their heads in and no more, or you will have either constantly running refill systems or a lot of bucket running to do. Geese will fuss away any water they can get access to. Tom places a board over the water trough so the geese can get a drink and do no more.

The only serious health problems Tom and Ruth have encountered with their geese in all the years of production was during one hot, wet summer when the pasture filled with moderate-sized stagnant puddles. The vet they called in after hundreds of birds had died determined they had ponds filled with fowl cholera bacteria. Since then they have followed the rule of "either a little water or a lot of water"—don't allow shallow pools of non-moving water to collect where geese can wade and share bacteria. Ponds are fine and watering systems are fine--just nothing in between.

Watch out!
The evening I talked to Tom and Ruth I was ready to murder my small flock of geese--they had spent the day chewing on my house and barn. Tom laughed as we traded stories on what all the resident honkers had stripped clean. Left to their devices, they will strip a wall of paint, a tractor of all its wiring and a trailer of all its tire valve stems. His advice is to leave a goose in a pasture with nothing in it except pasture. If you have equipment or buildings you prize, keep the geese away.

Processing
One place you will see a major divergence between goose and chicken production is in the processing. A goose processor will need a very different set-up than a chicken processor. If you don't plan to process your geese yourselves, you will want to check around your area to find a processor who will handle them before you get the first goslings. Goose processors are not common.

Although valuable, goose feathers and down are five times harder to get off than chicken feathers. If you wish to harvest the feathers and down (see below), the birds must be mostly hand plucked. Scalding is done much as in chicken processing, but the goose must be scalded for three times as long as a chicken (at 145 degrees.) Once scalded, and hand plucked or put in a plucker, 99% of the feathers will be gone, but then the carcass must be waxed to take off the final layer of down. To wax a goose, it must be dipped in hot wax two to three times, lifting into the air between each dunking. After the final dunk the bird is dipped into cold water. The wax layer can then be peeled off with the final one percent of the down imbedded in it. Evisceration is the same as with other poultry.

The Neurbergers have set up a certified poultry processing plant inside of a semi-trailer box on their farm. Their original intent was to have the processing plant be moveable, but they have not yet ever moved it. The story of their plant is a tale in itself, which will have to wait for another day.

Feather Harvesting
Two of the fine products of goose production are feathers and down. A goose actually has three types of feathers—breast feathers, underwing feathers and rough feathers. If you are collecting feathers you will need to harvest each of these types separately. Once the goose has been scalded, you can pick the curly breast feathers (which are used for pillows) and the pure down that is found under the wings (for comforters). Then the goose can be put in a mechanical plucker, which will take off the rough feathers of the back, legs and neck.

APPPA PRODUCER PROFILE   *Neuberger*

For those wanting to try something a little wild, how about "feather farming"? I had heard about this and asked Tom for the details. He explained that you can actually pluck down from live birds. The down will regrow and you can then return to the bird for a second and third crop (or more if you wish to overwinter the geese.)

To "farm feathers" you begin when a goose is about ten weeks old; at that time it is growing out of its first set of real feathers. You will see a lot of loose feathers around the pen area. At this point the goose is "prime" and you can pull feathers without hurting the bird. Catch the goose and hold it with the breast up and the wings between your legs. Strip the breast feathers and the down under the wings. In seven more weeks you can do this again, and every seven weeks after until cold is coming. Each prime period lasts about a week when you can harvest the down. If you have breeding or overwintering birds you can start up again the next year and plan on about three harvests per summer. For meat birds, you can harvest the gosling's feathers at ten weeks and again at 17 weeks, but then you will want the feathers to regrow so that you can harvest mature feathers at slaughter.

Goose down is used in manufacturing for bedding or clothing. The rough feathers are used for upholstery and fishing lures, etc. Buyers of small quantities may be very hard to find (try an internet search for "feather sales;") you may want to try home processing of finished products before you give up the feather business.

Marketing

When they first started in the goose business, the Neubergers had the support of the South Dakota Goose Association, which helped with marketing, processing and distribution of the birds throughout rural South Dakota. With few cities of any size in the state, South Dakota has a unique and extremely rural population. By working hard with direct sales, word-of-mouth and publicity using the Goosemobile, the Neubergers were able to develop a vast clientele of farm families and small town residents. Even now they sell to only one grocery and no restaurants. They sell a goose for $2.25/lb., with even the feet available for sale at $3.00/lb., with a vast majority of sales occurring during the winter holiday period. Tom and Ruth have built their success on high quality, good-tasting products with an emphasis on grassfed health benefits.

Years ago, every farm family coveted a goose for Christmas dinner. With today's "fat-free" consumer and micro-wave cook, the smell of browning goose is something few have experienced. The Neubergers say that the primary reason their production numbers have dropped so dramatically over the years is a drop in consumer demand as younger generations gain control in the kitchens of South Dakota. This frustrates them, as cooked the right way (breast down for the first hour, Ruth swears,) goose is very tasty and less fatty than either beef or pork. Emphasizing the qualities of grassfed goose as a fine dining product could be the basis of a strong market in areas that have restaurants, etc., supported by a denser population.

CHAPTER NINE
Poultry Nutrition and Health

Many people like to keep poultry around just because they like them. But others are raising poultry for the purpose of making a profit. In order to make money raising poultry, you must pay very close attention to the bird's diet and the quality and quantity of feed and supplements you give them. Avian health issues can also be a concern--even though birds on pasture tend to be, overall, very healthy, occasional health issues can crop up that, if left untreated, will devastate your flock.

The "Pasture" in Pastured Poultry: An Oregon View
Aaron Silverman Spring 2000

We raise about 2,000 pastured broilers and three acres of mixed produce and cut flowers. Chickens are run on an acre of the market garden from May through the middle of June, then moved to pasture for the rest of the season. The acre the birds are run on is seeded with a cover crop tailored for grazing.

We have experimented with several seed mixes, including straight white clover, oats/annual rye/field peas, oats/cereal rye/annual rye, and a mix of all the above. Cover crops must be chosen that are palatable to the birds and can withstand (are even enhanced by) mowing. Pure clover had the best results for the chickens, requiring no mowing and producing lots of succulent growth. However, for a short-season cover crop, clover is just too expensive at $80-$90 per 50 lb. bag. Peas and oats had great growth, and performed well in the beginning of the season, but were difficult to keep at grazing height. We thought the birds would eat the succulent peas, but they merely tromped on them instead.

Layers Enjoy Diverse Pasture

For several reasons we've now settled on mixes of grains and annual rye. This type of mix can be either grazed by other animals early in the spring to maintain grazing height, or mowed. Either treatment may stimulate the grains to "tiller" out, enhancing root growth and additional leaf growth. Cereal rye has some alleleopathic effects on weed seeds, diminishing their ability to germinate. This type of mix may also be established much later in the season, an important aspect in a climate that often sees little precipitation during the growing season until well into October, when light levels are much diminished.

After the birds are moved onto the pasture, the garden ground is mowed and disked. A portion of the section may be planted with winter squash or pumpkins with little additional fertility (depending on overall fertility levels). The remaining ground is cultivated through July and early August, treating it as a bare fallow. Beds are formed and over-wintering coles (brussel sprouts, kale, etc.) and garlic are planted in the remaining area. Recent research has suggested that microbial activity is enhanced if cover crops are allowed to go to a slightly more carbonaceous phase than normally grown. After the birds are removed each year, a small section of chicken ground is allowed to go to seed, disked and rolled later in the fall than the fallowed ground. This has been sufficient to reseed the strip, often with a more lush and thicker stand than the rest of the field. As we bring more ground into cultivation, a larger section may be allowed to reseed each year.

The "Pasture" in Pastured Poultry, Continued...
Aaron Silverman Winter 2001

In the Spring 2000 article previous, we related our broiler forage/cover cropping experiences over our first three seasons raising pastured poultry. Two seasons later, significant expansion of our farm has led us to refine our choices of pasture forages.

Creative Growers is a diversified farm in western Oregon, located 15 miles due west of Eugene, Oregon. We raise about nine acres of mixed produce and cut flowers for restaurants and a small subscription program. This past year, as member of a new producers' cooperative, Greener Pastures Poultry®, we raised about 5,000 pastured chickens. The chickens are closely integrated with our row crops, providing a large portion of our whole farm fertility program. Creative Growers is managed by two couples and had two additional full time employees in 2001.

Establishing pastures for pastured broiler production requires thoughtful evaluation. The choices available are overwhelming, and specific forages selected will vary throughout the country, based on region's climate and the needs of the individual farm system. Very generally, forages are divided into two groups of species: grasses and legumes.

Grasses may be annual or perennial and have two main growth habits. Bunch-type grasses develop at the point of seed germination; space between plants are often open or filled with other plants. Examples are orchard grass and timothy. Sod-forming grasses spread from the point of seed germination, either by rhizomes or stolons (types of underground stems or shoots). Examples of sod-forming grasses are Kentucky bluegrass, perennial ryegrass, and reed canary grass. Tall fescue, commonly found throughout the country, is a bunch grass that can form a dense sod if grazed closely due to its short underground stems. Grains such as oats or rye are other grass species that may be used as forages. Grasses tend to have wider latitudes for establishment conditions than legumes (timing, soil fertility, etc.), and usually have lower seed prices. However, grasses are not nearly as attractive for chickens as legumes and other broadleaf plants, and will never be fully utilized.

Legumes are unique in the plant world due to their symbiotic relationship with rhizobacteria that allow them to "fix" atmospheric nitrogen in the soil. The most commonly used forage legumes in pastures are clovers. The two types seen most often are subterranean (sub) clovers and white clovers. Sub clovers are cool season annuals. In the South and West Coast (Hardiness Zone 7 and warmer), subclovers are fall-seeded, and self-reseed readily. In regions colder than Zone 7, subclovers are spring seeded. Subclovers are highly palatable during their growing season and are well-suited to pasture mixtures for both irrigated and dryland pastures. Most white clovers are long-lived perennials. Wild white clover is a low-growing type found throughout the country. Although well-suited for grazing, wild white clover does go dormant in regions with dry summers. Dutch White and New Zealand White are common varieties of intermediate growing types. The intermediate types grow to about 12 in. high, yet have the ability to thrive under heavy grazing pressures. While highly palatable and digestible, white clover does pose a bloat risk in ruminants without careful management, so be careful with it if you practice multi-species grazing. A third legume commonly found in pastures is birdsfoot trefoil. Trefoil does well in adverse soil conditions, especially waterlogged areas with heavy soils and low fertility. Looking like an elongated clover with yellow flowers, trefoil is palatable if kept grazed or mowed under 10 inches tall. Vetch is a commonly used legume for green manuring crop ground, but vetch seeds are poisonous to chickens.

When choosing between forages (or mixes) to establish in our pastures, we try to balance the following:

Goals and Limitations: what role will your pasture have in your farm system? What are the limitations present for pasture establishment and production (knowledge, equipment, water, fencing, etc.)?

Climate: is there adequate precipitation throughout the growing season to provide growth throughout the summer? If not, will pastures be irrigated? Non-irrigated forages require deep rooting systems; irrigation or adequate precipitation allows for greater flexibility in choosing composition.

Purpose: how will the ground be used after the cover crop is established? Assuming you'll be using the ground for grazing animals, what animals? For how long? What are the needs/restrictions of the grazers?

Future Use: what is to happen to the ground after grazing? Is this permanent pasture, or future ground for row crops? If rotating ground between different uses, what time frame is needed for the transition to be successful?

While the specific forages we use at Creative Growers will not apply to all areas of the country, a walk through our pastures should illustrate the process we employ for deciding the forages used for each pasture type.

We run our broilers on three different types of ground: unimproved permanent pasture, improved permanent pasture, and integrated row crop ground. The primary goal for our pasture is providing quality forage for the chickens, while not conflicting with the needs of the other grazing animals. Our farming system requires us to choose forages that may be established in the fall. Pastures are seeded by broadcasting onto the soil surface, requiring higher seeding rates than seeding by grain drill. Most of our fields do not have access to irrigation, requiring forages that thrive under dryland conditions.

The climate of western Oregon is best described as "Mediterranean." Unlike many regions of the country, our weather patterns seem almost schizophrenic—wet and dry. Winters are typically mild, with average temperatures around 36° F, low light levels, and lots and lots of rain. The tap turns off during the summer, with daily highs in the 80s, and nighttime lows in the 40s, and virtually no rainfall from July through September.

We have retained a small area of unimproved permanent pasture as a control. After several years of running both chickens and sheep, this ground has become much more healthy and lush, while still retaining much of its original composition. Primarily consisting of native annuals--bent grass, subclovers, and various broadleaf "weeds"—this pasture is very lush April through mid-June, then dry July into September. In areas with slightly heavier soils the pasture may contain fescue, with reed canary grass in low lying regions. While these forages do stay greener longer, they are not very palatable to the chickens.

Improved permanent pasture is used for the bulk of our pastured poultry production and provides the bulk of material for our farm composting system. We remove the lush spring growth early with a silage chopper, to allow for adequate regrowth by July for summer grazing. Pasture composition must be palatable to the chickens, be enhanced by mowing and grazing, and have the ability to develop deep root systems for summer growth without irrigation. In our experience, chickens aren't great "grazers" in the traditional sense, preferring broadleaf plants over grass. Multiple uses require a diverse mixture of plants—those that readily produce high levels of leaf matter (for compost), broadleaf plants for best poultry forage, and a deep root system for non-irrigated summer growth/greenery. For this type of pasture we have settled on a balanced mixture of orchardgrass, perennial ryegrass, tall fescue, annual ryegrass, subclover, and New Zealand white clover. In our region, this mix is suitable for either irrigated or dryland pasture and for both grazing and haying.

Selecting forages for use on ground that will be rotated into row crops must take future use into account much more acutely than selecting for permanent pastures. In addition to providing quality forage for our birds, the main concern is a smooth transition from "pasture" to crop ground. Grasses and grains tend to produce wide, fibrous root systems which are carbonaceous and difficult to break down; allowed to go to seed, they can also become persistent weeds. We run our chickens on one-acre strips in our field crop ground from May through June; the ground is then worked and planted into fall and overwintering crops (garlic, leeks, brassicas, etc.). After spending several years with trial mixes of grasses, grains, and legumes, we have settled on pure stands of New Zealand white clover for this type of ground. Although expensive (generally $90/50lb. bag rhizocoated seed, enough for two to three acres depending on seeding rate), its advantages far outweigh seed cost. In addition to fixing nitrogen, white clover is highly palatable to chickens, who literally run about grabbing mouthfuls. White clover is easily turned under, providing plenty of organic matter to bind the fertility of the chicken manure, without being overly carbonaceous. It is established in the fall as well. White clover's slow winter growth allows other plants to establish themselves in moderate numbers, providing a diverse forage mix even in a heavily seeded stand.

No matter what forages you choose for either permanent pasture or integrated crop ground, management remains key in the ultimate success of your pasture. Allow for adequate establishment prior to grazing, especially if establishing deep-rooted forages for dryland pasture. Graze or mow (think of mowing as mock-grazing) pastures so that the pasture stays about six to eight inches tall. In addition to being ideal height for chickens to forage rather than flatten, this ideal height will delay the plants' accumulation of carbon. As plants begin to store carbon in their cells, usually in preparation for seed formation, their palatability and feed quality decline. While unavoidable, careful management will delay this development and provide quality feed for your pastured chickens throughout much of the growing season.

When deciding between forages and mixes, remember that there is no single "right" answer. Continual experimentation and observation will allow you to develop the best combination of forages for your farm - those that work within the limitations of your climate and system while meeting the needs and goals of your farm.

Grass Conversion Rates by Poultry
Joel Salatin Winter 2001

We can't ask for the same feed conversion rates as the confinement poultry industry, since their birds receive grease, fat, oils and drugs (both hormones and pharmaceuticals) to artificially stimulate weight gain. That is one reason why our birds are dense and theirs are soft. The industry uses live weight to compute feed conversion. We average about two lbs. of feed per each pound of live-weight. Dress out should average 75- 78%, which pushes conversion to about a three to one ratio for a processed bird.

How much of a chicken's diet is obtained from the grass they eat on pasture? I have never guaranteed 30%, but I have gotten that high when compared with birds not on pasture eating identical feed at the identical time of year. A friend in Kansas got more than 30% feed savings on tender, freshly-planted ryegrass pastures. Even the old pre-1950 poultry books tout 15% on non-rotated yards. The conversion factor will vary significantly depending on a number of variables.

1. Bird genetics. We have seen dramatic differences in foraging aggressiveness from hatchery to hatchery and even batch to batch within the same hatchery. We don't know all the whys and wherefores, but each batch will exhibit a little different grazing ability. Tim Shell's Pastured Peepers are superior in this regard. (See page 22.)

2. Time of year. The climatic conditions of the five-week pastured period alters the quality, speciation, and desirability of the forage. Heat or cold have an effect: cold reduces consumption because the birds want more energy, which is coming out of the feed trough.

3. Palatability of the forage. This is one most people underestimate. White clover is far more desirable than lignified fescue. Often it takes years to get agronomic fertility levels high enough to support strong stands of white clover. One key component of palatability is length of the forage. Chickens like vegetation shorter than four inches and that takes tremendous management to do it with grazing stock instead of mechanically. Mowing shreds blade tips, offering the birds wounds instead of succulent morsels. The birds always graze more aggressively in the spring on tender plants than they do later on more mature tissue. This has to do with succulence and brix readings.

4. Exercise. If we drop below 50 birds in a pen, we find that our feed conversion drops, apparently because the birds are running off all the calories. We walk a fine line between average daily gain and pasture consumption. If they consume too much pasture and run off all their feed trough calories, the conversion rate will be dismal. This is one reason to keep the birds confined in the pen rather than letting them free range. Reducing dressed weight by merely half a pound per bird will dramatically change the feed conversion rate because most of the calories are going to body maintenance rather than growth. Growth only occurs after all maintenance is met.

5. Moving time of day. The birds' most aggressive grazing period is two hours pre-daylight, which occurs long before the sun rises. Every quarter hour we wait to move pens after daylight reduces the grazing time period. As the dew comes off and the day gets warmer, the birds begin lounging not because they have grazed their fill, but because physiologically they demand a rest period. If a producer lounges around in bed for a half hour or more, the birds' grazing window shrinks. No matter how good the forage, if it is only offered when the birds naturally want to rest, it will be wasted.

6. Other goodies. Remember, we are not just grazing pasture, but everything in and on that pasture. Grasshoppers and crickets certainly reduce feed consumption as much as anything, but high densities of proteinaceous insects require a dense sward. These insects will not aggressively invade a thin sward, which is typical on a freshly-planted forage that has not yet covered the soil surface.

7. Grit. Many pastured birds do not receive enough grit to aggressively eat hard-to-grind grasses and legumes. They will go to the easier ground or pelletized feed to make it easier on their gizzards. If they are working too hard to grind up material, they will choose one easier to digest.

8. Feed wastage. More often than not, when I visit another pastured poultry producer, I see huge amounts of feed around the feeder. This may not look like much, but it adds up fast. A half-inch spread around a feeder can easily be 3 lbs. of material. Multiply that by 35 days on pasture and you have 105 lbs. per pen, divided by 70 birds equals an additional pound and a half. Feeding smaller portions more frequently, or being more careful on feeder design can help. Brower's polyethylene feeders are the least waste design I've seen.

9. Brooding. As with all animals and plants, the infant period sets the stage for performance far into the future. Chicks that are too cold, too hot, too crowded, too dirty or nutritionally deprived during the brooding stage will never gain like a properly brooded chick. This is perhaps the most common area where feed conversion gets lost.

10. Physical aspects of the feed. Grinding into too fine a mash or not grinding enough can both make ingesting the right amount of feed difficult. Finely ground mash is too powdery and makes swallowing difficult. Bulky feed requires too much gizzard grinding and that energy takes away from growth performance.

Obviously, many variables exist in the pastured poultry model. Success is out there waiting to be discovered!

On Farm Research and Snake Oil
By Joel Salatin Summer 1997

Each year new difficulties arise that sometimes make us wonder how long we can continue to raise industrial birds on pasture. As the industry selects for certain genetic traits for a production model going to 180 degrees different than ours, we must be more creative in order to get satisfactory performance.

In the last couple of years we have actually wondered how long we can continue to use regular hatchery birds and the "normal" feedstuffs. The quality of both feed and birds has deteriorated dramatically in the nearly 15 years we've been raising these broilers. The industry selects for birds that can gain faster on higher calorie feeds. To compensate for the physiological strain that puts on bones, tendons, organs and nerves, the industry develops ever stronger medications, hormones, arsenicals and vaccines.

Each year the birds need higher octane and more engine fine-tuning in order to perform but our paradigm calls for leaded, stodgy fuel and no mechanics to tweak the engine every few hours. As we and the industry move farther and farther apart, we find it nearly impossible to get good performance from the ration we have always used.

In addition, when we started, people weren't feeding dead cows to cows, and you could get rendered animal proteins from a fairly pure source—all cow, for example. But over the years, pure-species animal protein has become virtually impossible to get. Everyone on the market contains some sort of poultry by-products, which pushes us into feeding industrial dead chickens back to our chickens.

Not only is that unnatural, and philosophically reprehensible, it is the slippery slope that has lead to mad cow disease in Britain and who knows how many animal diseases.

We finally realized last year that we could no longer feed just "natural" things to these chickens and have them perform; we needed to upgrade the octane with some high-tech biologicals. Obviously, we didn't want to use medications and synthetics but there are some high-powered biological extracts, some genuine hyped-up nonsynthetic snake oil. We simply could not get enough vitamins and minerals into these birds from natural feedstuffs to get decent performance because these birds were completely different than the ones we had a decade ago.

As I travel around the country, I ask a lot of questions to see what I can learn from folks who are creatively refining this pastured poultry model. In addition, most conferences have trade shows with every kind of nutritional product and promise you can imagine. The problem is that all of these product sales reps show charts that compare their product to nothing.

It is always "ours" compared to "nothing." But you and I don't make decisions that way. We need to decide, among several options, which is best. But these products don't compare themselves to each other; the control group is always "nothing." So which snake oil should I buy?

We've tried numerous things over the years. We tried Nutri-Carb and nearly killed half a batch of chickens. We tried hydrogen pyroxide with no results. Over the years, when someone promised a great solution, we'd try it. That's how we came up with the Conklin Fastrack probiotic--others were just glorified minerals with enough bugs thrown in to be called a probiotic. The Fastrack really gave us results in reduced late-growth heart attacks.

We resolved, therefore, to run trials this year comparing these competing snake oils with each other, dividing our flocks and running these products head to head to see which would give us the best performance. We measure performance first in terms of mortality and secondly in weight gain. Since we don't mask sickness with medications or vaccines, health is real and not a charade.

The first batch of 1,300 this spring were divided into four groups of 325. One received our regular old ration. Group two received the regular old ration plus Immuno-Boost, a water supplement. The third group received the regular old ration plus Willard's Water, a catalyst altered water supplement. The fourth group received no water supplements but rather a ration and feed supplement (Nutri-Balancer) containing no animal protein developed by Fertrell, an organic soil amendment company that has been in business for nearly 30 years.

Very shortly we noticed that all three supplemented batches spun circles around the control. This was enough to prove that our hunch about needing to look at some snake oils was well founded. All three of these groups performed extremely well. Imagine how elated we were to find the Fertrell Nutri-Balancer group doing so well without any animal protein except some low-heat fishmeal (Sea-Lac).

As we went to the field, things rocked along but the Immuno-Boost clogged up the waterers terribly and we abandoned that product. With the Fertrell birds doing so well, we finally switched everything to that and did not run the trial completely to the end since we are in the profit business, not the pure research business.

Incredibly impressed with what we were seeing from the Willard's Water, we broke the next batch into a group of Fertrell only and Fertrell plus Willard's. Of course, we joked about having 10 pound birds in two weeks and all the stuff that goes with experimenting with snake oils. Interestingly, adding the Willard's Water to the Fertrell did not make an iota of difference. We determined that we could not make the birds "weller." Once they are well, they are well and that's that. Now we've abandoned the Willard's water and intend to try some other snake oils to see if we can get any benefits.

We are pleased enough with the Fertrell supplement that we have put all layers and broilers on it as our new "control" and will begin running trials from this new benchmark. We also want to give some competitors a shot at the ration, specifically Leland Taylor and his Clodbuster stuff, as well as Dyna-min and others. Do you have ideas?

I really believe that this type of research is sorely needed, but it cannot be funded by companies or institutions or it can be biased. All I want to know plain and simple is what performs. I have purchased, at full retail price, every ounce of every product that I've used. Any trial that uses "free" product should be questioned for integrity. I am not trying to hurt or help anybody except to share our experiences with practitioners who are smart enough to make their own decisions.

We do not ever intend to go back to animal proteins. The theory is that the minerals cause the body to metabolize the proteins in the grains. As we've reduced the minerals in our soils through the use of chemical fertilizers, we've reduced the vitamin-mineral content of our grains and hence the body's ability to extract the nutrients that are there.

By proper supplementing with minerals, it makes everything kick in like it should. Pretty fascinating. To think that the whole protein issue is really a mineral weakness really makes you shake your head. It's also fascinating that the other two water supplements, on our old ration containing the meat and bone meal, had incredibly dramatic results, almost miraculous. Certainly if a person were using conventional rations I have no doubt that using either of these products would solve a multitude of problems and pay for themselves many times over. But finding the crutch that fits is not as good as healing the broken leg. I think the ideal is to get completely away from the industry as much as possible; a good place to start is with getting away from meat and bone meal.

Two weeks ago we took our second batch to the field, 1,053 birds. We could not find one cripple or a single gimpy, weak bird. Since then we've had several frosts and have not lost a single bird. They are clean and white, with the pinkest skin I've ever seen—wonderful color. The chicks have clean rear ends—they look almost shampooed and airbrushed instead of manure smeared as is common. Of all the things we tried, I certainly did not expect an animal protein-less feed to work, but it has made a believer out of me. We no longer need a hospital pen.

Why Do You Mix Your Own Feed?
By Jeff Mattocks Fall 2001

Many folks call or write and ask about how to mix their own feed. Some of these folk ask, "what's the difference between their mix and what the feed stores sell?" I can think of one good reason for mixing your own feed. YOU control the ingredients and will always know exactly what your chickens are eating. I will guarantee that all of the large feed manufacturers change their ingredients at least weekly. They do this to save money and increase profits. The "Big Guys" use a corn and soy-based diet similar to what you would mix. However, once they have met the basic guaranteed analysis on the label, look out! This is when the wheat midds, floor sweeping, and grain dust gets incorporated. What I am saying is that basic nutrition is met with 800 lbs. of corn, 400 lbs. of 48% soy meal, 400 lbs. of wheat midds, 100 lbs. of animal by-products (meat and bone, feather, poultry, or blood meal), 50 lbs. of fat (soy, com, canola, animal, or re-claimed restaurant oils), 30 lbs. of calcium-phosphate mineral, 15 lbs. of feeding limestone, 6 lbs. of vitamin pack, and 7 lbs. of salt. These are the amounts needed to formulate a basic broiler feed. The total comes to 1,808 lbs. What do you think makes the other 10% or 1,921 bs. in your ton? You can bet that they are not probiotics, kelp meal, extra vitamins, trace minerals, alfalfa meal, or anything extra you may feel necessary. Okay, that's enough about the "Big Guys."

Basic ration ingredients	Protein%	What they do for you.
Corn, ground or cracked	9%	Primary energy source--adds a delicious yellow color without using color additives (marigold extract)
Roasted Soybeans or Extruded Soybean Meal	38%	Primary protein plus energy--not a soy by-product.
Oats	10%	Fiber, protein, energy reducer--this is how we slow down the ration.
Fishmeal	63%	Protein essential amino acids--SeaLac (special low heat process)
Calcium (predigested)	0%	Proper bone and metabolic development--aragonite or Oyster shell flour is easier to digest.
Poultry Nutri-Balancer (available from Fertrell Company)	0%	Complete blend of phosphorous, vitamins, trace minerals, kelp meal, unprocessed salt, probiotics.

What do all these ingredients do? Well here we go!

This is the quick run down of what they are and what they do. There are, however, hundreds or thousands of different ration possibilities. It all looks fairly easy, right? (Ha-Ha!) Did you know that if you feed over 15% oats and barley in combination, 30% wheat, 25% peas, 10% flax seed, or 5% fishmeal, YOU WILL HAVE PROBLEMS!!!!!! Each of these cause side affects or require added enzymes for proper digestion. Even soybeans must undergo a heating process to breakdown the Trypsin inhibitor. I am glad to see folks making their own feeds although I get real nervous when folks design their own feed rations.

Some of the side affects of feeding too much are:
1. Excess oats or barley—poor digestion, runny manure, wet litter, slowed growth rates—requires added B-Glucanase Enzyme for proper digestion.
2. Excess wheat contains excess Pentosan which causes runny manure, wet litter and overall diet digestion reduction--requires added Xylanase Enzyme for proper digestion.
3. Excess peas may contain tannins. Tannins decrease protein digestibility 6% digestibility reduction for each 1% of tannins. There is NO corrective measure.
4. Excess flax will pass on a "paint like" smell or taste to the finished poultry product. There is NO corrective measure.
5. Excess fishmeal will pass on a "fishy" flavor to the finished poultry product. There is NO corrective measure.

As you can see there are many hidden stumbling blocks to designing a poultry ration. These are some of the most commonly used or misused ingredients that I have seen. Please design your poultry rations carefully.

Reference: *Commercial Poultry Nutrition*, Second Edition, S. Leeson and JD. Summers, University Books.

Nutritional Tips
By Jeff Mattocks Spring 1998

Consider moving the pen twice per day in the beginning of the fifth week as pasture and labor allow. At this age, the broilers will be able to consume more forage than one move will provide.

¤ To help reduce stress during the hot summer temperatures use vinegar in the water. Many people believe that this helps because vinegar is a natural blood thinner. This would allow the circulatory system to work more efficiently, thus reducing the possibility of heart attacks/adult mortality. The vinegar can be used throughout the growing cycle at the rate of 1/2 oz. per gallon of water. Vinegar may also be used on an as-needed basis at the rate of 1 oz. per gallon of water.

¤ Size of mash grind. A kernel of corn once ground should be four to five pieces, not dust. The same size for the soybeans will be just fine (enough). Don't be afraid that the small chicks will not be able to eat these size particles. At birth they have an instinct to peck for the largest pieces first. You will probably be surprised at what those 2 oz. birds will be able to consume. The mash feed size can generally be modified by using a roller mill or cracker mill.

¤ The preferred calcium source is oyster shells. The oyster has already done most of the work for the chicken. They absorbed the naturally occurring calcium and purified it so the chicken doesn't have to work as hard to digest and absorb it.

¤ An outstanding suggestion I recently heard: Russell Groves suggested to me that the brooder house/pen would condition the chicks better if they could get out into fresh air and fresh forage. The proposal is to provide a doorway allowing the chicks out into a wire mesh area, where they could get fresh air and forage growing up through the bottom mesh for them to eat while growing up. He and I feel that this would transition the chicks better for their change to the moveable pens (a head start).

Poultry Feed Ration Update
By Jeff Mattocks Summer 2001

Here are the rations we recommend currently. Last year we did some field trials with feeds. The Hamilton family of Canada fed a 19.5% protein ration from day one through finish. They got equal or higher weight gains with fewer mortalities. We have discovered that by starting the chick a little slower and finishing a little faster, the internal organs and immune system have a chance to keep up with the growth rate. Thus we encountered less Ascites and fewer heart attacks. This year we have some experimental ration formulas that do not contain soy products. I will let you know how they work as the growing year goes on.

19% Broiler Grower:

Shelled Corn	1015	lbs.
Roasted Soybeans	625	lbs.
Oats	200	lbs.
Fishmeal, 60%	75	lbs.
Aragonite (calcium)	25	lbs.
Poultry Nutri-Balancer	60	lbs.
Total	2000	lbs.

16% Pullet Grower:

Shelled Corn	1215	lbs.
Roasted Soybeans	450	lbs.
Oats	200	lbs.
Fishmeal, 60%	50	lbs.
Aragonite (calcium)	25	lbs.
Poultry Nutri-Balancer	60	lbs.
Total	2000	lbs.

17% Layer Ration

Shelled Corn	965	lbs.
Roasted Soybeans	600	bs.
Oats	200	lbs.
Aragonite (calcium)	175	lbs.
Poultry Nutri-Balancer	60	lbs.
Total	2000	lbs.

All Rations should be coarse ground or rolled.

You may add 2 lbs. of fishmeal to 20 lbs. (5-gallon pail) of Broiler Grower 19%. This addition will provide a 21% protein mix for Chick Starter. This would be for chicks while in the brooder.

The Chick Starter ration may be slightly altered to feed other species of fowl.

You may add 4 lbs. of fishmeal to 20 lbs. (5-gallon pail) of Broiler Grower 19%. This addition will provide a 26% protein mix for turkey and game bird starter, to be fed from day 1 thru day 28.

You may add 2 lbs. of fishmeal to 20 lbs. (5-gallon pail) of Broiler Grower 19%. This addition will provide a 21% protein mix for Turkey Grower #1, to be fed from day 29 thru day 56. Once turkeys are out on pasture they should receive regular 19% Broiler Grower until slaughter.

Nutrition News For 2002
Jeff Mattocks Spring 2002

The best pastured poultry producers are continually looking for improvements in their production models. Given that feed is one of the variables to consider, I would like to offer suggestions for ration mixes to those producers who are willing to try new ideas or techniques.

During the past year we have experimented with several different options and additions, most of which turned out to be "Magic Fu-Fu Dust" or "Smoke and Mirrors." Even still, I've found that if you do enough experimenting, you may catch a winner now and then. Well, this year I believe the one single thing which will make a noticeable difference to you is alfalfa meal.

Alfalfa meal will provide the following enhancements to your poultry:
1. Extra Vitamin A, to promote a stronger immune system.
2. Contains recognized but unidentified (by the commercial poultry industry) growth factors that aid and stimulate additional growth naturally.
3. Adds Xanthophylls to the diet to enhance broiler skin and egg yolk coloration. Adds appeal to general diet appearance which will mask any minor feed fluctuations.

Of course, as with most things, overfeeding alfalfa meal will have negative effects. Some of the negative affects you see are:

1. Alfalfa contains higher levels of Saponins and Phenolic Acid.
   ¤ Saponins are a sugar group linked to a steroid group. This is probably one of the unexplained growth factors that contribute to better growth and health performance. We should keep in mind that too much of a good thing is bad.
   ¤ Not much is known about Phenolic Acids. The little bits of information available link Phenolic Acid with a newly discovered antioxidant, Gallic Acid. This is being used to treat cancer patients. This information leads me to believe that Phenolic acid in small quantities is supporting better health.

2. Too much fiber in the diet.

3. Too much color to skin or yolks.

The following chart shows my recommendations for those willing to try alfalfa meal as part of their ration. Keep in mind the negatives, and keep me posted of any observations you have in trying the new mixes.

Ration Suggestions for 2002

19% Broiler Starter/Grower:

Shelled Corn	1015	lbs.
Roasted Soybeans	625	lbs.
Oats	100	lbs.
Alfalfa Meal	100	lbs.
Fishmeal, 60%	75	lbs.
Aragonite (calcium)	25	lbs.
Poultry Nutri-Balancer	60	lbs.
	2000	lbs.

16% Pullet Grower:

Shelled Corn	1215	lbs.
Roasted Soybeans	450	lbs.
Oats	100	lbs.
Alfalfa Meal	100	lbs.
Fishmeal, 60%	50	lbs.
Aragonite (calcium)	25	lbs.
Poultry Nutri-Balancer	60	lbs.
	2000	lbs

17% Layer Ration:

Shelled Corn	965	lbs.
Roasted Soybeans	600	lbs.
Oats	100	lbs.
Alfalfa Meal	100	lbs.
Aragonite (calcium)	175	lbs.
Poultry Nutri-Balancer	60	lbs.
	2000	lbs

These rations should be processed using medium grind, half ground/half cracked or roll processed. NOT DUST—powdery feeds are harder to digest.

Looking for a Feed Mill?
Jeff Mattocks   Fall 2000

I grew up working in a small-town feed mill. I started at the age of eight, working on Saturdays. I would carry out bags for retired customers or whomever. People were always welcome to look around the store, ask questions and generally get that "warm fuzzy feeling." You should expect the same thing today.

When a person goes in search of a feed mill, they should look for the following:

1. Friendly, personable staff.

2. Ask for a tour or look around. Look at the overall housekeeping and upkeep.

3. Ask to see the exact commodities from the storage bins used to prepare your feed. Ask if they have test weights for the grains and perhaps test samples. When you look at the grain, ensure that it is whole, unground without too many splits, cracks, and pieces.

4. Ask for a guarantee that commodities will not be interchanged or replaced without your consent.

5. Take your feed mill owner a chicken to sample. Once he has eaten your chicken he will be more likely to do what you ask. This act of friendship will last a long time with someone in the agricultural business.

These are some of the things to look for. My customers make these requests of me and Fertrell. I have no problems with guaranteeing our products.

We offer and provide tours of our facilities upon request. This is not too much to ask from an honest businessman.

Remember: the CUSTOMER is ALWAYS RIGHT!!!!! This includes you when you buy and the customers who buy from you.

Summer Ration to Winter Ration Changes for Pastured Layers
By Jeff Mattocks   Fall 2000

During the summer, pastured layers have all that wonderful access to grass. But what about winter? What should you do? I propose that during the winter months you should change your layer ration as follows:

Basic Summer Ration:

Coarse Ground Corn	990	lbs.
Roasted Soybeans, Ground	600	lbs.
Oats, crimped or ground	200	lbs.
Calcium	150	lbs.
Poultry Nutri-Balancer	60	lbs.
Total	2000	lbs.

Basic Winter Ration:

Coarse Ground Corn	1040	lbs.
Roasted Soybeans, ground	550	lbs.
Alfalfa meal	200	lbs.
Calcium	150	lbs.
Poultry Nutri-Balancer	60	lbs.
Total	2000	lbs.

These changes will help keep beneficial nutrients derived from forage in the layers' diet even after the grass dies out. The winter changes will have no significant impact on basic nutritional levels. However, these changes will keep the layers healthy and egg yolks will stay nice and yellow.

During the summer months with good pasture, protein levels will tend to be higher than the layer may need. With oats in the diet, the extra fiber will tend to act as binder and buffer. During winter months, the soluble proteins from legumes and insects are not available; therefore the use of alfalfa meal makes good sense.

APPPA BUSINESS MEMBER PROFILE
The Fertrell Company, Inc.

The Fertrell Company was first started in Baltimore, Maryland, in 1946. The founder was interested in roses, and frustrated with commercially available fertilizers, so he worked in his backyard to develop a natural nutrient mix that got the superior results he was looking for. Friends encouraged him to share his product, and the Fertrell Company was born.

Dave Mattocks joined the company and, in 1991, he and two partners bought controlling interest in the company. Since then they have been growing 20-25% per year, and Fertrell products are now available nation-wide through a series of distributors.

Fertrell distributors are hand chosen by Dave and his son Jeff (the Fertrell Poultry expert and member of the APPPA Board). When they see a farmer who is really doing well and excited by the changes Fertrell products have brought to his or her farm, Dave and Jeff will work closely with that farmer to help him further understand the products available and how to work with other farmers. Distributors will then set up satellite dealers, making the product and support available over a very wide geography. Several APPPA members are also Fertrell Distributors. You can find a distributor in your area by calling Fertrell at 800-347-1566 or going to their webpage at www.fertrell.com. Many Fertrell products, including the Poultry Nutri-Balancer, are allowed in certified organic operations.

All Fertrell formulations are made at their mixing plant in Lancaster County, Pennsylvania. The 16 on-site employees are very proud of what they do, and Dave stresses with them the value the farmers they serve are gaining from the personal integrity each employee puts into their product.

"Our number one goal is to bring opportunity to the farm community for sustainability," Dave explains. He goes on to say that "our second goal is to move more farmers to producing certified organic products by supporting the strengths of each farmer." The Fertrell Company is committed to working with farmers to understand the needs of their soils, livestock and overall systems and finding the appropriate ways to supply those needs. The Company's main focus is on working with the soil to grow better crops and feed for the livestock. They also supply livestock supplements to improve health and growth.

Dave tells me that livestock, including pastured poultry, are most efficient at getting the nutrients they need to thrive from the plants that make up their food. "You will see an 85% better efficiency from naturally available nutrients over those that are added to feed post-harvest." This means that soil-building to improve feed quality is by and far the best way to improve the health and vigor of animals. Fertrell sells many soil-building products, including mineral mixes and fertilizer formulations, and will work with each farmer individually to understand the needs of their soils and livestock. Not all nutrients are available from all soils, and not all farmers are able to raise the feed they need, and so Fertrell developed several livestock nutrient mixes, including the "Poultry Nutri-Blancer" which is very popular among pastured poultry producers. The Fertrell Poultry web page explains the Fertrell philosophy that "health comes first and performance will follow."

Dave closes by telling me "The Fertrell Company believes there is a very strong future for alternative agriculture with a natural connection. The Company and all its employees want to affect the world and touch as many lives as possible- our biggest desire is for the farmer to be successful."

Contact Information: 800-347-1566 www.fertrell.com, Bainbridge PA.

Little Lessons Learned
Diane Kaufmann Summer 1998

"The difference is, mostly dead is not all dead."
  Miracle Max in *The Princess Bride*

This June, on the second night after my two-week-old chicks went out to pasture pens, we had a sudden weather change from hot and humid to cool winds and a summer thunderstorm. I expected some problems the next morning, but I'm not sure what could have prepared me for finding 20-30 dead chicks in each pen, victims of trying to get warm and dry by piling on top of each other. In total, I lost 92 chicks that night. It could have been many more.

As I was tossing all those little bodies into a bucket I began to notice a difference—some bodies were already stiff, but some were blue with cold and very limp. I assumed those chicks were recently dead and hadn't had time to get into the rigor mortis thing. But my eye happened to catch a glimpse of a beak slowly open and close on one of the chicks in the dead pile. As I sat watching I began to see a few other chicks doing the same thing —albeit very infrequently. Since there didn't seem to be anything to lose except some time, I started gathering up all the chicks who were just limp, not stiff. I took them back to the brooder and laid them under the heat lamps. They still looked lifeless and I was convinced it was a waste of time. I had a meeting for the rest of the day so I didn't get back for chicken chores. When I got home that evening, my daughter wanted to know why there were all those chickens running around in the brooder—she thought they were supposed to be in the outside pens. Amazingly all but two brought back to the brooder warmed up and survived! A lesson learned worth sharing: Mostly dead is not all dead!

Which Direction are They Growing?
Sometimes you just can't win. Or maybe you have to redefine winning. I'm choosing to redefine! Our farm is participating in a two-year study of pastured poultry sponsored by a SARE grant and administered through the Center for Integrated Agricultural Systems at the University of Wisconsin-Madison. We're one of five farms tracking things like on-farm production facts and figures (death loss, feed consumption, profitability, time management, etc.—all the good things, ya know!). Well, this year has been the most disastrous by far of the seven years we've been raising pastured poultry. Its embarrassing to be in the study. Why? Of our first batch of 300 we were "able" to process 140 at 10 weeks of age averaging about 2 pounds. Yep—re-read that sentence and there's not one typo in it. Even with a death loss of 10% we still should have processed 270 at 8 weeks averaging 3.8 pounds. So what went wrong?

We've also had problems with cannibalism like we've never seen before. For the faint of heart, please skip the next sentence. I've taken to recording the daily DQ's (drawn and quartered); have you ever watched chickens slaughter another live chicken—down to the bones? It's not pretty.

I wanted to blame the hatchery or the feed additive company whose product we were using. Both of these were used for the first time this year. But knowing that when you point a finger at someone else there are three fingers pointing back at you, I decided to be a little more rational about it and assume the problem was with management (me). I called the hatchery first—anything different about these birds this year? Nope, no one else having problems like this. Ok. Called the feed salesman. We play telephone tag for several days and then we connect. I explain the problem, we go over the ration. Hmm, everything seems to be in correct amounts–other than don't use wheat with young birds. What about the soybean meal? Well, its not soybean meal, its raw splits. What! The salesman tells me raw soybeans contain an enzyme that acts as a GROWTH INHIBITOR! The enzyme in raw soybeans is a trypsin inhibitor, which actually prevents the complete breakdown of proteins in the small intestine by the pancreatic enzyme—thereby slowing overall growth. Gadzooks. Ignorance is not bliss but I feel better. I now know how to solve the problem of no growth and cannibalism—use roasted or extruded soybeans. And I correctly diagnosed the source of the problem—me! An expensive lesson learned but valuable; I sincerely hope you all learn from my experience. (The bright side is also nine roaster pans made into soup stock and casserole meat for winter.) Batches started since switching to soybean meal are doing MUCH better and now growing in the right direction!

Ed note: in 2005, with high fuel prices, I experienced similar problems with soybeans that were roasted, but not roasted ENOUGH! Be very careful when feeding soy. Be sure your feed mill understands the concern.

APPPA PRODUCER PROFILE

Jonathon and Ellie Coulimore

By Aaron Silverman Spring 2002

Jonathan and Ellie Coulimore have raised pastured poultry since 1994 in Vancouver, Washington, 20 miles north of Portland, Oregon. They plan on producing nearly 2,000 broilers and just under 200 layers in 2002. While the whole family is involved in the poultry and other farm enterprises, the farm does not provide all the family's annual income. Continual innovation and trials are the hallmarks of the Coulimores' system.

Broiler Production System

Jonathan plans on raising five batches of broilers in 2002, from May through October, with birds arriving every four weeks or so. He purchases his chicks from a hatchery in Eastern Washington (Dunlop Hatchery), and has tried numerous strains over the years. He currently raises "Peterson" strain of Ross Arbor Acres. His typical harvest age is eight weeks, three days, and he obtains an average dressed weight of 5-5.5lbs without giblets. The brooder system has undergone significant modification for this season. Previously, Jonathan brooded his birds in old baggage carts, obtained from his work with Delta Airlines. While this unique innovation allowed him to brood small batches of chicks and deliver them directly to their field pens, it limited the size of his batches. He has begun a transition to a more open style of brooding with the construction of a 15 x 30 ft. greenhouse, covered with white plastic. The white plastic acts similarly to shadecloth, providing diffuse light that moderates temperature spikes in the greenhouse. Batches of 400 chicks are brooded three to four weeks in a 12 x 18 ft. plywood box. The brooder's heat source is also in transition, with an old-style electric brooder inside one box, and a propane pancake heater in another. The birds have access to an outside run after two weeks. One of Jonathan's concerns is the higher cost of propane versus electricity, which may be offset by the ability to raise a greater number of chicks with a single heater.

Jonathan uses a traditional Salatin-style field pen. They are 10 x 10 ft., covered with aluminum sheeting, and moved every day.

Feed

One of the unique aspects of Jonathan's production system is his feed. He has set up several grain bins in an old garage and has a roller mill and mixer. By producing his own feed, he is able to trial various ingredients and rations, and has settled on an incredibly diverse ration. His current ration contains: corn, roasted soybeans, wheat, oats, alfalfa meal, corn gluten meal, lysine, Redmond conditioner, azomite, aragonite, Fertrell Nutribalancer, fishmeal, molasses, and vinegar. His ration is about 21% protein, and he mixes this with 20% whole oats and wheat at about 5.5 weeks. He notes that molasses is added to reduce the dust in the feed, and lysine is added to counter the minimal lysine level of the corn gluten meal.

Processing

The Coulimores do not operate a state-inspected plant, but have created an on-farm processing system that would be acceptable by most health departments. While killing, scalding, and plucking occurs outside, evisceration is done inside on a stainless steel evisceration table. Birds are collected early in the morning, and butchering begins about 5 a.m. By 1-2 p.m., all birds are bagged, stapled, and stored layered in ice in an old freezer. About 2,500 lbs. of ice is used throughout the system for 400 birds. Jonathan admits this is probably more than is necessary, as he loads up the freezer for storing the birds up to three days after butchering. All offal is composted on the farm. Customers pick up their order that afternoon.

APPPA PRODUCER PROFILE   *Coulimore*

Labor
Daily chores are handled primarily by Ellie. A group of neighboring home-school kids assist with butchering.

Marketing
The Coulimores rely on word-of-mouth for the majority of their marketing. They send out about 100 letters each spring, detailing the season's offerings and calendar. Most orders are sent in within the first three weeks after the letters are mailed; reminders are either mailed or phoned to sell the last 20-25%. Rarely do the Coulimores enter a butcher date with surplus product. Their first customers were through a home-schooling support group they were a part of while home-schooling their children. They experience about a 25% annual customer turnover rate and rely on customer references to pick up new customers.

One aspect of the Coulimores marketing that works well for them is their tiered price structure. Prices are based on the number of birds purchased at a single butcher date. The volumes are broken down by " one to five," "six to nine," and "10+." Jonathan attributes his high average sale of 15-20 birds to this pricing structure. No deposit is taken for orders.

Record-Keeping
As with many homestead-scale producers, Jonathan admits his record-keeping is not as accurate or complete as he'd like. His records focus mainly on end-weights and mortalities. He keeps a production log to record mortalities throughout the season. His main health issue has been ascites, which he has found to fluctuate throughout the season, peaking during periods of cool, wet weather. He does keep track of how much is sold, and a list of customers and their orders.

The Bottom Line
Jonathan estimates that he makes about $4 per broiler, before accounting for family labor. He attributes this to his small-scale, minimal marketing costs, and availability of both family and home-school labor.

Layer Production
The Coulimores also raise about 200 layers. Replacement pullets are brooded in a similar fashion to the broilers, and go to the field between four and five weeks old. The hens are housed in a raised field house, with nests that have Shenandoah automatic egg rollout tracks. Local home-schoolers help wash both eggs and the rollout tracks. The Coulimores deliver eggs to a school in Portland, as well as to their local church. Jonathan grinds a similar ration to the one used for the broilers, with a lower protein content and increased whole grains.

Unfair Advantages
Jonathan attributes much of their success to the close-knit network of home-schoolers in his area. He also has used his knowledge of the freight system to become the Northwest distributor for Fertrell (Nutri-Balancer, Sea-Lac fishmeal, etc.) Each season brings new innovations and trials, something Jonathan feels is critical to continued success. His ability to store and grind his own feed is the keystone to his ability to conduct ingredient and ration trials.

Mortality: Common Causes
By Jeff Mattocks Summer 2002

The most common causes for pastured poultry mortalities are ascites, air quality, temperature control, coccidiosis, enteritis, heart attacks, curly toe, and spraddle leg.

Ascites

Ascites (water belly, pulmonary hypertension) is the number one cause for broiler mortality in commercial and pastured poultry production. Incidence rates as high as 25% in commercial broilers and 15% in pasture poultry broilers. Ascites is responsible for 20-30% of all male broiler deaths. Ascites is contributed primarily to the superior growth characteristics coupled with under-developing internal organs, primarily the lungs and heart. The problem is a lack of oxygen or a higher demand for oxygen, which the heart and lungs may not be able to deliver to meet the metabolic processes. Ascites is most often triggered by stressful conditions, usually dampness and cool temperature; the brain sends a signal requiring more oxygen to the digestive tract to generate more heat.

When the lungs can't meet the demand for oxygen it sends a signal to the brain to increase blood flow. The heart then begins to beat faster. However, the right ventricle is undersized and the heart cannot handle the return pressure. The back press then causes an over pressure in the liver, which in turn starts to seep plasma into the body cavity (i.e. water belly.) Therefore, either the liver shuts down or the heart quits, causing death.

Primary preventative measures to relieve Ascites are increased quality air flow and the prevention of night time cold stress. Other alternatives to fight Ascites are to decrease dietary energy by feeding mash diets rather than pelleted diets, skip-a-day feeding or limit feed access. There are side affects to skip-a-day feeding and limited feed access. Both methods will cause frantic feeding behaviors, which will lead to skin wounds from the birds jumping on each other to get to the feed. Skin wounds provide prime opportunity for E-coliform bacteria infection. This is known as I.P.. or infectious process.

Air Quality

I receive many calls per year about mild to severe respiratory problems. I start my diagnosis by asking a lot of questions regarding the symptoms and conditions of the poultry. Most of the cases that I try to diagnose throughout the year are related to air quality, and most of these occur in the brooder.

Many folk, particularly beginners in pastured poultry, treat their chicks like their infant children. Everyone is cautious about drafts and chills and these are things to be aware of. The downfall to being overcautious is the tendency to seal up the entire brooder so that NO fresh air can get in. Chicks require a minimum 100% air exchange six times in a 24-hour period. This doesn't include drafts that come in at floor level. It also does not include chilling the chicks with a rush of very cool air. It does include a subtle, continuous movement of air in the brooder or any controlled environment.

The problem most often encountered with poor air movement is sinusitis. Sinusitis is a direct result of excess humidity and ammonia release from manure. The ammonia will cause an irritated respiratory tract, which causes tissue scarring, which decreases oxygen absorption to the blood stream, which accentuates Ascites. (My grammar is not really that bad. I wanted you to see the domino affect of a bad condition.) Both sinusitis and ammonia scarring will retard growth weights, if they don't kill the birds first.

Both sinusitis and ammonia build up can be controlled with air quality management.

Temperature Control

We all know that temperature control is critical. I experienced this first hand this spring when brooding chicks in February when our day time and night time temperatures swing at least 20 degrees. I know that I don't deal well with 20 degree temperature swings; how well do we think a day-old chick can deal with these types of temperature changes? I had to adjust my heat lamp distance a minimum of four times a day to try and maintain some sort of constant temperature.

Chicks need to regulate their own temperature. By this I mean that we provide enough area with supplied heat that the chick can find it when it needs it and get away from it when they don't. Amazingly, the chicks are really smarter than I give them credit for. I only thought they were stupid because they didn't do what I thought they should.

Leg Problem Flow Chart

Curled Toes → How many?
- 5% or more? → Riboflavin Deficiency
  - Check feed formulation for adequate riboflavin.
  - Quick cure feed brewer's yeast on top of feed.
- 1-5% or less? → Genetic Abnormality
  - May not be curable.
  - Supplement added Vitamin E 50 IU per chick per day

Splayed or Spraddle Leg → How many?
- 5% or more? → Manganese Deficiency
  - Check feed formulation for adequate manganese.
  - Quick cure supplement manganese on top of feed.
- 1-5% or less? → Genetic Abnormality
  - Once tendon has slipped there is no cure.

Other Leg Disorders → Walks strange or lays around. Does not want to walk? → Possible Calcium Deficiency → Verify dietary Calcium content.
- If calcium is low adjust calcium in ration.
- Supplement Oyster shells until new feed is ready.
- If calcium is OK → Probably just a Lazy Chicken Breed

I get a couple of calls each year where the chicks are very irritable and even cannibalistic. Asking my twenty questions, I find the problem is almost always excessive heat.

Coccidiosis and Necrotic Enteritis

Coccidiosis and necrotic enteritis are often confused, as the symptoms are similar. The symptoms include pasty butts, diarrhea, lifelessness, excess water consumption, and eating shavings. The difference will be blood spots in the manure. Blood spots are a clear indication of coccidiosis. The good news is that both problems can be treated the same way. MANAGE YOUR LITTER! That was simple. Whenever you see clumped litter (generally around the feeder or waterers) you have harmful bacteria and/or coccidiosis. Actually there are several precautionary steps that can be taken. First, keep the clumped litter removed. Second, raise the feeders and waterers so that the lip of the feeder and waterers are level with the average birds' back. Third, periodically apply thin layers of new shavings on top of the old. Fourth, maintain a good AIR FLOW.

Most occurrences of coccidiosis and enteritis will occur in the brooder. Generally symptoms will become noticeable around day ten. The mortality will peak between day 14 and day 21. Then the deaths will slowly reduce because the remaining chicks have built their own immunity to coccidiosis. If either of these problems have gotten out of hand and you are in the middle of a crisis, feed whole raw milk to the chicks for seven days. This will coat the stomach and soothe the pain so they can continue to eat and drink normally while the immune system kicks in and protects the chick. This is the easiest method.

The other method is a *"break glass-in case of emergency" type of solution ONLY!*

A small mixture of copper sulfate into the water has an adverse affect on harmful bacteria and coccidia organisms. Copper sulfate has a stiffening affect on the digestive track, thus making an unfavorable environment for feeding and reproduction of bacteria. The application for this method is: 1 oz. copper sulfate mixed with one-gallon distilled water. This solution is your concentration or stock solution. Add 28 oz. of concentrate to five-gallons of drinking water. Treat for three days ONLY! If necessary you can treat again after three days of no treatment.

Heart Attacks

Heart attacks are also referred to as SDS or sudden

death syndrome. The symptom for SDS is real easy to identify. The chicken will be on its back with its feet straight up in the air. What causes it? Well the university experts are still guessing. There are ideas that it is heat-related, which causes an electrolyte imbalance; this, in turn, causes left ventricular fibrillation. That was a mouthful; I got it from *Commercial Poultry Nutrition* written by Leeson and Summers, University of Guelph, Ontario.

During heat stress times I recommend feeding apple cider vinegar at the rate of 1 oz. per gallon of drinking water. Some people really think it works and some people say they have no results. Try it! It's cheap and you won't hurt anything. Then let me know what you think. I also encourage reducing stocking density, especially during the mid-summer to allow better air flow. Increasing the amount of available oxygen will reduce stress and unnecessary heart pumping. I am also fond of moving the broilers during the midday. By doing this you will remove them from their old manure, improving air quality. Also, new grass is cooler from evaporation and it isn't exchanging nitrogen from manure with its carbon matter. This technique won't work for everyone because not everyone is home during the midday. If you can try it, you'll like it.

Curly Toe and Spraddle Leg

These are the common leg problems that most pastured poultry producers experience. Each is fairly simple to explain. Curly toe is either a riboflavin deficiency or it is a genetically transposed weakness. If you have a riboflavin deficiency, you will have 10% or more afflictions. If you have two out of 200, it is a mistake of nature. If you have 20 out of 200, you have a problem. Check your feed recipe to ensure you have adequate amounts of riboflavin. By the way, adequate is 5.5 PPM in finished feed or 11,000 PPM per mix. Immediate corrective action is to feed diced or ground raw liver or to supplement Brewers Yeast onto the top of their feed.

Spraddle leg, otherwise known as slip joint tendon, is either an injury or manganese deficiency. What happens is the achilles tendon has slipped out of the joint between the foot and drumstick. The leg will stick out perpendicular to the bodyline. As with curly toe, if you have two out of 200, it is mostly likely a result of an injury. If you have 20 out of 200, you have a problem. There is NO corrective action once the tendon has slipped. Again you need to ensure that your manganese requirements are being met. It should be 70 PPM per pound of feed. For immediate treatment, you can just sprinkle some manganese supplement onto the top of the feed.

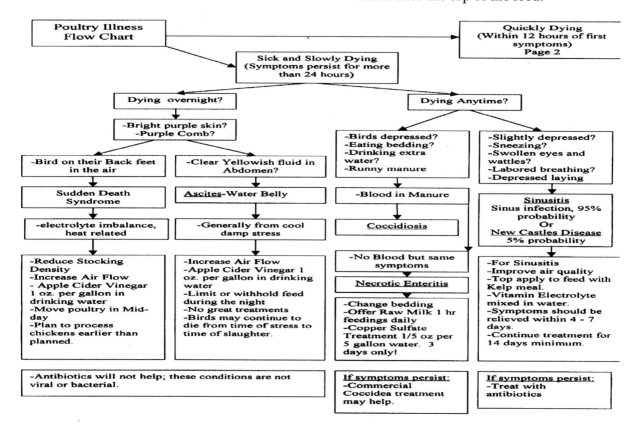

APPPA BUSINESS MEMBER PROFILE

Acadian Seaplants Limited
By Jody Padgham Summer 2004

Acadian Seaplants Limited has been harvesting and processing marine plants off the coast of Nova Scotia in Eastern Canada for over 35 years. Acadian Seaplants operates five major manufacturing facilities strategically located close to cultivation and harvesting sites throughout Atlantic Canada. Approximately 300 employees (including full-time and seasonal) work for the company. Acadia Seaplants' head office is in Dartmouth, N.S., across the harbour from the Port of Halifax. The region is endowed with several internationally renowned centers for Ocean Research and Technology. Acadian Seaplants exports internationally to over 65 countries and is a fully integrated company from marine plant cultivation and hand harvesting of pure seaweeds to production and application, development, manufacturing and technical customer support.

Acadian Seaplants is proud of the use of diversified, technology-based manufacturing to produce products for all sectors of agriculture. Their products include fertilizers, feed, food, food ingredients and brewery supplies, all derived from select species of marine plants. Their sustainable resource management is award- winning and a vital priority for the Acadian Seaplant experienced team of seaweed resource managers. They understand that as the world's largest independent manufacturer of these marine products, the natural resource is vital to their business and the businesses that they support internationally.

Acadian Seaplants' Kelp Meal is a natural source of iodine. This has made it a very popular organic source of that mineral. Many customers that use Kelp Meal in egg layer diets do so as a means to increase iodine levels of the eggs. At the same time the pigments in the kelp have also typically helped give egg yolks the deeper colour sought by many producers. This has been especially useful in the free-range and organic markets where more natural looking eggs are preferred. In layers and turkeys, the addition of kelp to the ration can result in reduced feather picking and cannibalism. Kelp meal in layer and turkey feed has also been shown to increase the overall health of the animal.

Acadian Seaplants Limited is dedicated to the science of marine plants and to the development, application, management, research and technical support that their customers need to sustain and thrive in today's ever increasingly aware marketplace. The aforementioned attributes combined with gentle manufacturing and quality management programs have given Acadian Seaplants the understanding of marine plant physiology. They have developed advanced systems for processing the seaweeds that are harvested and cultivated into the superior products businesses people expect. Acadian Seaplants' most important mandate is to meet the needs, and exceed the expectations of, their customers and the people that they support.

Acadian Seaplants Kelp Meal is offered in bulk and in special feed formulations through The Fertrell Company. If you have any questions about their minerals or their products that contain Acadia Seaplants Kelp Meal, please do not hesitate to contact the Fertrell Company at 800-347-1566 (www.fertrell.com) or Acadian Seaplant.

Contact Information: 800-575-9100 www.acadianseaplants.com

Dealing with Coccidiosis in Pastured Poultry
Timothy Shell Summer 2001

Big problems with coccidia are generally the result of a breakdown in the poultry producer's management which results in stress to the birds. The stress, in turn, breaks down their immune system's ability to handle what would otherwise be a normal health challenge. Small problems occur even in good operations.

Coccidia are ubiquitous. They are carried in the feces of almost all wild birds. Farm poultry will always be exposed and, under normal conditions, develop an immunity to the coccidia. You can compromise the "norm" by several things.

Coccidia problems are generally a symptom of poor management on the part of the producer due to violating some of the following prerequisites for healthy poultry. You need to correct the problem if this is your case, not just medicate. On the other hand there will always be the few that have a predisposition to get sick in any group of livestock even when the best of management has occurred and optimum conditions are provided. Nature normally dispatches these creatures to prevent passing on the tendency to future generations. It is generally unprofitable to keep what nature has discarded.

1. Clean water. Be sure that the birds have access to water that you would be willing to drink yourself. This is not necessary once the birds are older and have a proper immunity developed. However, in early life you must not allow for dirty water. The big offender here is brooder bedding (which contains feces) being scratched into the waterer or fecal/bedding dust collecting in/on the waterer. Clean your waterers very regularly. Be sure to eliminate any wet or dirty places/puddles that the birds might have access to. They do not understand not to poop in and drink out of the same puddle. Mud is dangerous if they have access to the outdoors.

2. Clean bedding. Be sure the birds have bedding to lounge in clean enough that you would be willing to kneel in it in nice work clothes and show the peeps to your best friend's children. Wet, dirty bedding causes a hygiene overload for the birds in excess of the pathogen tolerance threshold for their immune system.

All livestock have specific tolerance thresholds for specific pathogens. Above that level they get sick. Below that level they do not get sick. Exposure to a disease-causing organism below a certain level of colony-forming units does not cause disease even though the pathogen is in the bird's system, and you never know it because it is dealing with it as planned. The fact that many of us have gotten away with allowing some horribly dirty conditions for our poultry is not a tribute to our skills as producers but to the wonder of the bird's incredible immune system.

Poultry exhibit a hygiene behavior that includes sliding their beak over their feathers to remove dirt. Their system is designed to handle ingesting a certain amount of dirt each day. If you give them more than they can handle, they may get sick. If the dirt is from overly soiled bedding they will be ingesting their feces via preening. You will notice that the coccidia-affected birds will look dirty. They "know" somehow to stop preening until their body can handle the crud. Your other birds would look that dirty too, they just clean up every day. The amount of dirt on the unpreened ones lets you know how much the others are cleaning off themselves in a short time. You must add bedding in whatever amount needed to deal with problem spots in the brooder (like around the waterers and in the nightly sleeping spot).

3. The DOUBLE WHAMMY. The busy person easily falls prey sooner or later to the fault of letting the feed and/or water run out for their poultry. The combined effect of hunger and/or thirst on the birds is to encourage them to sort through the ground/bedding searching for particles of food or moisture. In a dirty environment, hungry and/or thirsty birds will almost certainly exceed the safe threshold of tolerance for pathogens as they ingest soiled material in search of food and water. This exposure level, coupled with the stress caused on their system by the hunger and thirst, creates a situation ripe for disease to set in.

4. The TRIPLE WHAMMY. Poultry love sunlight. They love to sunbathe. This is a great benefit to them. If the weather turns cloudy and damp/rainy and/or the birds have no access to direct sunlight in the brooder in early stages of life, they may be at a dis-

advantage for proper hygiene. The sunlight is a disinfectant and a therapeutic tonic in the birds' world. Doing without can contribute to outbreaks of coccidia as well as other diseases. The combined affect of all of the above can be disastrous.

Symptoms

Birds faced with an overwhelming infection of coccidia will look dirty and unkempt. They will be weak and listless, hunkered down in a corner and not moving much. They do not look healthy one day and just drop dead the next. You can tell several days ahead which ones are on the way out. They can have bloody manure from the bleeding of the intestine caused by the irritation of the coccidia on the papillae. Severe infections will have foamy, yellow, mustard like manure. If you have birds in this condition you have already experienced significant losses in the productivity of the rest of the flock.

Treatment

Deal with dirty bedding and water. Use a plywood circle under the waterer large enough to keep the birds from scratching feces into it. Elevate the platform three to four inches above the bedding. Use drink cups or nipple waterers to provide sanitary water. If you have had a severe infection in the brooder, clean it out and disinfect and re-bed with clean bedding.

Supplement with water-based probiotics in the waterer. Jeff Mattocks of Fertrell recommends fresh raw cow or goat milk to me as a supplement in or on the feed or fed free choice as a successful remedy for coccidosis. The milk is mucus-forming and coats the intestinal track. It also has beneficial bacteria and enzymes in the raw form. I used it on the feed about 2.5-gallons of milk per 5-gallon bucket (25 to 30 lb.) of feed, well mixed. Putting it in the feed makes sure the birds all get a dose. Usually the birds turn around in 48-hours. An old poultryman told me they used to use milk products to treat coccidia before the medications came out.

Remember, poorly managed pasture models can lead to exposure levels to feces and other stressors equal to or greater than that of modern confinement poultry facilities.

APPPA BUSINESS MEMBER PROFILE

Dotson Farm and Feed
By Karen Wynne Fall 2005

Location: Lafayette, Indiana
Product line: Fertrell poultry feed, supplements and fertilizers and certified organic hay
Years in operation: Since 1967

Contact information:
2929 N. 9th Street Rd.
Lafayette, IN 47904
Phone 765-742-5111
Fax 765-429-5601

Business Philosophy:
Integrity is first, then fast efficient service and fair pricing.

When I spoke to Gordon Dotson this fall, he had completed his last cutting of organic hay for the season. In central Indiana where they are located, the Dotsons can usually get four good cuttings off their hay fields. While a lot of the mixed timothy, bromegrass and red clover hay is distributed locally, the bulk heads east to Pennsylvania where organic dairies put it to good use, especially in the winter months. A droughty summer made a big impact on the yield, but has allowed for good quality hay.

Gordon and Luita Dotson's business, Dotson Farm and Feed, has evolved over time, starting as a landscape maintenance business in the late 1960's and adding 80 acres of hay in the next decade. Increasing hay production balanced out the loss of landscaping clients in the recession in the late 1980's; at the same time they began supplying Fertrell products to local farms. Their son Steven came to work a few years later and the three have been working together to provide their customers with quality products. "Integrity is first, then fast efficient service and fair pricing on products, " Gordon says of the Dotson's business philosophy.

They also work hard to provide their customers with any information they need. "If we don't know, we get an answer," Gordon says. The Dotsons can share organic production information from years of experience. Back three or four decades ago when there were not a lot of organic resources available, Gordon relied on books by Ruth Stout and Rodale to guide him. Now he is happy to share that information with other producers; certified organic growers in the area are still relatively few.

What do the Dotsons have to offer pastured poultry producers in the area? In addition to certified organic hay, they provide local producers with a wide variety of Fertrell fertilizers and poultry feed and supplements. They also do soil testing (always good to do before you buy fertilizer!). Feed supplements include Poultry Nutribalancer, Sea-Lac fishmeal, feed calcium (aragonite), kelp meal, Probios, Direct Feed Microbials, and booster packs. These are all formulated to provide balanced nutrition for feed mixes or supplemental nutrition for stressed birds. If you don't know what you need, the Dotsons are always available to answer your questions. If they don't already know, you know they'll find out.

Introduction to Tenosynovitis (Viral Arthritis)

By Don Brubaker Fall 2003

I would like to take a moment to introduce myself. I grew up on a 28-acre poultry farm in the heart of the Pennsylvania Dutch Country in Strasburg, PA. After years of raising corn, soybean, vegetables, hogs, steers, and broilers the chemical way, my family's thought patterns began to change. After the birth of a son with disabilities, we wondered what went wrong. To make a long story short, that experience brought us to a natural and organic lifestyle that we enjoy today.

On June 2, 2003, I started with The Fertrell Company and I'm enjoying working with the natural and organic farming community. In this article, I would like to talk about viral arthritis (VA). This virus was first reported in 1957 and from time to time has created difficulty for poultry producers. VA is a reoviral infection that infects primarily meat-type chickens. There have been several serotypes identified. The virus is passed through the droppings of infected chickens and can also be transmitted through dirty egg shells. The virus will not last long in the respiratory and digestive areas of the chicken's body; only in the sheath of tendons will the virus survive for any length of time. In mild cases you will see swollen joints and reddening of the legs. In the more severe cases the bird's legs become so sore that they can no longer walk. The tendons will swell and actually tear and the bird will become a cull.

Prevention

Once the damage is done there is no treatment option for the infected birds, so it is very important to do all you can to keep the virus off your farm.

Getting the material antibodies built up in the parent stock is the first thing you have to do. Unfortunately, this may be the most difficult, because as small producers we do not have control of the breeder flocks. The one thing we all have going for us is that there are many other producers buying eggs and chicks from the same hatcheries, so they need to do a good job in all aspects of the breeder flock. You can communicate with your hatchery or egg supplier to see if they have a program in place to build up the titers (antibodies) in their breeder flocks for reovirus or, for that matter, any other disease.

If you still have a problem after you find out that the breeder flocks have good titer levels, you can look into a good vaccination program for your farm. Because of the small numbers of birds many of you start with, day old vaccination at the hatchery is the best move. On-farm vaccines usually come in 5,000- to 10,000-dose vials. Many of you may not believe in the use of vaccines, so it is very important that you, a pastured poultry producer, do your homework and find out what method of control works best for you.

Farm Management

Day-old chicks can be infected with VA but usually will not show signs of the virus till they are six to seven week-old chickens. The virus is resistant to chemicals and heat. So lets start with the hatching-- some of you hatch your own chicks. Eggs need to be cleaned as soon as possible after the hens lay their eggs. Clean the eggs with warm water and a cleaner (that is accepted by your certifier if you are organic). The incubator should also be cleaned after each use. If you move the eggs to hatching trays for the last three days, don't forget to clean those trays also. Once your chicks are hatched, they need to be moved to a clean and warm brood area within 24-hours. It might sound like I'm a clean freak, but prevention is easier than dealing with the virus.

If you are having problems with VA you should come up with a sound bio-security plan. When caring for your birds, always start with the youngest first and work towards the oldest. It would be even better to have a separate person care for the new chicks as they came onto the farm. In any case you should keep yourself cleaned up. You should wear washable boots and clean clothes every time you care for the birds. All tools and equipment used to care for the birds in a given day should be kept clean. Whenever possible it is a good idea to have a ten- to 14-day break between groups of birds. This will help to prevent the build-up of this disease and many others. It may seem like I have spent a lot of time on breeder flock health and cleanliness of the farm, but I see no other way to prevent VA from being a problem on the farm.

What is Blackhead or Why You Shouldn't Raise Chickens and Turkeys Together
By Karen Wynne Summer 2005

Blackhead is a poultry disease caused by a protozoan parasite (remember that one-celled creature from junior high biology?). It mainly affects the liver and cecae, two pouches located off the large intestine. Blackhead causes stunted growth, poor utilization of feed and death. The disease is common where poultry are raised, but it is most lethal in the production of turkeys and game birds. Chickens are normally more resistant, although incidences may be increasing.

The Culprit
The parasite responsible for blackhead is Histomonas Meleagridis, which is spread through poultry feces. While this parasite does not survive on its own for long, it can survive for years in the eggs of worms that inhabit the cecae. These worm eggs are passed by the bird and can infect other birds if ingested by them. Earthworms can also ingest and carry the protozoan. Chickens can be infected by the disease without showing any symptoms; turkeys can often be infected when they come in contact with equipment, housing, and pasture previously used for chickens. Younger birds from six to sixteen weeks are more susceptible to death from blackhead. Normally, mortality rates stay below 15%, but they can reach 100% in turkeys if uncontrolled.

Symptoms
External symptoms of blackhead include loss of appetite, increased thirst, droopiness, drowsiness, darkening of the facial regions (hence the term blackhead) and diarrhea. Generally the disease is identified by yellow-green circular lesions on the liver. In addition, the cecae and liver are swollen and hardened. Laboratory tests can also be used to confirm the diagnosis.

Control and Prevention
Blackhead is best controlled by keeping different species of birds separated. Many pastured chicken producers raise their turkeys at entirely different farms or at least reserve separate pastures for the different species of birds they raise. In addition, turkeys should not be pastured on an area where chickens ranged unless several years have elapsed. (My dictionary defines several as more than two, but fewer than many, if that clears anything up!) Rotating ranges within a species is still a good idea.

Also, managing the cecal worms will help reduce blackhead incidence. This is difficult to do when animals are on pasture; the worm eggs can survive for long periods of time under less-than-ideal environmental conditions. In permanent housing, raised floors with slats or wire can reduce the opportunity for the birds to ingest the eggs. Some chemicals are approved for cecal worm control, but nothing is available that affects the protozoan itself.

Information from the Mississippi State Department of Poultry Science www.msstate.edu/dept/poultry/disproto.htm

---

Roost Mite Control 101
By Robert Plamondon

Roost mites live in cracks and crevices and emerge to suck blood from your chickens. They are almost too small to see unless they've been feeding, at which point they swell up with stolen blood. If you feel invisible bugs crawling up your arms after collecting eggs or dealing with the chickens, you've got roost mites. If you see little red or red-brown dots on just one side of eggs, you have roost mites (they were crushed when the egg was laid). If you pull out a handful of nest-box litter and it moves on its own, you have roost mites. They're very common. I'll be talking about them like a broken record until the weather gets cold again. After predators, they're the biggest threat to your hens.

The quick fix for roost mites in the nest boxes is 5% Malathion dust in the nest box (Pyrethrin is the organic allowable equivalent). Scraping roosts and wooden nest boxes clean and then painting them with an oil (preferably a non-drying or slow-drying oil) gives longer-lasting protection. Used motor oil thinned with kerosene is traditional; linseed oil thinned with kerosene smells a lot better. I mean to try linseed oil thinned with turpentine for an old-timey, slower-drying effect.

Whitewash works, too, I'm told. I haven't tried it. So does lime-sulfur spray. It gives a longer-lasting effect than Malathion, but not as long as oil. It's very safe. It smells like rotten eggs, though.

Avian Influenza in Poultry
By J. P. Jacob, G.D. Butcher, F. B. Mather, and R.D. Miles

Ed Note: At the time we are compiling this book, the threat of Avian Influenza is in all the news. It is not yet a problem for pastured poultry producers, but, particularly due to poultry politics, may become one over time. See the following article for more information on that issue.

Avian influenza is a viral disease affecting the respiratory, digestive and/or nervous system of many species of birds. Avian influenza virus infection can occur in most, if not all, species of birds, both domestic and wild. Influenza viruses vary widely in their ability to cause disease (pathogenicity) and in their ability to spread among birds.

Wild species of birds usually do not develop clinical disease, but some influenza viruses cause severe illness or -- death in chickens, turkeys and guinea fowl.

Clinical Signs
The severity of the disease ranges from inapparent (mild) to rapidly fatal. Lethal strains of the virus can strike so quickly, particularly in young chickens, that there may be no clinical signs other than sudden death.

Avian influenza viruses of low to moderate pathogenicity are identified regularly in the United States in the domestic poultry populations. Avian influenza virus is reintroduced into domestic poultry by migratory waterfowl, which are carriers of the influenza virus.

Clinical signs vary greatly and depend on many factors including the age and species of poultry affected, husbandry practices, and the inherent pathogenicity of the influenza virus strain. Clinical signs may include:
- ruffled feathers
- soft-shelled eggs
- depression and droopiness
- sudden drop in egg production
- loss of appetite
- cyanosis (purplish-blue coloring) of wattles and comb
- edema and swelling of head, eye-lids, comb, wattles, and hocks
- diarrhea
- blood-tinged discharge from nostrils
- incoordination, including loss of ability to walk and stand
- pin-point hemorrhages (most easily seen on the feet and shanks)
- respiratory distress
- increased death losses in a flock

The clinical signs of avian influenza are similar to those of other avian diseases. Avian influenza may be confused with infectious bronchitis, infectious laryngotracheitis, fowl cholera, and the various forms of Newcastle disease.

Typical history, signs, and lesions may be suggestive of mild forms of avian influenza. Confirmation of a diagnosis is by serologic testing and virus isolation and identification. Because virulent strains of avian influenza are considered to be exotic to the United States, they are reportable to the USDA. Virulence level is evaluated by virus isolation and controlled laboratory challenge of experimental chickens.

Postmortem Lesions
Lesions vary greatly depending on pathogenicity of the virus, age of the bird, type of poultry, etc. Lesions may include swelling of the face and area below the beak. Removing skin from the carcass will show a clear straw-colored fluid in the subcutaneous tissues. Blood vessels are usually engorged. Hemorrhage may be seen in the trachea, proventriculus, beneath the lining of the gizzard, and throughout the intestines. The lining of the gizzard may be easily removed.

Other areas likely to show swelling and hemorrhages include the muscle along the breastbone as well as in the heart, gizzard fat, and abdominal fat.

Young broilers may show signs of severe dehydration with other lesions less pronounced or absent entirely.

Serotypes
There are many different strains (serotypes) of the avian influenza virus. Some of the highly virulent

strains evolved from milder strains following repeated chicken to chicken passages. The avian influenza virus has been shown to mutate at an extremely high rate as it serially infects poultry. Chickens are not the normal host for avian influenza, so the virus they pick up from other birds has a tendency to mutate and become pathogenic. In 1994, an avian influenza outbreak in Mexico c started out mildly, but mutated into a "killer" virus that decimated many poultry flocks. This same scenario had occurred in the northeastern United States in the mid-1980s. Today, extreme biosecurity precautions prevent spread of the virus to the United States and neighboring countries in Central America. Current research efforts on c avian influenza are directed toward understanding why and how mildly pathogenic viruses become highly pathogenic.

Transmission

Infected birds shed the virus in fecal and oculo-nasal discharges. Even though recovered flocks shed less virus than clinically ill flocks, recovered flocks will intermittently shed and should be considered infected for life.

Waterfowl (wild and domesticated) are the primary natural reservoir of influenza viruses. Wild waterfowl usually do not show clinical signs, but they can excrete the virus for long periods of time. In addition, waterfowl can be infected with more than one type of influenza virus. Detection is further complicated by the fact that they often do not develop a detectable antibody response after exposure to the virus.

Influenza virus has been recovered from water and organic material from lakes and ponds utilized by infected ducks. Co-mingling of these birds with range-reared flocks is a factor in some outbreaks.

The avian influenza virus can remain viable for long periods of time at moderate temperatures, and can survive indefinitely in frozen material. As a result, the disease can be spread through improper disposal of infected carcasses, manure, or poultry by-products.

The disease also can be easily spread by people and equipment contaminated with avian influenza virus. Avian influenza viruses can be transmitted on contaminated shoes, clothing, crates, egg flats, egg cases, vehicles, and other equipment. Any object located on an infected poultry farm must be considered contaminated and should be completely cleaned and disinfected before it is moved from that premises. Clothing worn on an infected farm should be laundered.

Insects and rodents may mechanically carry the virus from infected to susceptible poultry.

Influenza virus has been isolated from turkey eggs suggesting vertical transmission, although typically the virus kills the embryo. There is little or no evidence of egg-borne infection of poults. However, eggshell surfaces can be contaminated with the influenza virus, and thus are a means of transmission.

Avian influenza viruses have frequently been isolated from clinically normal, imported exotic birds. These infected birds are a potential threat to cage birds, wild birds, and poultry. Live-bird markets are a reservoir of infection. Such markets serve as a focal point for gathering and housing many species of bird. These facilities are rarely cleaned or disinfected.

Treatment

There is no effective treatment for avian influenza. However, good husbandry, proper nutrition, and broad-spectrum antibiotics may reduce losses from secondary infections. It must be remembered that recovered flocks continue to intermittently shed the virus. All buildings should be cleaned and disinfected after an infected flock is removed. The poultry litter or manure should be composted before application to cultivated lands.

Prevention

A vaccination program, in conjunction with strict quarantine, has been used to control mild forms of the disease in commercial chicken and turkey flocks. With the more lethal forms of the disease, however, strict quarantine and rapid depopulation of infected flocks remains the only effective methods of stopping avian influenza. The success of such a program depends, of course, on the full cooperation and support of the poultry and allied industries.

With the realization that there is a reservoir of influenza virus in wild waterfowl, every effort must be made to prevent direct or indirect contact between domestic poultry and wild waterfowl. Persons handling wild game (especially water- fowl) must change clothes completely and bathe prior to entering poultry houses.

It is very important to prevent the spread of this disease into the United States. It is very easy to spread avian influenza on clothing and through human contact. Do not visit or go near any poultry flocks unless proper biosecurity actions are taken.

Conclusions

Specialty or hobby-type flocks have an increased risk for direct or indirect exposure to avian influenza because of their contact with wild birds and other poultry. These flocks are commonly mixed and marketed through a live auction market distribution system where proper sanitation is not always practiced. This system mixes various types of stressed poultry and has been a key link to avian influenza outbreaks in commercial flocks.

The poultry owner is the first line of defense in identifying outbreaks of avian influenza. If birds develop signs of avian influenza, or if exposure is suspected, immediately notify your state poultry officials.

From PS38, one of a series of the Animal Science Department, Florida Co-operative Extension Service, Institute of Food and Agricultural Sciences, University of Florida. Original publication date April 1998. Reviewed June 2003. Visit the EDIS Web Site at http:// edis.ijas.uj/.edu.

Bird Flu Editorial: A Common Sense Approach
By Jean Nick Fall 2005

First, let's talk about Avian Influenza, aka "bird flu."

Bird flu is a family of diseases caused by versions of the influenza virus. Some are more serious than others and some spread better than others. New versions are always appearing and very few are much of a threat outside the area they originate in.

Same family of diseases that cause human flu but a version people can get is rarely a problem for birds and visa versa.

There are three possible levels of threat with any bird flu outbreak:

1. All versions of bird flu infect birds, but very few can infect non-bird animals (including humans). Bird-only versions are a threat to birds (some are more lethal than others and/or more lethal to some types of birds than others), and to the lively-hoods of those who raise such birds, but no threat at all to human health.

2. Every once in a while a version of the bird flu changes enough in a very specific way (influenza viruses are renowned for being constantly changing and evolving) so that birds can pass it to humans and those humans get sick (and perhaps even die, depending on the version and the care they receive). This is cause for concern to folks who come in direct contact with infected birds, but not a threat to the general public.

3. And once in a very rare while a version of bird flu changes enough more so that humans can not only catch it from birds but can pass it to other humans as well. If the symptoms the version causes are extreme
enough this is a huge public health problem. Remember: This last step in evolution of a flu is VERY RARE.

The current version of bird flu everyone is talking about is in stage 2. Stage 3 hasn't yet happened and MAY NEVER HAPPEN with the current version, even if it makes it to North America.

Experts pretty much agree that "someday" the current version of Avian (bird) flu, or another version yet to appear, will get to North America.

But "someday" could be next week (unlikely), or 20, or 80 years from now. No one has any way of knowing when it will happen or if this version is the one.

So what will happen if this version (or some other version) makes it to North America?

First: Remember it HASN'T HAPPENED YET with the current version.

But IF a stage 1 or stage 2 type of Avian flu makes it to North America (most likely via migratory birds who are carriers, but don't tend to get sick enough to die, but it could also come in with imported live poultry that are not quarantined or tested properly) what do we need to worry about?

1. The government may tell producers in a certain area or areas (such as those where migratory birds tend to land) that they can't keep chickens (or other poultry) outdoors—perhaps for a certain period of time. There is no evidence this will be of any help in slowing the
spread of bird flu, but government officials will be under a lot of pressure to DO SOMETHING.

You may decide keeping your poultry inside sounds like a prudent thing to do in certain situations even without a government order. It is always a good idea to think about the whats, whens, and hows before the situation occurs.

2. If a Bird Flu is actually found in an area (either in wild birds or in domestic poultry) the government may decide to kill each and every domestic bird residing within a certain number of miles of the outbreak, including many that are NOT infected and even entire flocks without a single infected bird -- to protect all the birds outside of the target area. This is traditional containment of an agricultural disease and has saved countless billions of dollars of loss in many types of animals (whether it is the best way to do it is open to discussion). Tough on a few producers, lifesaving for the industry in general. I'm not sure that any person has the power to stop federal or state from killing domestic birds if the proper orders have been issued; and whether the wording of order might provide exemptions for birds that have no contact with the outside or not. So, yes, authorities might show up on your farm and bag all your birds and take them all away. Any eggs you have should be exempt, so perhaps you could recover that way?

Having a plan for such an eventuality may be worth considering, especially if poultry are a big part of you lively-hood.

3. If your birds get sick/die from flu or something you think might be flu you needless-to-say have a big farm problem and may be at risk of getting sick yourself.

Knowing what symptoms to look for, how to keep yourself well, how to get the professionals in fast, and having the resolve to do it are probalby important things to find out and think through BEFORE anything happens -- and good general advice since there are other disease that can strike poultry once in a blue moon that are very serious.

If a version of bird flu gets to stage 3 anywhere in the world it WILL be in North America before you even hear about it and we are all in serious trouble.

Keeping your immune system in top notch shape, perhaps exploring alternative treatments for the flu, and being prepared to stay at home and away from crowds for a period of days or weeks may increase your chances of getting through it.

Getting a "flu shot" will NOT help prevent you from getting the bird flu.

Why are authorities telling various groups of people to get a flu shot then?

So that people who are more likely to catch human flu will have less of a chance of getting run-of-the-mill human flu AND thereby there will be fewer chances of a person suffering from people flu coming in contact with a bird with bird flu or a human with bird flu caught from a bird. Scientists think that last situation MIGHT allow the two types of germs to get together and make baby germs that would be a stage 3 bird flu.

So how worried am I?

Not real worried. This year's Bird Flu is still in stage 2 and probably will not make it to stage 3. I will be prepared to care for my family as best I can if it does make the jump, but don't plan to worry about something I have no control over.

The stage 2 version has yet to make it to North America and may well not make it here at all. We will monitor it's movement and keep calm and encourage others to do the same.

If it stays stage 2, but does make it to North America, I'm still optimistic. My chickens are in top notch health and have strong immune systems. We feed a high-quality feed that includes probiotics (Fertrell Nutribalancer). Our farm isn't high risk as it doesn't attract wild waterfowl to land most of the year (we might consider fencing to keep the wild geese that come to a neighbor's fields for a few weeks in the winter off our land).

Predators: Thieves in the Night
By Jody Padgham Summer 2004

Predator Identification Key

Clue	Possible Predator
Several birds killed	
a. Birds mauled, but not eaten	Dogs
b. Birds killed by small bites on body and neatly piled.	Mink or Weasel
c. Heads and crops eaten on several birds.	Raccoon
One or two birds killed	
a. Birds mauled, abdomen eaten.	Opossum
b. Deep marks on head and neck, some meat eaten.	Owl
One bird gone--feathers remain.	Fox or Coyote
Several birds gone--no clues.	Human

Dogs. A dog usually kills chickens for sport. Several dead birds with much mauling of the carcasses is usually evidence of a dog. Dogs usually visit the chicken pen during the daylight hours rather than at night.

Mink-Weasel. Birds usually show signs of attack on the sides of the head if a mink or weasel has visited the poultry house. With these predators, several birds will probably be killed and piled neatly together. The back of the head and neck are frequently the only parts of the carcass consumed.

Opossum. The opossum generally attacks only one bird at each visit. Usually, the bird's abdomen has been eaten. Eggs may also be the object of the opossum's raid.

Owl. The only likely culprit here is the great horned owl, which does sometimes attack poultry. One or two birds are usually killed, with the talons being used to pierce the brain. The owl will usually eat only the head and neck. Feathers found on a fence-post near the chicken pen may provide an additional clue.

Fox-Coyote. The old sayings about the sly fox were not by accident. The fox and the coyote are very smart and difficult to catch in the act of raiding the flock. Since birds are frequently carried away with little evidence left behind, the only way of determining losses may be a head count. Visits from these predators will usually be very early in the morning. Keeping birds in a secure pen or house until late morning is good insurance against losses from a fox, raccoon or coyote.

Determining the identity of the predators is essential in preventing repeat visits. Once identification has been made, appropriate steps can be taken. Eliminating the point of entry is the first deterrent and eliminating the source of the problem by trapping or other means is the second. Trapping should be done properly to minimize the chances of catching an innocent animal. Seek the advice of a wildlife specialist if you have no experience with trapping.

Prevention is the best solution to the predator problem. Properly construct houses with no access points and maintain fences to keep predators at bay. Use chicken wire or hardware cloth to discourage chewing. Rats and mice are not usually a problem with grown birds, but they can encourage entry by other predators by gnawing holes in wooden pens or burrowing under pens and fences.

Modified from publication F-824 produced by the Oklahoma Cooperative Extension Service • Division of Agricultural Sciences and Natural Resources

You Are the Key to Preventing Avian Disease
adapted from USDA-APHIS (1980)

No matter what kinds of birds or poultry you own, you are the key to preventing infectious avian disease.

People can introduce disease. Anyone going near a flock should wear sanitized clothing and footwear.

Vehicles can introduce disease. All cars and trucks that can't be kept off the premises should be cleaned and disinfected first.

Equipment can harbor disease agents. Houses, cages, and all other items should be washed and disinfected between flocks.

All poultry should come from official NPIP-tested flocks.

Buy replacement stock from proven disease-free hatcheries and breeders.

Fighting cocks, pigeons, ducks, pheasants, fancy fowl and cage birds all need the same disease-prevention care as commercial chickens and turkeys.

Follow a plan of preventative medicine, including timely vaccinations with USDA-licensed products.

Isolation is important to avian health—put as much distance as possible between your birds and other birds, vermin, and traffic.

If unusual health problems develop, take samples of blood, organs or birds to your veterinarian or State avian disease diagnostic laboratory.

Laboratory tests can unmask disease conditions before a devastating epidemic can get out of hand.

Quick action in the face of an epidemic can save a flock or an entire industry. Remember, you are the key.

Because of the heavy losses suffered from contagious disease outbreaks . . . USDA veterinarians strongly urge poultry producers, fancy poultry breeders, and exotic bird breeders to improve their sanitation in every phase of husbandry.

The suggested procedures are applicable in the prevention of any disease that can be transmitted by the movement of people, birds, and equipment.

What can you do?

Find out where your current or potential disease problems are. Take a good look at the movement of people, equipment, and birds on your farm.

Analyze your situation. Get expert advice from your veterinarian, county agent, or State or Federal animal health officials.

CHAPTER TEN
Processing

Some folks say that processing is what makes or breaks a beginning pastured poultry producer. Most people can find a way to get a home flock butchered—if not getting through it on your own, then perhaps finding a neighbor to take over the task. But those who wish to process on anything but a family level will soon get frustrated unless they create an efficient processing set-up. Most states regulate the number of birds you can process on-farm and sell. Find out from your local Department of Agriculture what the laws are in your state before you make any investments in equipment. To sell at a farmers market or to a store or restaurant you will probably have to go to a state or federally inspected processing plant. These may be uncommon in your area, so check before you make your business decisions. Asking other pastured poultry producers in your state where they have their birds processed saves research time for you.

How to Process a Chicken at Home
By Jody Padgham Summer 2005

If you are raising chickens to eat, you will need to kill them as a last step of the production process. Many who are starting out try home butchering. It is a process that almost anyone can do at home with very common household equipment. However, you will quickly learn that butchering a chicken using the canning kettle as a scalder, like Grandma did, certainly works, but gets old quickly if you are processing more than a few birds. We give you here instructions on how to set up a functional, yet extremely inefficient butchering operation. Once you get the hang of it, you may want to abandon the process entirely and seek out a local processor, or you may want to upgrade and purchase time—and energy-saving home processing equipment. If you decide to go that route, see the following article on "Small-Scale Processing Equipment."

Food safety is quite a catch-word these days. Those of us who produce our own food at home can be lulled into thinking that because we "aren't the big guys" our food is safe. Let me warn you that home-raised birds often carry the same toxins that we read about in the news—E. coli, salmonella, etc. In fact, one of the contributors to this book did quality tests on their finished birds and were surprised to see high E. coli levels. (See page 197.) So don't feel that you can ignore food safety rules and still have a healthy product. If you are careful with sanitation and keep birds at the proper temperatures, it is very likely that your end product WILL be much safer than anything similar found in the store, but ONLY if you follow recommended guidelines.

Cleanliness in home processing is critical. Clothing and hands must be clean. Many that process significant numbers of birds at home choose to wear latex gloves while butchering. Strong bleach water is an important disinfectant for hands and all surfaces (rinse with fresh water before touching birds). Ensure your processing area is free from flies. Process birds using plenty of fresh, pure water.

Most critical to food safety is that birds maintain proper temperatures at all points of the processing process. With a high metabolism, a live bird has a temperature higher than a human. When you kill the bird, you will be dunking it in hot water (140-155° F) and then placing it on a table or in shackles during the time that it takes to pluck and gut the bird. (And given we are raising poultry on pasture, it is often HOT outside while we are doing this.) It is critical that the bird get to a 40° F temperature as soon as possible after its death. This means that you must work quickly, and put the bird into very cold ice water as soon as you are done handling it so that it can chill down as quickly as possible. Pathogens that are harmful to human health flourish between 50 and 100° F, we must get our carefully raised bird out of this danger zone as quickly as possible.

Once chilled, the bird must be handled to ensure that it stays cold. We can't stress enough how important these considerations are. If you don't think you can follow them, strongly consider taking your first batch of birds to anyone else you can find that will process them, as your and your family's health can be at severe risk.

We include these instructions primarily for the family that plans to raise birds for their own consumption. Most states limit the ability of those who home-process to market their birds. Many states allow producers to sell up to a certain maximum limit through on-farm sales. (For example in Wisconsin we can home-butcher 1,000 birds per year and sell them direct to consumers from the farm.) If you wish to do on-farm processing and sell to others, contact you local Department of Agriculture to find out what the rules are in your state.

Equipment Needed:
- A way to contain chickens so they can be caught easily (pen, back of pick-up, crate)
- A scalder set-up—a kettle or heated water bath that contains at least two gallons of water that can be kept at a steady temp of 140-155° F for the entire butchering process. A propane turkey fryer works well.
- Clean, washable surface at comfortable height for evisceration
- Fresh running water
- Buckets to contain offal
- Very sharp cutting knives
- Pliers
- Knife sharpener
- Killing cones or axe
- Water bath with ice
- Small bowl with ice if you are keeping giblets
- Clean, heavy food-grade bags for finished product

The Butchering Process:
Prep
- 12-16 hours before butchering withdraw food from chickens to be processed. Allow them to have water.
- Catch and transfer birds to holding area. Chickens are 99% easier to catch after they have gone to bed at night, so plan to catch after dark and hold till processing. Catch them gently, grabbing each bird around the body and wings, not by the legs. This prevents stress and bruising. I find using heavy gloves for both the catching and later putting into cones saves on scars.
- Heat scald water to 140-155° F. Use a thermometer to closely monitor the temperature.
- Clean all surfaces and equipment with a strong bleach solution. Rinse.

Killing
- Place a bird in a killing cone, head down. Draw a very sharp knife across the side of the throat from the outside just behind the jaw. You want to cut both the large vein and the cross vein so the bird will bleed freely, but not cut the esophagus or windpipe (to reduce contamination). Hold the head firmly as you cut, hold for a few seconds longer to be sure the bird is bleeding well. The bird will flop for a few minutes as systems shut down. You know they are dead when you stop seeing pulse in their anal area/vent. The dead birds will start to stiffen within 15-20 minutes, so don't kill more than your family or helpers can pluck in that time.
- You may also kill by chopping the head with an axe on a stump or using a knife while the bird is tied to a clothesline or suspended from shackles. Axe slaughter is the old fashioned, but messier method. A throat cut allows greater bleed-out of the bird and a higher quality finished product.

Scalding
- Dipping the dead bird in scalding water for several seconds allows the feathers to be pulled more easily. It is very difficult to pull feathers from a dry bird.
- To scald, take a bird that is freshly dead and, holding the legs, dip up and down in water that has been heated to 140-155° F. Generally 10 complete dips is sufficient. Maintaining an exact temperature is fairly critical. If your water is cool, you will spend much too much time yanking feathers. If the water is on the hot end of the temperature range, you can dunk fewer times, but if it is TOO hot, you will damage the skin and end up with an inferior product. If you are looking for a start-up scalder set-up, one of those outdoor propane turkey fryers works well.
- Once the bird is scalded, put it through a mechanical plucker or set it on an evisceration table and pluck the feathers. Pull wing and tail feathers first, as they can be the toughest. Use plyers to pull hard nubs. Rub your fingernails against the feather grain to dislodge feathers most effectively. Rinse well with fresh water.
- If you are processing any kind of waterfowl (ducks, geese), you will find the plucking process very challenging. Most waterfowl processors wax the carcasses in order to remove feathers. That is

the subject of another article. If you have waterfowl and wish to have someone else do the processing, be aware that it can be difficult to find a processor. You will want to find someone who agrees to process them BEFORE you buy the birds.

Evisceration

¤ Turn the plucked bird on its back. Cut off the feet at the elbow joint with a kitchen shears or pruning shears. (Dogs love to eat the feet, though they are also prized by some humans for soup.)

¤ Pull or cut off the head.

¤ Make a cut through the skin at the base of the neck. Reach to the birds right side and feel for the sac of the crop. It should be empty if you have withheld feed. Pull that out the best you can. Most of the esophagus and trachea should come with it.

¤ Between the legs: ½ in. above the vent, make a horizontal cut with your knife just through the skin. Be careful not to cut into the intestines. Reach into the cavity with your hand and displace all of the guts. Pull the mass carefully outside of the body cavity. Reach with fingers back inside to find the lungs (hard against the backbone), heart and gonads.

¤ Once guts are out of the body, make careful cuts on either side of and below the vent to remove the entire digestive system from the bird. If there is any expulsion from the vent, be careful it does not get on any part of the bird and wash it away quickly with water. Dispose of guts into pail.

¤ If you wish to save the gizzards, you will have to release the heart, liver and gizzard from other viscera. The elongated, jelly bean-shaped gallbladder is attached to the liver, and will release a shamrock green ink if you are not careful in cutting it away. The gizzard must be cut open and cleaned of all stones, etc. Toss the giblets in a bowl or ice cream tub with ice and save the cleaning tasks till you are done processing all your birds. They can be rejoined with their giblets later.

¤ Turn bird onto its chest and cut the oil gland from the top of the tail. Remove the neck if desired. Check to be sure the windpipe is gone.

¤ Return to back and, if desired, cut another horizontal slit parallel to the initial cavity cut. Tuck the legs into the flap of skin created. Tucking the legs makes it easier to bag the birds when they are chilled.

¤ Thoroughly rinse the bird, inside and out.

¤ Place bird into ice cold water bath.

¤ Start with another! As a solo processor, I usually kill one bird; when that one is dead and ready to go to plucking, I kill another so that it has time to die and bleed out while I am cleaning the first. Those with larger crews can plan their timing accordingly.

Chilling

¤ Keep birds in a cold water bath until their internal temp is 40° F. They will be stiff and feel very cool to the touch.

¤ I like to chill my birds for 24-hours before freezing. Not everyone does this, but some say it improves tenderness (kind of like "hanging" beef, pork or venison before cutting). I take the birds from the ice bath, drain and place into heavy plastic bags. I don't tie the bags shut, but twist them closed and place upright in the fridge. After 24-hours, I take each bag out, drain and tie closed with a tie. The birds are then ready for sale or freezing. They may be cut into pieces at this time.

Cleanup

¤ Clean all equipment immediately with soap and water. Use bleach solution to sterilize.

¤ Compost feathers and entrails in pile with hay, leaves or other carbon source, protected from dogs and predators.

¤ Don't plan on having chicken dinner for at least a week. The smell will linger on your hands and thoughts and discourage full appreciation of the final product until a few days have passed....

Homemade plucker with turkey fryer scalder

Small-Scale Poultry Processing Equipment
By Jody Padgham Summer 2004

I started out on my processing equipment exploration thinking I'd talk to several pastured poultry producers to find out what they liked and didn't like about their processing equipment. But, like every other time I've tried to do something like this, my explorations took me in a direction I didn't anticipate. It may be no surprise to some of you that when I ask folks what they like about their particular brand of equipment, they all say "it's what is right for me and my scale of operation." I found that all of the companies that sell processing equipment to people like us produce very high quality equipment that has been time-tested and works well. The major difference separating the different brands of equipment is the scale of operation that it is intended for, or very minor differences in design. Everyone I talked to had been happy with all the equipment that they used, and only made changes as their production grew or changed.

So, instead of a lot of testimonials, I thought it more useful for all who may be out looking for the perfect processing set-up to know who is out there and what they are selling. Almost all mentioned here are APPPA Business Members and are always happy to serve APPPA producers.

I'll start by reminding you that if you are just starting out, you may not want to jump in and buy new processing equipment right away. It is possible (though we don't want to think about it) that pastured poultry won't be a good long-term fit for you or your family. Give yourself a break and try things out for a year or two before you spend the big bucks to get that perfect processing set-up.

So what do you do if you have birds on the ground and are wondering just how they will be processed eight weeks from now? If you live in a traditionally rural area, as I do, you may be surprised how many serviceable pluckers are sitting in the back of someone's barn or shed. You may get lucky, as I did, in that my neighbor's brother had raised birds for 10 years, built a home-made plucker, and then gotten "out of the business." I was able to first borrow, and then buy for a very reasonable price, my first "no-frills" machine. No, it isn't efficient or fancy, but it is giving me the opportunity to try out my bird raising, marketing and processing skills before I decide to take the plunge and invest in better processing equipment. I'd recommend that anyone starting out put an ad in the local shopper for a "used chicken plucker" and see what comes. Whatever you get may drive you crazy, but at least you haven't invested $1,000 before you find out that you hate chickens! For those "do-it-your-selfers," there is always the option of building your own equipment. Herrick Kimball has a great book available that describes in great detail how to build his "wiz-bang" plucker for less than $500. Contact Herrick at hckimball@baldcom.net or 315-497-9618 to buy a copy of his book. Don't forget that APPPA members can put a classified ad in the GRIT for only $5.00 per issue. There are often ads for equipment for sale. Try putting in a "wanted" ad and see what you get.

For those who have made it past the "two-year test," here is what I found as far as farm-scale processing equipment. There are basically five companies that sell equipment relevant to us "pp" types:

(Please note that prices DO NOT include shipping.) **These prices were current as of 6-10-04.** With steel prices fluctuating, you will want to call to confirm prices before you finalize your decision. Contact information for the companies can be found following the chart.

Small-Scale Processing Plant Equipment
Those ready to get into processing at a larger scale also have several companies to choose from. All offer high quality equipment, and all will work with you to custom design a system that is just right for you. Contact the folks listed in the table on the next page to ask about a whole system for your small processing plant.

| Small-Scale POULTRY PROCESSING EQUIPMENT ||||||
Name	Location	Product	Birds/ batch	Price	Notes
Pluckers					
David Schafer	MO	Featherman Pro Plucker	5	$975	Food grade plastic, 22" drum, spray kit avail.
Eli "Poultryman" Reiff	PA	Mechanical Plucker	5	$1525	Stainless, 27" tub
Pickwick-Zuber	MN	Pickwick Mini-Spin Pik Picker	3, up to 30 lbs. total	$2500	Stainless, 24" tub
Pickwick-Zuber	MN	Pickwick Spin Pik Picker	5, up to 80 lbs. total	$5600	Stainless, 30" tub, timer, auto discharge. Will do geese, turkeys, suckling pigs
Brower	IA	BP 25SS Spin Picker	3-6 chicken 1 turkey	$2310	Stainless, 25" tub
Brower	IA	Tabletop Drum Picker	1 bird at a time	$472	Galvanized, you hold bird
Brower	IA	Tabletop Drum Picker with legs	1 bird at a time	$755	Galvanized, you hold bird
Pickwick-Zuber	MN	Pickwick Drum Picker, 3 sizes	1 bird at a time	$370-925	Galvanized, table top or with legs, with or without motors
Scalders					
Pickwick-Zuber	MN	PKES Electric Scalder	3	$190-210	Hand dunk, fiberglass 30 gal.
Eli "Poultryman" Reiff	PA	Manual Scalder	5	$692	Stainless, 42 gal, hand dunk
Eli "Poultryman" Reiff	PA	Gas Rotary Scalder	5	$2149	Stainless, 42 gal
Brower	IA	Manual Scalder	3	$1770	Galvanized, 38 gal, dipping basket

Distributor Addresses:

Ashley Machine; Jim Israel, owner; 901 Carver St, Greensburg, IN, 47240; 812-663-2180 (small processing plants)

Brower; Houghton, IA; 800-552-1791 (also small plant equipment) www.browerequip.com

Pickwick-Zuber, Inc. (Formerly Pickwick-Zesco); 7887 Fuller Rd, Suite 116, Eden Prairie, MN; 55344; 800-808-3335; sales@zesco-inc.com (also small plant equipment) www.zuberinc.com

PoultryMan; Eli Reiff; 922 Conley Rd, Mifflinburg, PA ; 17844; 570-966-0769

Schafer Natural Meats; David Schafer; 760 SW 55th Ave, Jamesport, MO; 64648; 660-684-6035; dna76@grm.net

Build your Own: Killing Cones and Transport Boxes
By Jody Padgham Summer 2003

Here it is, close to the time of year when we remember we really were thinking of replacing a few pieces of equipment last year and probably never got around to it. Whether time got out of hand or economizing is an important factor in your operation, these patterns may be just the thing to fill in the equipment gaps. Randy Anderson and Vince Maro of northwest Wisconsin recently shared how to quickly and inexpensively build your own killing cones and transport pens.

Killing Cone:

Randy had purchased some used cones, but soon found that the cones were too small for his 4-5 lb. broilers. This simple pattern can be adapted for larger chickens or turkeys:

Buy 16"-wide roof flashing from the local lumber store.

Measure 34" on one edge. Measure 17" on the opposite edge, centered over the previous measurement. Connect the ends of these two measurements and cut the edges, which will be evenly angled, with a tin snip or metal blade on a power saw.

Make a cone by bringing the cut edges together, overlapping 1 in. Drill three holes along the seam through both pieces. Insert rivets and smash flat, if you can find something to support the cone while you are smashing. (If you don't have the ability to rivet things, I think short, flat-headed bolts will do. Just be sure the nuts are on the outside of the cone so the bird's skin doesn't get caught.) The finished cone will be about 4 in. in diameter on the bottom and 9 in. in diameter on the top.

If you want a smooth finish on the top take the long side before riveting the cut edges and bend ½ inch over, using a vice grips or vice and pounding flat with a hammer.

Punch a hole in the top side for hanging. Flatten the back a bit if desired.

Transport Crate:

For about $14.25 you can easily make this sturdy transport crate, which is designed after a typical wood poultry crate. We didn't get a chance to weigh it, but I'd guess it weighed about 15 lbs. This crate can be used to carry 8-12 chickens. (Note: Vince decided to make this crate extra tall so he could carry turkeys. A standard broiler crate would typically only be 10-11 in. high. If you wish to make this crate shorter, cut the 2 x 2 in. pine to the length you wish the crate height to be.)

Materials:
- Two pieces ½ in. plywood cut 24 x 36 in. (two 4 x 8 ft. sheets will make five crates)
- 2 x 2 in. pine cut into ten 12 in. pieces (two 8 ft. will be plenty)
- Wood lathe: Eight pieces cut 24 in. long, ten pieces cut 36 in. long. One cut 16 in. long.
- Two narrow 2 in. hinges
- One latch. Can use a toggle latch, a barrel latch, or your favorite. As flat as possible for eventual stacking of crates.
- 1 in. deck screws
- 2 in. deck screws

Cut the plywood pieces to size. In the top piece, cut a 14 x 14 in. square hole, centered from each side.
Screw both hinges to one of the sides of the cut door, and attach to the plywood top. Attach the latch on the opposite edge. Turn the top over. Using 1 in. deck screws, attach the 16 in. long lath parallel to and slightly over the cut with the latch on the other side as a stop for the door.

Lay the uncut plywood piece on a flat surface. Use four of the 12 in. long 2 x 2s as upright corner supports and balance the top with door on top. Use 2 in. deck screws to go through the top to attach the supports in each corner.
Place two additional 2 x 2 in. supports on each long side, 1 ft. apart. Attach to top with 2 in. screws.
Place one additional 2 x 2 in. support on each short side and screw in.
With a helper, turn the entire crate over. Place the loose bottom on top of the 2 x 2 in. braces, lining up the corners. Screw each brace to bottom using 2 in. screws. Turn upright again.
Attach the lath to the sides. We ended up drilling the lath and then screwing with 1 in. deck screws. Place four pieces of lath on each side, matching the lengths with the sides. Space lath evenly and match edges. Leave ½ in. of open space on bottom to allow water to run out during cleaning.
Final step: attach last pieces of lath to long top edges as supports for stacking. Be sure the lath is thick enough to accommodate the latch you have chosen.

That's it--pretty easy! Very sturdy, not too heavy, and CHEAP! These crates should last several seasons if well cared for. One participant noted that if the bottom plywood was painted, it might last longer and be easier to wash. These guys just created this design this year, so there surely will be refinements--think of your own as you experiment and customize to your own operation.

Eviscerating Poultry the Easy and Sanitary Way
By Tom Neuberger Spring 1999

Anyone who has ever eviscerated poultry on an eviscerating table for very long knows back and neck pain. Plus, how about the struggle to keep the work surface sanitary?

The solution to allowing an eviscerator to work with good posture and to reducing contamination is a shackle system. These bird holders are, without exception used in every major poultry processing plant.

I imagine many small operators don't consider shackles for they feel it would be terribly expensive. This would be true if one considered dozens of shackles on a motorized line. But, this isn't necessary for a small operator. In fact, a few shackles hanging from a tubular steel stand would be considerably less costly than a stainless steel evisceration table.

A shackle has three slots with which to position a bird into a three point hang: the two feet and the head. This will position the bird with the back down and the posterior facing the operator.

If the shackle is approximately nose high the eviscerator can now stand erect and look directly into the cavity so as to plainly see into the cavity as he or she works. All the time the shackle is holding the bird firmly so the eviscerator can use both hands.

The shackle should be stainless steel for ease of cleaning. A good stainless shackle costs $20-50 depending on size and configuration. It would be best to get universal shackles that will hold small pheasants, large turkeys and anything in between.

The stand to hold the shackles should allow a garbage can or something below the shackles in which to place the entrails and waste. A handy accessory is a pressurized hose hooked on the frame of the stand to spray down a finished bird or to quickly wash off a contaminated bird.

A good efficiency idea would be to make a bigger rack with more shackles. This will allow room for the person doing the picking to move the birds from the picker to the shackles without bothering the eviscerators. A longer rack could also give room for more eviscerators.

You should be able to buy eviscerating racks for around $400 new, less for used. They are so simple to make that almost anyone with a welder can make one.

We guarantee that anyone who eviscerates with the birds in a shackle will be able to double their rate as compared to using a table, without the back and neck discomfort. Plus, you'll have cleaner birds!

Aaron Silverman Operating His Shackle System

Just 100 Chicks!
By David Schafer Summer 1999

Alice swore she would never, ever process a chicken. Ever. Final word. She was serious. With Alice, that means I had to resort to a flank attack—a frontal confrontation would have been suicidal.

I had to employ the subtle approach, such as leaving alluring gross margin analyses lying on top of the desk; having loud conversations with grazing buddies who were also working on their more stubborn half; and seating her in the front row of as many Salatin Pastured Poultry Revivals as possible! Hallelluja Brother!

After employing these devious tactics, that one moment of weakness I was patiently waiting for finally occurred: "Well, maybe just 100 chicks," she said. Yes! Victory was mine! That was several thousand chickens ago.

Despite the proven track record of success, many grass farmers (and their practical wives) are still reluctant to enter the pastured poultry business. The hang-up for nearly everyone? Processing. Yuck! Who wants to do that? Precisely! That is why plumbers make $50 an hour! Who wants to do that?

The obstacle which has been a close second to the physical aspect of processing is the availability and cost of the processing equipment. Well, things have greatly improved on that score. Several hundred home-made chicken pluckers have been built for under $400 based on Ernie Kauffman's prototype built in 1995. (Ernie's plucker is still going strong at 5,000 per year and others have reportedly plucked up to 15,000 birds per year!)

Ernie's picker utilizes a plastic 55-gallon drum mounted over a rotating disc, both of which are peppered with fingers. A small lawn mower engine powers it.

Kenneth King of Hutchinson, Kansas, read the writing on the wall and introduced a scalder and picker to the growing market several years ago. In 1998, Brower, the Iowa based poultry processing company, came out with a reduced-cost line of equipment for the small-scale producer, a clear signal that small-scale, particularly pastured poultry, production is here for keeps. Their advertisement is also within.

Alice and I recently had the opportunity to be in Hong Kong and visit the huge downtown market. Unlike our local markets, the crowds don't show up until mid-morning. We were, therefore, able to witness the chicken processing which occurred on site earlier in the morning.

The birds were removed from plastic crates and tied together by their feed—five in a bundle. Then they were taken to the communal kill/scald room where they were hung over a stainless trough—as a batch—and bled. Being tied together restricted their movement considerably.

Still tied together they were tossed into a cauldron of hot water and stirred with a big paddle. No dunking. There were four other big pots of water over wood fires to keep enough hot water ready. After scalding, the bundles were thrown on a cart. When the cart was full they were wheeled out to individual market stalls, untied and plucked.

All stalls had their own picking units. These pickers measure 2 x 2 x 3 ft. high. They are belt driven by a quarter-horse motor and made of aluminum. The fingers are bigger and stiffer than what we have seen here in the States. A feather chute funnels feathers into a container on the ground. After picking, feet were lopped off with a cleaver and the birds eviscerated and checked for remaining feathers. Heads stayed on.

We were impressed enough with the performance and price of these pluckers to begin the process of importing them to the U.S. We expect the price to be somewhere between $800 and $1,000 and hope to have them available by September. (For an update see page 155.)

APPPA BUSINESS MEMBER PROFILE

The Poultry Man- Eli Reiff
By Brian Moyer Summer 2004

About three years ago, I attended a pasture walk at a farm in north central Pennsylvania where they raised sheep and pastured poultry. At some point in the tour, the farmer introduced one of the attendees. "This is Eli Reiff. He has a USDA-approved processing facility at his farm. I can tell you that without Eli, many of us wouldn't be doing pastured poultry." One afternoon a few weeks later, I made the long drive from my farm to Eli's place in the town of Mifflinburg in central PA to pick up some crates he had for sale. When I got to his farm, I pulled up to the processing building where I found Eli leaning against the door. I said, "How are ya, Eli?" and he said, "Well, if I get up in the morning and I got chickens to dress, it's gonna be a good day." I thought to myself, "I gotta bring birds to this guy," and Eli and his family have been dressing my birds ever since.

Eli Reiff has been processing poultry for 24 years. Eli, along with two other family members, does the processing. They generally schedule most of the work for the morning, leaving afternoons to do other farm work. They process five days a week after Memorial Day and dress 200 to 225 birds per day. They process 40,000 chickens per year, along with 1,200 turkeys that get scheduled starting after Labor Day.

While Eli and family where busy dressing my birds this last time, I picked his brain about poultry processing and pastured poultry. Eli scalds at 147 °F for 40 seconds. Older birds a little longer, layers for 60 seconds and turkeys for 50 seconds. Eli can tell the difference between pastured and indoor birds the minute they come to the door. "They smell different. Indoor birds can smell of ammonia or manure and the skin is thinner and can be damaged in the scalder." He tells me. "The steam from the scalder brings out the smell. Indoor birds have more fat and the fat is greasier and runnier then pasture birds. Pasture birds have firmer fat. They are brighter looking. They have more white in their feathers and more yellow in their legs." Common problems that he looks for in birds are things like inflammation in the knees. This is seen mostly in indoor birds and is due to fluid build up in the legs. When this happens, Eli removes the leg from the carcass. He also sees ascites, or fluid in the belly, which Eli says can be caused by high stress. Bruising on wings is common; it often happens when loading the birds into the crates.

Reiff's Poultry Dressing business is a USDA-approved custom facility, which means once a year an inspector will review his shop. In Pennsylvania regulations fall to the USDA. There is no state inspection. They charge $1.50 per bird, or $1.25 per bird if it is under nine weeks of age. Older birds take longer and not as many can be scalded or plucked at once, so are a little more expensive.

Since Eli sees and talks to so many pastured poultry producers, I asked him his thoughts on where he thinks pastured poultry is going. " We're not even half way to where it can go." He tells me. He is amazed that so many new people are still reading Joel Salatin's book and then coming to him for processing. People from all different backgrounds. I've seen people pull up to his place with chickens in the back of a Volvo wagon!

I asked Eli what tips he would give growers to help them in raising high quality birds. "Don't move them to pasture too soon. If the weather is questionable, it is better to hold them in the brooder longer. If you wouldn't sleep out there, don't expect your birds to. And don't over-crowd your crates when taking them to slaughter. At eight weeks of age you should have no more then 10 birds per crate. Less if it is hot weather."

In 2001, Eli started constructing his own processing equipment to sell. He calls this business "The Poultry Man" processing equipment and has sold over 200 pieces all over the country and in Bermuda and the Bahamas. Their mission is to "Help farmers to be able to do what they want to do at an affordable price."

APPPA BUSINESS MEMBER PROFILE

I take birds three times a month to the Reiffs, and although the trip is a long one, I always find the Reiff family cheerful and happy to see me. Eli is always willing to help educate growers and processors. He has hosted a field day at his place and has been invited to conferences to speak about poultry processing. When I asked Eli if he could quarter some of my birds for me today he said, "Hey, we put the 'c in custom."

Contact Information: Poultry Man LLC, 922 Conley Rd, Mifflinburg PA 17844. 570-966-0769

Schafer Farms Natural Meats (Featherman Pluckers)
By Jody Padgham Fall 2004

David Schafer is the owner of Schafer Farms Natural Meats (Featherman Pluckers), which has been in operation for 15 years. Four people run the business from Jamesport, Missouri. Featherman pluckers are sold all over North America, and can also be found in the Caribbean, South America and Africa.

Brief History of ownership: My Amish friend, Ernie Kauffman, and I toured a defunct pheasant plant and observed their equipment. From there Ernie designed our first homemade tub plucker in 1994. Through the rest of the '90s I mailed out crude plans for build-your-own. In 1999, my wife and I saw two-bird pluckers in use in the Hong Kong farmer's market and later shipped some of these to the states. Those were the Featherman Jr. In 2000, another Amish neighbor, Syl Graber, and I developed a larger version made with a plastic 55-gallon tub – the Featherman. In December of 2003, Syl and I introduced the Featherman PRO, a lighter, more professional looking, molded plastic model (no 55-gallon barrel). Syl purchased a sawmill so production transferred to Carl and Floyd Miller and Abe Kurtz. We now offer Featherman PRO, Featherman Quail, and Featherman Non-electric (both models) for pastured-poultry operations. In addition, besides being a "chicken plucker factory," as Abe calls himself, Abe Kurtz is also a fine tarp maker and can make hoop house tarps of any size.

Our mission statement reads: "To improve the quality of the food supply, encourage local entrepreneurship, and increase the viability of the small family farm by providing low-cost poultry processing equipment." To that I'd like to add that I credit pastured poultry with liberating us from a 540-acre family farm where the work was never done to a 64-acre farm where we just raise meats for direct market. It was the addition of poultry to our beef and lamb that caused our customer base to increase to the point we could make a living on a smaller piece of land and jettison some burdensome enterprises. I've personally seen what pastured poultry can do and know the bottleneck is not production or marketing–it's processing.

Some Featherman Plucker highlights:
1. Gentle Pick--slow RPM pluck, the softest fingers available allows for a generous margin of error in scalding;
2. Clean--feather chute voids feathers for a tidy work area;
3. Time saving–a spray ring liberates the operator from holding a hose;
4. Powerful--a direct drive train through a 1 hp motor and speed reducer increases torque plus there are no belts to tighten or replace;
5. Portable–fits in my pick up (MPU) or car trunk at 125 pounds;
6. Economically priced–using food grade, HDE plastic instead of stainless steel cuts the price in half.

Before the advent of the home-made pluckers, it took almost $4,000 to buy a commercially made tub plucker. Kenny King smashed that barrier when he came out with his JAKO line of pluckers and scalders. We pulled the price barrier back even further. It is immensely gratifying to have stumbled into the business of making a machine that allows folks to more easily do something that had such a huge impact on our lives. We feel very fortunate to have this opportunity.

Contact Information: Schafer Natural Meats; David Schafer; 760 SW 55th Ave, Jamesport, MO; 64648; 660-684-6035; dna76@grm.net

APPPA BUSINESS MEMBER PROFILE

Pickwick Zesco Poultry and Meat Processing Equipment
By Jody Padgham Fall 2003

Chan Zuber, General Manager of Pickwick-Zesco, told me recently that he got into the business of poultry processing equipment back in the 1940s when he was in high school. He started out cleaning and re-assembling used processing equipment that his Dad, a Pickwick dealer, wanted prepped for resale. He never thought he'd still be working with processing equipment this many years later, but is quite proud of the business that Pickwick-Zesco has become today.

Now located in Eden Prairie, MN, just outside of Minneapolis-St Paul, Pickwick-Zesco is a national distributor of processing equipment for poultry and meat processing plants of all sizes. Zesco and Pickwick came together in late 1999 when Zesco agreed to assemble and sell Pickwick equipment.

Offering a full-line of processing machinery and supplies, Chan and his brother and Dad also have expertise in designing processing plants from large to small scale. They have designed over 400 small plants that are still operating around the U.S. They are willing to work with anyone interested in putting up a plant today, though inconsistencies between states in law interpretation put a larger design and regulation burden on the plant owner than ever before. They even offer a free processing plant planning guide on the Pickwick-Zesco website: http://www.pickwick-zesco.com/guide.htm

For poultry producers, Pickwick-Zesco features chicken scalders, pickers, dunkers, evisceration lines, processing tables, knives and other equipment needed for small and medium sized processing facilities. In November of this year they are rolling out a new line of equipment that is specifically designed for small-scale and home processors. They hope to have prototypes of that equipment at the Small Farm Conference in Missouri; look for them there for a first hand look.

When asked about Pickwick-Zesco's business philosophy, Chan tells me that they strive to "leave the client better off than before they met us." He goes on to say that "even if we have to go so far as to recommend one of our competitors, we want our customers to have their needs met."

When asked what product among their new line of small processing equipment he was most excited about, Chan said that he really liked their new Electro-Static Spraying System. This is a crop spraying system that can be custom-fit for all size operations–from hand-held units to pull-behind for the tractor. When spraying any liquid, this system puts an electrical charge on the particles, which when sprayed will coat the entire leaves of plants (rather than just the top side). This revolutionary system will allow for much better coverage of any fertilizer or control product using less product. Look for the Electro-Static Spraying System on the Pickwick-Zesco website or ask about it when you call.

Chan concludes our conversation by telling me that before Zesco bought Pickwick in 1999, he never knew how many small farmers there were in the U.S. He now sees "a good future for small farms in the U.S."

Contact Information: Pickwick, 7887 Fuller Rd, Suite 116, Eden Prairie, MN 55344; 800-808-3335 ; www.zuber-inc.com

Cadillac Mobile Processing Plant
By Tom Neuberger Winter 2001

In 1985 we installed 100 bird-per-hour poultry processing equipment in a big, two story building in a nearby town. Although we had the first opportunity to purchase, it was more building than we needed. This situation proved to be a blessing for it forced us to accomplish a vision we had had for some time: a mobile poultry processing facility on our farm, where the poultry are raised, which would save much time and expense.

Although we had no need for something mobile, the idea was practical in that when we became tired or retired we would have something that could be marketed and transported anywhere in the world. We thought about retrofitting cargo trailers, stock trailers and mobile homes that could be pulled by a pickup.

One day when visiting neighbors who had purchased two used reefer trailers for a modest sum to use as pig nurseries, I saw what we needed. Eight ft. by 48 ft. reefer trailers have plenty of floor space for the equipment necessary to process hundreds of chickens a day and, most importantly, they are lined with white glass board for ease of pressure spray cleaning.

Some models of reefers have side doors. This is a must for the workers to get in and out. We installed a side door and a ramp for the workers to get in and out.

A hundred miles away there is a trucking industry consignment auction every few months. Reefer trailers that are 10-15 years old and are still roadworthy, with working freezer units, are available for $3-7,000. These older units are available at a modest price as the trucking industry upgrades to bigger and more efficient equipment.

The only preparation needed prior to placing equipment in a reefer unit is to improve the floor and install electric and water lines. The floor of a reefer has aluminum rails about 3 in. apart. This is unacceptable for cleaning. We poured concrete to cover the rails. A reefer trailer has drain holes at the four corners. We have the trailer several inches lower at the front. Water runs to the front and drains into a tank placed under the front edge of the reefer. This water is then pumped by a sump pump into a tank on a truck. The bloody wastewater is then spread on our pastures. It is great fertilizer!

Reefers without a working freezer unit are much cheaper. But, the freezer units are a good investment to use as air conditioners for the workers during hot weather and as a means of chasing out flies. Although we have screen doors, the pests get in but leave in a hurry when the freezer units are turned on.

The poultry enter via the back door of the unit. This is perhaps the only serious disadvantage of a reefer mobile unit. The crates need to be raised four feet to enter. We use forklift forks on our farm loader tractor.

From the back to the front we have the following equipment in the unit: kill tunnel, scalder, picker, eviscerating line and chill tank. There is plenty of room to accommodate at least six workers.

After the birds are chilled and the unit is cleaned, we hang the birds back on the eviscerating line to drip dry and to place them in plastic bags. From there the birds are transported to our further processing facility which is the made over garage on our house. Here the birds are weighed, sorted, cut up, ground, smoked, etc. and eventually placed in the outdoor walk-in freezer.

I have often thought a more efficient and perhaps more economical avenue for further processing and freezer storage would be to purchase another reefer and place it right beside the processing unit. After the birds are chilled they could be moved into the further processing unit. The front end of the unit could be sectioned off for freezing and the back end used for cutting, grinding, etc.

Cost? We have $25,000 plus labor in our unit. This number will vary a lot depending on the cost of equipment. There is quite a range of new and used equipment available. One piece of equipment we wouldn't compromise on is the eviscerating line with shackles. A line is much more sanitary and comfortable for the works than tables. A line also increases capacity. A line can be motorized or stationary. Either way they are a must!

We operate under a federal producer exemption (PL 90-492). I would think such a unit could meet state or federal inspection. I don't know how you could be more sanitary.

WI Mobile Processing Unit
By Jan Nitz Spring 1999

The new Meadowland Poultry Co-op, located in west central Wisconsin, has put together a state-of-the-art mobile processing unit. This is the fifth MPU in the country that has been built to provide local farmers with low cost, on-farm poultry processing. Months of hard work, planning and anticipation came together this spring with the debut of the Meadowland MPU in Chippewa Falls, WI.

The equipment is "awesome" according to John Mower, chief designer of the unit. We started with a new 18 ft. trailer, donated stainless steel counters, commercial dunker, scalder and picker and a whole lot of ingenuity. Thanks to many people we were able to put together a remarkable mobile processing unit, complete with hot water pressure wash system, all weather canopy and stainless steel tables.

Poultry producers will pay a onetime membership fee to become co-op members. After that they will rent the equipment for 65 cents per bird processed. Since they are co-op members and the co-op owns the equipment, they will actually be using their own equipment. The fee charged will cover mileage, maintenance and loan repayment.

The producer will be responsible for providing a number of things when the MPU arrives at the farm, including:
- an outside garden hose type water hook-up;
- 110-volt electric hook-up;
- location with good drainage and accessibility for the 18 ft. trailer;
- one pound of ice for each bird processed. A quick cool down is essential for safe handling of poutry, therefore, for liability reasons, the MPU custodian will not set up the unit if an adequate amount of ice is not on hand;
- an experienced processing crew;
- paying the custodian for the rent of the equipment upon arrival at the farm;
- having birds crated and in the processing area.

All of the offal (entrails, feathers, heads and feet) will be composted by local farmers. A custodian will be traveling with the unit to deliver, set-up, oversee equipment, and supervise cleaning. He or she will not provide labor to process birds. The farmer is responsible for the labor. The custodian will be responsible for:
- delivering the MPU to your farm at the scheduled time
- checking for the proper amount of ice;
- collecting payment;
- setting up the MPU;
- arranging barrels to collect offal and chill birds;
- supervising use of equipment;
- offer advice if there is a glitch in the system;
- oversee clean-up;
- inspect unit after clean-up;
- gathering and loading equipment (hoses, extension cords, knives, canopy, barrels) when cleanup is complete.

We are currently working with the Wisconsin Department of Agriculture, Trade and Consumer Protection to be sure we are meeting or exceeding all requirements and regulations. Using the MPU will not exempt farmers from the State of Wisconsin imposed limit of 1,000 birds per farm. The MPU is not part of a marketing plan, farmers using the MPU will be responsible for using or direct marketing their own birds. This unit is simply a way to give farmers access to commercial equipment at a reasonable cost.

2005 Update: The Meadowlands MPU is now privately owned and available on a rental basis in the Bloomer, WI area.

Mobile Processing Units (MPUs)
By Steve Muntz Spring 1999

So.... you want to raise pastured poultry, and there aren't any processing plants within a hundred miles who will process the birds for you. You don't want to invest in a new processing plant. What do you do? Well, if there are several other producers in your region, the answer may be a mobile processing unit (MPU). MPUs are beginning to show up all over the country as producers work out solutions to the chicken processing challenge. I've visited with the folks who operate several different MPUs in the country and have brought the information they provided me here to the GRIT! To start off, here's a comparative table of some different MPUs.

Ed note: It is especially important that you note this article was written in 1999. We include it here, as it gives you a good idea of how diverse MPUs can be and what some of their features are. However, some of these units are no longer operating, or the situations around them have radically changed. For a more recent update, see the following article, Revisiting the MPU, written in the summer of 2003. Although MPUs seem to many to be the perfect solution to the challenge of finding processing, several states have, unfortunately, developed very hostile stances against them, for a variety of regulatory reasons. Be sure to check regulations closely before you get very far on developing your own area MPU.

Four Different Mobile Processing Units

	New York	Kansas	Minnesota	Nebraska
How did it get started?	South Central NY R,C&D saw the need and decided to help out. Some grant assistance provided.	A couple of people were interested in pastured poultry and wanted to share equipment. Kellogg grant got them started.	Members of the state Sustainable Farming Association were interested in processing chickens and turkeys	An Extension Agent thought the MPU would be a good idea for local farmers. The farmers got some partial grant money to purchase the unit.
Description	20 ft. trailer, pulled by pickup truck	16 x 8 ft. trailer pulled by pickup truck	Set up in an old converted school bus	Fifth wheel enclosed trailer, 16 x 18 x 7 ft. Purchased used and ready to go. Partition between plucking room and evisceration room.
Plucker	Homemade drum picker	Homemade out of an old washing machine	Used 5-6 bird drum plucker	Electric drum plucker
Scalder	Homemade propane scalder	Homemade out of hot water heater	Used multi-bird dunker/plucker system- propane	Propane scalder with dunking apparatus
Chill System	Galvanized tanks with ice water	Plastic drum with cold circulating well water	Stainless steel tanks with ice water	Stainless steel and polyethylene tanks with ice water
People Required	two to four	four to five	four to five	four to five
Max Birds /Hour	35	75-80	40	90-100
Site Preparations and Requirements	Fairly level spot, 110 electric, water	Slightly sloped site floor for water drainage, 110 and 220 (40 amp), water	Slightly sloped site for water drainage, 220 outlet, water. Bulletin on site prep provided,	Well drained area, gravel preferred, near furrow in field for waste water. 110 electric, water
Legal Status	Non-inspected, operates under federal/state exemptions	Non-inspected, operates under federal/state exemptions	Non-inspected, operates under federal/state exemptions	Non-inspected, operates under federal/state exemptions

	New York	Kansas	Minnesota	Nebraska
Training Required	First time users undergo training	Help an existing processor first to qualify	First time users undergo training	Training provided by main operators
User Fees	Members $25 per use plus $0.25 per bird over 100. $0.31 per mile pickup and delivery. $50 charge if not returned clean.	Free to members- originally set up through Kellogg funding.	Any member can use for $0.50 per bird with 100 bird minimum. 30 mile travel included in price, $1.00 per mile over 30.	No charge to the three farmers who put their time and money into the MPU and truck that hauls it. The unit is rented out at times.
Maintenance	Maintained by NY RC&D council with fees collected.	Maintained by Ken King who produces Jako equipment and uses the MPU to test new products.	Maintained by one person, using fees collected.	Maintained by the loose association of three farmers who got it going.

All the Mobile Processing Unit (MPU) coordinators interviewed said that the biggest bottleneck of the system was the gutting. One person killing, scalding, and plucking can generally keep three or four busy at the evisceration and quality control area. Finding a person who can rapidly gut a bird and move on to the next in line will help get the hourly bird rates up. Most of the operators of these systems are processing around 150-200 birds per use of the MPU, though the Nebraska group will usually run about 400 chickens. None of the MPUs have any freezing/cooling facility to hold the birds for any length of time, as most are sold right off the farm on the day of processing. Maintaining proper cleanup and sanitation procedures were a common complaint among the MPU coordinators interviewed. Some of them charge a fee if the unit is not returned properly cleaned. Offal from the processing is being handled differently on each farm. Many people compost the offal but some bury it and some will take it off into the woods.

Inspected Mobile Processing Units

None of the four MPUs described above are inspected facilities. All the farmers using these MPUs are operating under federal and state exemptions for on-farm processing of a limited number of poultry and are not required to be inspected. What about getting a federally inspected MPU? There is one MPU in operation in South Dakota which undergoes inspection, but not every time that processing occurs. That is because this MPU is operating on a different level of federal exemption that allows processing of up to 20,000 chickens on the farm. The coordinator of that MPU is Tom Neuberger. Tom did run a local processing plant in town but the plant was sold out from under him and so Tom decided to build something he had dreamed of for quite some time - a mobile processing plant. This MPU was built in a 48-ft. refrigerated semi-trailer. It runs most efficiently with three people and they will usually process about 300 chickens in a day. It has plenty of space, waterproof paneling and a built in pressure washer for rapid cleanup. The trailer is on a slight pitch so that all water used in the operation moves to a forward drain which then fills a tank outside the trailer. That tank is pumped into another tank on a sprayer truck and the water is sprayed on the farm ground as a low concentrate fertilizer. Tom has never used more than about 1,000-gallons of water for one processing day. Chickens are cooled down in small, used, stainless steel milk tanks. These old tanks are wonderful for chilling since they are insulated and hold down the temperature of the water for a very long time. The refrigeration aspect of the truck is not used for storing the processed chickens; but for air conditioning the unit in the heat of the summer. Tom's plant requires a 220 electric hookup for the plucker. The scalder is a propane model. Currently the Neubergers are the only ones using this MPU. (See page 157.)

The option of a federally inspected facility is currently being seriously considered in Kentucky. Several different organizations are working cooperatively to see if a farmer friendly, federally inspected MPU can be developed for use in the state. These organizations recently learned that they will be receiving grant money to assist with the development process. Something like Tom Neuberger's model may be what they are looking for, if it can be built on a little smaller scale that would make it easier to move from farm to farm. The other challenge will be in keeping the cost to a reasonable level.

More and more, MPUs are helping to meet the processing needs of small-scale pastured poultry producers in the country. If you are interested in building an MPU in your area, check the legal requirements for your state first.

Revisiting the MPU
By Jim McLaughlin Summer 2003

In the Summer of 1999, Steve Muntz wrote an article on Mobile Processing Units (MPUs). APPPA felt it was time to revisit the topic and look at what was new in the MPU arena.

The development of MPUs were a direct offshoot of the pastured poultry movement. Many producers who wanted to raise pastured poultry either for their own consumption or to sell had no access to commercial poultry processing plants. Producers found processing equipment hard to come by or very expensive, and therefore were limited in efficient ways of processing their birds. They also found that if they could process poultry on farm they would save a considerable amount of money. So, the idea of putting processing equipment on a trailer and making it available to producers was born. Many of the early units were the result of several producers getting together to meet a common need. These producer groups received funding from organizations such as Heifer International, Sustainable Agricultural Research and Education (SARE), and others. In the last few years individuals have been building units for themselves and making them available for rent to other producers.

Missouri
David Schaffer from Missouri had one of the first mobile units in the mid '90s. The unit had all the processing equipment loaded onto a trailer and was moved from site to site. The producer would take the equipment off the trailer and place it under a shelter for processing then back onto the trailer to be moved to the next site. Unfortunately, the weight of the equipment made it difficult to load and unload and people stopped using the unit. Today David sells low cost processing equipment under the Feather Man label.

New York
Central NY Resource Conservation and Development Project, with funding from Heifer International, built a MPU expressly for the purpose of providing new producers processing equipment to use without the expensive costs associated with purchasing new equipment. In 1995 the Pastured Poultry Association of Central New York was formed. Part of their purpose was to oversee the use of the MPU. The association also helped new producers get started in pastured poultry. The association has disbanded but the unit is still in use today, seven years later. The unit has a homemade picker and scalder and is mounted on a 20 ft. trailer frame. The original cost to build it was $3,000. The unit is used from June to October and producers use it to process 2,500 to 3,000 birds a year. While the unit has had some minor upgrades over the years, maintenance costs have been very low. For more information on the MPU go to: http://www.norwich.net/socnyrcd/mobile.html

Another New York group, the North East Pastured Poultry Association, had a MPU built for their producer members. Heifer International also helped fund the construction of this unit. The unit is used to process 2,000-3,000 birds each year by 10-12 producers. This unit purchased new equipment and cost around $8,000 to build. Scheduling and maintenance is provided by volunteer labor. In the winter of 2003 the group had major renovations done to the wiring, plumbing and steel work. Heifer International has also contributed funds to help provide for office and administration expenses.

Nebraska
Dave Boslee in Hasting NE had one of the first MPU's in the country. He started with a burned out 5th wheel chassis and mounted an 8 x 20 ft. container on the frame. The box is divided inside to separate the killing area from the evisceration area. The scalder and picker, manufactured by Pickwick, has provided good service to the producers. Each new producer must go through training by an experienced processor before they are able to use the MPU by themselves. The trailer is rented on a per-bird basis, which covers the cost of propane and use of the towing vehicle. Producers are asked to refill the gas tanks on the truck after use. The current cost is 70¢ per bird. Dave estimates that 5-6 producers can process 75 birds per hour.

Minnesota
The North East Chapter of the Sustainable Farming Association of Minnesota has a MPU. Joel Rosen is responsible for repairs of the unit. The MPU is actually an old school bus that has been converted to

a poultry processing area. The unit is completely plumbed and wired for 220-volt operation. Users pay an annual $25 fee to belong to the local chapter, and pay a 50¢ per bird fee with a 100 bird minimum per processing. This fee includes propane and up to 30 miles of travel. Each farmer is responsible for control of all wastewater and offal removal. The unit, which has been in operation since 1995, has had only minor upgrades and repairs over the years. There are approximately 25 producers who use the unit, processing between 3,000 and 5,000 chickens and turkeys per year. The producers who use the unit operate under the Minnesota 1000 bird exemption. However, Rosen states, the unit has been inspected by the Department of Health. There are no major problems with the unit although the group has to replace the current hot water heater. They are considering installing an on-demand type heater to provide continuous hot water.

Inspected Units

Kentucky

In 1997 Heifer International's Steve Muntz began planning a MPU to be used in the state of Kentucky. At the time there was no place for a small producer to process poultry and no provision in Kentucky state laws to allow for on-farm processing of poultry. Steve contacted the USDA to see if a federally inspected MPU was possible.

The USDA has two distinctions of inspection: federally inspected and custom exempt. Unless a plant is processing more then 20,000 birds per year, the USDA basically turns all governing of poultry processing over to the state, which is called custom exempt. When a plant exceeds the 20,000 birds per year the USDA requires that the plant is inspected and licensed. The USDA inspects both the physical plant as well as the animal slaughtering and handling process, this is called federally inspected. Some states allow individuals to process up to 20,000 birds with only a state licensed facility. In most states allowing custom exempt status, the process is not inspected but the plant is inspected for cleanliness. (There are exceptions to this, some states do inspect the birds as well.)

In some states there is an exemption from both the inspection of birds and the licensing of the plant. These states provide an exemption allowing a producer to process up to 1,000 birds per year on their farm without any state or federal license or inspection. This is the exemption many pastured poultry producers have taken advantage of. There has been much said about the exemption stating that it is a 1,000 bird "per farmer" exemption as opposed to "per farm." Producers have thus considered each person working on the farm a farmer and deem that each one can do 1,000 birds each. So, they believe, a family of five that works on the farm could process 5,000 under exemption. I don't believe the "per farmer" argument would stand up in court, but I also don't know if it has been argued in court. However, even under exemption, both the USDA and the state has the right to come on site and inspect and can force an operation to cease and desist. This, unfortunately, has happened to several producers across the country.

Muntz's problem in Kentucky started when no one in the state would sign off on building a MPU. The USDA wasn't interested because he wasn't planning to exceed the 20,000 bird limit. The state said it had to be inspected and the USDA said it was the state's problem. So after much rigmarole, both the USDA and state said they would oversee construction and provide exempt status. Muntz and his team did a great job on building a state of the art unit. Unique to the unit is that it is also designed to process shrimp.

The MPU is a 20 ft. gooseneck trailer that is housed at Kentucky State University and may be picked up with a properly equipped pick-up truck. The trailer has a collapsible canopy that extends on one side to make a screened area for killing, scalding, and picking. The carcass is then passed inside the trailer through a small door for evisceration. The trailer is air-conditioned inside and has hot and cold water. It has a capacity of 500 birds per day. Problematic to Kentucky's unit is that the unit must be set up at an approved docking station. The docking station must have a concrete pad, potable water supply, electric connections, composting station and easy access to a municipal sewer supply or a septic tank system with a grease trap. The cost of establishing a docking station for the MPU is estimated at $4,000-$6,000. Specifications for the docking station are still under development and differing specifications may be applied to different species or locations. Trained operators of the MPU will be able to rent the facility for a fee and cleaning deposit. They will also be required to sign a user agreement specifying liability issues and responsibilities for operation.

Heifer International is the actual owner of the KY MPU, but the use of the unit will be guided by a management team consisting of representatives from the organizations responsible for the development of the unit.

A great deal of time and $70,000 has been spent in building the KY MPU. This includes the development of one docking station (a site where the MPU can be hooked up and used), which has been built at Kentucky State University. A second docking station is now in the planning stages. A well-trained crew should be able to process 400-500 birds in an 8-hour day with this facility.

To date there are 15 certified Facility Managers. These are individuals who have gone through a intensive two-day training program. Training not only includes proper use of the equipment and process, but in-depth procedures on sanitation, food safety and proper handling of the finished food product. Heifer International recently received a grant to hire a facility manager to operate the unit. This should ease the necessity of having each user go through the mandatory two-day training. Each producer will now be assisting the facility manager, who is responsible for the unit. Several thousand birds have been processed to date and the group expects the number to increase as new docking stations are located across the state.

In addition, two small bumper-hitch trailers with dual chest freezers and a small generator have been built to assist with the storage/transportation of finished product. These trailers will be available from the University of Kentucky and Berea College.

For more information the website is: http://www.kyagr.com/mkt_promo/LPF/Livestock/meatmarketing/processing.htm

New Hampshire/Vermont

New Hampshire also has a MPU that is "open to USDA sanitary inspection." It is operated under federal exemption as New Hampshire does not have a meat inspection program. Ray Garcia of Cabin View Farm in Littleton, NH owns and operates the unit and does all the processing himself instead of renting the unit out. In addition to providing service in New Hampshire, Garcia is also allowed to operate in nearby Vermont. He has been able to consider this his full time job, although he say's it is a little slow in the winter. To fill in the slow time he now also offers lamb processing. Garcia began processing in 1993 and has a self-contained enclosed trailer that houses the equipment. The killing, scalding and picking is done outside with a Pickwick plucker and a manual, propane fired large-pot scalder. Evisceration takes place inside the trailer.

Garcia is able to process 35-40 broilers per hour by himself. He is looking to add more people in the busy season and is also planning for a new, completely enclosed trailer. As with all state or federal inspection the killing, scalding, and picking has to be done in a separate room which will require a larger trailer. He currently charges by the bird and mileage one way. For 25-35 birds it is $3.50 per bird, 36-125 birds are $2.95 per bird and for 126-200 birds he charges $2.85 per bird. If you want the necks, heart, gizzards or livers there is an additional charge. This fee includes processing, ice, and bagging. The producer is responsible for removal of all waste. When a customer makes an appointment Ray sends a Processing Agreement explaining that the producer is responsible for capturing the birds and transporting them to the MPU, provide electric and potable water. If a farm has an old electric service Garcia has a generator on board to power his equipment. In addition to broilers he also processes turkeys, ducks, geese and lamb. He is considering a brick-and-mortar federally inspected plant to process hogs and beef.

Washington

Terry Swagerty with the Washington State Extension Service is the overseer of a new fully enclosed MPU in Colville, WA, in the eastern part of Washington State. The idea began in 2001 when Terry saw a need to assist local produces with an inexpensive way to have poultry slaughtered. Swagerty secured a $30,000 grant to construct the unit. Much like Steve Muntz, he soon found the idea wasn't as well received by the state as by the producers. As of this writing the unit has only been used on a trial basis, but this year (2003) he hopes to see several hundred birds going through the unit. Producers this year will operate under the 1000 bird exemption, in future years producers will have the option of processing up to a 20,000 limit under a state license. As with the Kentucky MPU all users must be trained in GMP, SOP, SSOP, and HACCP to insure proper handling of the consumer product.

This unit is an 8 x 20 ft. van trailer with a killing area

separate from the eviscerating area. A three-bay wash sink, hand wash sink, and washable interior are all part of the requirement of WA State Department of Agriculture, the regulating agency in charge of processing. In WA state poultry is not included in the "meat" reg's but under a separate category of "food" so the health department oversees the MPU. An Ashley scalder and Brower picker are the main components in the unit, with chill tanks to provide an ice bath for the finished birds. One thing Swagerty did was to build the components to operate on 110-volt systems to accommodate some of the older farmsteads in the region. Future plans include a truck-mounted cooler unit to hold and transport carcasses.

The Island Grown Farmers Cooperative in Lopez Island, WA has the first USDA licensed MPU. However the unit is limited to red meat and not available for processing poultry. The first of its kind in the U.S., the Mobile Meat Processing Unit is the result of several years of patience and hard work on the part of local livestock producers, Washington State University Cooperative Extension, Lopez Community Land Trust (LCLT) staff and numerous supporters. Funded and developed by the LCLT, the Unit will be operated by the Co-op, traveling throughout San Juan County to slaughter beef, lamb, pigs and other meat right on the farm. The USDA inspection will enable local farmers to directly sell their meats by the piece to local residents, stores and restaurants. For more information on the Lopez county unit visit their website is: http://www.lopezclt.org/sard/mpu.html

Sioux

The Cheyenne River Sioux Tribe built a MPU to process buffalo. The unit is designed and used strictly for buffalo. The mobile processing unit was designed to allow for field slaughter of buffalo. The idea came about from trying to avoid placing the buffalo in an unnatural situation in order to slaughter and process buffalo meat. The website is: http://www.basec.net/~crstgfp/MPU.htm

My research turned up more MPUs then we are able to report on in this article. There is a new one in the planning stage in New Jersey. The Finger Lakes Region of New York has a new one in its first year of operation. It is my understanding Wisconsin has at least one unit, Green Hills Farm Project in Missouri has a MPU that has been remodeled this year, and Nebraska has one more in addition to the one mentioned in this article.

CHAPTER ELEVEN
Marketing

If we were to organize this book presenting topics in the order of importance, or in the order of "what to do first," marketing would be the first chapter. Any business expert will tell you that it is critical that you "find your market first." A study of who is out there and what particular kind of product they want very well might dictate the fine details of your production. The way you present yourself and your products to your customers may make the difference between success and failure. In this section we offer numerous resources to help you prepare for understanding and serving your own particular market niche.

Direct Marketing Isn't About Selling, It's About Dreaming
By Brian Moyer   Spring 2003

"I'm no salesman" is what I hear from other farmers when the topic of direct marketing comes up. To me, "salesmanship" is not one of the key elements when it comes to direct marketing. There are many more important things to consider. Let's say you are selling at a farmers market. The very fact that there are customers walking around the market means you don't have to do any selling. They are already sold. They are sold on the very idea of a farmers market, or they wouldn't even be there. The only question that remains is: what are they going to buy from you? You don't have to stand there yelling," hey , have I got a tomato for you!" We know the customer definitely doesn't want that, but what do they want?

People emerge from their humble abodes to gain as much information, entertainment and perceived value as they can get in one trip. They generally don't go out to just buy a few veggies and that's it. They want to meet the person that grew this food. They want to know how to prepare it. They want to know why they should buy this heirloom variety instead of the conventional. They want to know if you are having a good season, or if the drought is giving you problems. They want to know because you know and they don't. Most people no longer know how food is produced. But they want to learn; they don't want to feel stupid. This applies to restaurants as well. Chefs can make a fabulous dinner but they don't necessarily know how the ingredients got to their door. Just because they work around food doesn't mean they understand it. A few years back I met a man who worked for a large supermarket chain and when he found out I was a farmer he said "I didn't think anybody did that any more."

It doesn't matter if you are selling at a farmers market, farm stand or at the back door of a restaurant. The key ingredient to direct marketing is enthusiasm. You must like what you do, or you probably wouldn't be doing it. If you are enthusiastic about what you grow and how you grow it, it becomes infectious. Customers want to share your enthusiasm because they want to feel confident in what they have purchased. Especially if it is a non-conventional item that you are trying to get top dollar for.

Photographs can help make a sale. We sell poultry at our farmers market, so it is challenging to display our product. Obviously, we can't have chickens sitting out on the table for people to look at, so we use things like photographs to help. We have a piece of colored poster board with some photos of our chicken production and an explanation of how we raise our birds. We also use colorful signs, a banner with a rooster on it, old wooden chicken crates and egg baskets; anything we can think of to draw people in. Produce farmers have an easier time because the product is right there, full of color and textures, but you can still put out some photos of your farm and fields full of vegetables.

Signage can be critical to making that sale. A sign explaining what the variety is, where it comes from, how to prepare it, and even when it is in season will do a world of good. If the public understands that something is only available for a short while, they may be inclined to stock up on it.

Photos and signs can also go under the broad heading of "entertainment." Entertainment doesn't have to be song and dance. Today, people's idea of entertainment includes things like information and education. Just look at the hot entertainment distractions today. Bike trail riding and rock climbing offer more than just an activity. They offer an experience. When people come to my farm I don't offer miniature golf or a petting zoo, but I do offer an on-farm experience.

On the average, when folks come to the farm it is in the afternoon and most have traveled a good distance to get here. When they come, I don't want to just hand them their items, take their money and say "thank you very much, see ya next month." I want to make sure they know what is going on (if they want to know). I want them to have a vested interest in the farm so that my farm has a value to them as a vital part of the community. If they are first time customers I give them a tour and explain what we are doing and why. If they are regular customers, they generally want an update of what is happening on the farm. Some even help with chores. Collecting eggs and feeding chicks is a great novelty to them and a help to me. We both benefit and it increases the value of their purchase. They can't get that experience in a supermarket.

There are two views of selling. One is just helping a customer part with their money. The other is helping to meet a customers need. Filling the customers' needs is what I'm talking about. If a customer passes by my stand without purchasing anything, it's only because I didn't have what they wanted or needed. Remember, they are already at your market, they are looking to buy something. They should at the very least leave your stand with information about your farm. That could be a business card, a brochure, or just plain talking to them. Heck, my wife Holley managed to get a vegan to try her goat cheese! All because of a discussion about our farm and Holley's enthusiasm for what she does. And believe me, no one was more surprised then me. Holley is not as "out going" as I. She is quiet and rather "laid back." But, Holley loves her goats and her cheese and it shows.

"The only thing I know about marketing is to call the truck driver." This is what one farmer told me in a phone conversation. He called me because he was frustrated in his efforts to sell his pastured poultry and even more so when he found out we are selling ours at a dollar more per pound then he is. He was trying to sell at a farmers market near a grocery store. He figured he wouldn't succeed because people would rather pay less at the store. But, that is not always the case. To begin with, if they came to the market they are clearly looking for something that they can't get at the grocery store. Pastured poultry is something they generally can't get at the local store. I suspect that this farmer wasn't offering any info about pastured poultry, the benefits of eating grassfed chicken, or even info about his farm. Let's face it, if I'm "selling" anything, it's our farm. That is, I'm selling the idea of our farm. That's why we have photos and info at our farmers market, farm tours on our farm, and newsletters to our customers.

Marketing needs to be present in every aspect of your farm. Ask your self these questions. How does your farm look? Do you have a farm sign? When you make deliveries to restaurants do you just drop the stuff off and run or do you deliver when the chef is there to make sure they are happy with your product? Do you use nice invoices or just scratch one out on a piece of paper? Can people write checks to your farm name or your name? Do you have a logo? If you sell some of your products through a CSA, do you ever hang out and meet the shareholders? Most CSAs will have special events during the season and we try to make an appearance to some of them. We also have a small demonstration pen that we can put on the back of the pickup along with a few chickens so the shareholders at the CSA can see how the chickens are raised. We also leave copies of our newsletter on pickup days so the shareholders can get to know us and our farm even better. This is the key, the relationship we have with our customers. We are on a first name basis with all of them. They feel perfectly comfortable calling us to order or ask questions or even to come and help out on the farm. Simply because they want to know and price is never a question when we have a relationship.

To get started in marketing it helps if you get to know farmers in your area that are doing a similar kind of farming. Even if they are doing the SAME thing as you, it's good to know them. We have increased our sales by building these relationships because we can refer customers to other farms. We have a neighbor who raises grassfed beef and pastured poultry. If our customers want beef, we send them to this farm and they send people to us for lamb. Building a network like this can make trips to the country more worthwhile for folks.

I hope this gives you some ideas. Marketing has to be a part of your farm plan, just like figuring out when to plant and harvest are. Try to take time each day to dream. Dream about what your farm could be and how to get it there. I know it's hard to find time. My time to do this is at the end of the day. When all of my chores are done and the sun gets low, I can sit by the doorway of the barn washing eggs or just petting the dog and let my mind wander. Try it. You'll be surprised at how your sales can increase. You're not selling. Just dreaming.

Steps to Successful Marketing
By Steve Bonney Spring 2000

Many farmers are turning to direct marketing as a way to receive higher prices for farm products. This is certainly a reasonable approach, but it is not easy. Marketing is not farming. It is a whole new business and must be treated as such. Businesses are built over time, according to a plan. There are usually no quick returns, but the long term rewards can be immense. Allow five years for any financial goals that include profits to be realized.

The first step to successful marketing is develop a marketing plan. This is a business plan that identifies all steps involved in developing your marketing business. It can be modified along the way but should initially contain a complete set of strategies for achieving your goals.

This means that you must first set goals. What do you and your farm family want for and from this business? Decide on the quality of life goals that you and your family want. Time for leisure and travel, time to engage in farming activities, opportunities to bring your children into the farm's operation are all examples of considerations that should be written into your plan. Is a family member going to manage the business? Are you going to employ labor? How much time will you spend on this business? Not all of these questions have easy answers, but must be considered in your plan. What financial goals do you want for this business? Do you want all farm products to be direct marketed? Can you process your farm products to add even more value and then direct market the processed product? Cheese, butter, yogurt and bottled milk are opportunities in dairy. Specialty meats, sausages, natural or organic meats are opportunities in livestock. Marketing includes all the steps necessary to sell your products. First determine what you will sell. Do you already have that product or is it time to change farming enterprises?

You already know that there are few options for selling commodities. The markets are restricted and controlled by others who are masters of low-cost buying and selling.

The same large corporations that control the commodity markets are also masters of adding value to commodities and mass marketing these products. These are not markets that you can, or want to, compete with on a small scale.

Nearly all opportunities for generating higher returns for farm products fall into the category of "niche marketing." There are thousands of niche opportunities that have been created by a very consolidated agricultural industry. In fact, the larger the corporations grow to take advantage of the economics of scale, the more niches that are created in the marketplace. Filling niches in the marketplace and achieving higher profits depends upon the ability to produce a differentiated product. The goal here is to provide a product for which there is consumer demand and for which there is less competition on the supply side. How do you differentiate your product? What do the low-price producers feed their animals--antibiotics, hormones, parasiticides, animal byproducts, grain grown with chemical fertilizers and pesticides? The further you remove your production practices from this conventional approach, the more your product is differentiated from the conventional market. That translates into less competition; hence, better prices. Organic and natural foods are currently undergoing the fastest growth rates in the food sector. While the category of organic food is well defined and highly regulated, the category of natural food can be defined by the farmer. Tell your customers exactly how your product differs from what they would find in a grocery store at a lower price. Give them reasons to buy your product.

Now the question becomes, "who are your buyers?" Are there customers who want such products in your area? If you can locate them, they are motivated buyers who emphasize quality over price. Those who want chemical free food and those who want locally produced food can become loyal customers. Where do you start to look? In your own community. This begins with networking. Who do you have access to? Relatives, friends, neighbors, members of your church congregation, associates of your club affiliations. This is also an opportunity to conduct market research while you are identifying customers. This market research is no more complicated that asking people what they want and if they will buy it if you produce it. You could do this orally or in written form. Keep it short, so that it takes very little of anyone's time.

For example:

1. Would you purchase (beef, pork, bottled milk, etc.) that contained (no hormones, no antibiotics, no animal byproducts, etc.) and was raised on grass?

2. What other considerations are important to you? (select all that apply). (Here you could list other differentiating practices that you don't currently follow: certified organic, etc.).

3. Is the ability to purchase locally grown food important to you?

4. Would you come to a farm to purchase food?

5. How many miles would you drive?

6. How much would you purchase in a six-month period? These are only examples. You ask what you want and need to know. This information guides you in the development of a product that has consumer demand.

Pricing is always the difficult step. You could match the retail price in stores, but that is still too low for a quality, differentiated product. Remember that if you compete with low prices you have gained very little. Education is a part of realizing reasonable profits.

Continually give customers reasons to pay higher prices for your products. Develop a brochure that defines your product and clearly states the advantages of your product.

Distribute these brochures anywhere there might be interest. AARP groups, health support groups, health food stores, buying clubs, field days, and county fairs are good choices. Brainstorm other possibilities with your marketing team. Giving away samples is a very effective marketing tool and is about as cost effective as any marketing strategy.

Have your customers give samples away to their friends. Be sure to replace the product that they hand out. Giving samples to chefs is a good way to establish sales to restaurants also.

In my experience, if you have a high quality product, you need to attract only a few good customers at first and then the news will spread by word of mouth. Start small and then let consumer demand drive your growth in production.

In summary: have a quality, differentiated product to sell. Develop a plan to conduct your marketing. Identify your target customers. Continually educate your customers and potential customers about the advantages of eating your product and always continue to differentiate and improve your product. Be persistent in your efforts, but patient with the development of your marketing business. Enjoy the path to success!

Direct Marketing: Seven Basic Roles
Summer 2001

Farmers thinking about direct marketing a food product can consider several basic work roles that are part of getting a food from the farm to the consumer.

Doing the necessary tasks for all these roles is likely to require teamwork, says Wright County educator Maribel Fernandez of the University of Minnesota Extension Service.

Fernandez cites seven basic work roles for persons doing direct marketing, as identified by consultant Pete Reese. They are:

¤ Vision and planning: Identify opportunities, draft a purpose statement, write clear long– and short-term goals, create a financial plan, recruit team members, lead an effort to come up with a brand. Also, help negotiate deals with suppliers and customers, build relationships with others involved in a similar enterprise, monitor and evaluate results and make adjustments or implement changes.

¤ Organization and administration: Assess strengths of each contributing person, create a task list, set up an office, keep financial records, help negotiate with suppliers and customers.

¤ Production and management: Turn goals into a realistic production flowchart, write resource requirements for major steps (people, machinery, facilities, supplies), recruit workers, make sure the work gets done, keep production records, look for ways to improve the product and make the process more efficient.

¤ Sales management and customer relations: Determine how to best spend sales energy and money, keep sales records, meet with large customers personally, supervise other people helping with sales and advertising, help negotiate major deals.

¤ Advertising and public relations: Identify how best to reach customers (flyers, signs, etc.), calculate a budget to support sales, implement development of a label with the brand and that meets legal requirements. Also, write sales materials and a telephone message, talk with local newspaper and newsletter editors to get articles written and published, keep a file of other people's sales ideas.

¤ Financial management: Support a financial planning process; select a financial institution and be the contact person; select a bookkeeper or accountant; manage the checkbook, savings account and investments; track and pay bills; be in charge of taxes, payroll and insurance; look for opportunities to save money.

¤ Regulatory compliance: Research labor, environmental, transportation, personnel and product safety and liability issues; build relationships with regulatory agencies; recruit legal or consulting help when needed; maintain all certification records, correspondence, and documentation.

" Most likely you won't be able or want to take care of all of these areas on your own," says Fernandez. "One person may take care of one or two areas, but in the long run it is better to allow people to do the work that fits them best."

Fernandez recommends writing down your purpose for getting into direct marketing and what you want to sell. Then write down all possible steps you can think of between the raw product from the farm and the final product on the customer's table.

Marketing Meats and Poultry at Farmers Markets
By Aaron Silverman Fall 2002

Greener Pastures Poultry (GPP) is a multi-farm collaborative in western Oregon. GPP provides market access for its members' pasture-raised poultry by operating a licensed processing facility. All its products are marketed under a unified brand label to restaurants and stores in Eugene, Corvallis, and Portland, and farmers markets in Eugene and Portland. GPP expects to process and market 20,000 birds in 2003.

Marketing Meats and Poultry at Farmers Markets
"Diversity" is a buzz word for both farmers and farmers market managers. Producers are advised to diversify their crop selection, possibly even venture into the realm of "value-added" products to strengthen their farm's profitability. Market managers, under pressure to attract a greater customer base for their vendors, increasingly look to diversity of products when having to weigh space constraints with the desire to increase participating vendors.

One visible aspect of vibrant farmers markets throughout the world is the presence of meat and poultry vendors. Direct marketing of meat and poultry has been far slower to appear in farmers markets in the U.S. This is due in part to the burden of regulations that encompass the processing of meat and poultry (see sidebar for resources pertaining to processing regulations on page 172). Although regulations vary widely between states, and even within states, nearly all states are consistent on one point: meat and poultry offered for sale must be processed in a licensed facility. For the purposes of this discussion, it is assumed that the products you are seeking to market have been processed in a licensed facility.

Marketing for the Producer
Marketing meats at farmers market is similar to marketing any value-added product, and very different than marketing vegetables or cut flowers. Few customers will be impressed by a bountiful pyramid of chickens, or tables laden with lamb chops. Purchasing meat requires a heightened level of trust by the customer, since no fondling or smelling is possible. Your ability to gain, and more importantly retain your customers' trust will determine your success in marketing meats at farmers markets.

We use three elements to gain our customers' trust. The first is the creation of a marketing brand—the customer's way of identifying our product outside of the farmers market. Brand identity is even more critical when processing and marketing is done on a collaborative basis. Your brand is more than just your name, it encompasses the "who" and "what" of your product, and in the case of most meats, the "how."

Central to building this brand identity is your story. Your product's story is crucial for distinguishing yourself to customers who are used to purchasing meat and poultry from a supermarket case. Pictures speak much louder than words. While you may get some comments about the "cute lambs" from the vegetarian crowd, you'll also get sales from the family whose five-year-old stops at your booth, enamored by the "cute chicks."

Consistency is the most crucial element in ensuring lasting success at your farmers market. When marketing meats at the farmers market, you are starting with many disadvantages. The greatest is the novelty of the product. People aren't used to purchasing their meats at farmers markets, even those that purchase the majority of their produce there. Purchasing meats doesn't fit as well in some customers' schedule, due to its highly perishable nature. If your product is high quality every time these inconveniences will be overlooked by your customers. You are not only competing against the products and pricing of supermarkets, you are competing against their convenience.

These three elements–brand identity, story, and consistency–apply to all marketing. The importance of these elements is heightened when marketing any value-added products, whether meats or jarred jams and pickles.

The physical aspects of marketing meats are also different than when marketing produce or cut flowers. Your ability to impress customers by sheer beauty and bounty is severely restricted by your product itself. Signage and pictures (i.e. your story) play a more prominent role in the initial "grab" of customers.

The challenge of designing a booth for marketing

meats lies in balancing the need for an attractive display with food safety requirements. A major decision in marketing meats is whether your products are fresh or frozen. Food safety guidelines require that perishable products remain under 40° F to minimize potential contamination by harmful bacteria. Frozen product is certainly easier to ensure adequate temperatures, and eliminates concerns of leakage. Frozen product, especially poultry, has a stigma for many customers. Fresh product requires more care to ensure adequate temperature control and appropriate packaging to prevent leakage. Fresh product may have greater acceptance by customers, and is often seen as being of better quality than frozen product.

Marketing Frozen Lamb

At several farmers markets in Oregon, there are vendors marketing frozen packages of lamb. Their lamb is processed and vacuum-packaged by a USDA-Inspected facility. Large chest coolers, each containing a single product, are arranged inside the booth in an inviting manner. Color photos and articles portray the happy, stress-free lives of the sheep. A large price board allows customers to make their decisions before entering the booth, facilitating a smooth exchange between the customer and the vendor. At the end of the market, each cooler is easily loaded back onto the truck and unsold-inventory is handled in an organized fashion. Sales can be tracked with some certainty, even in a harried market, by comparing the beginning and ending inventory of each cooler. While this doesn't alleviate the challenge with meats of uneven demand for some cuts over others, it does help ensure you have adequate supply of those cuts favored more by your customers.

Marketing Fresh Chickens

For the past two years, Greener Pastures Poultry (GPP) has marketed fresh chickens at a weekly farmers market in Eugene, Oregon. All poultry are processed in a licensed facility that GPP operates. Both whole birds and parts (boneless breast, hindquarters) are marketed, and packaged in tight, clear bags. Because products are fresh, electricity is needed to operate refrigeration. A vehicle is needed to hold the high volume of packages that are sold (between 150 and 200 packages per market) at the appropriate temperature (35° F). The door of the cooler acts as the backdrop to the booth. At the front of the booth sits a large sandwich board listing prices for each item, names of area restaurants and stores using GPP's products, and information about promotions or specials. Hanging from this price board is a box with brochures and a photo album showing "The True Lives of GPP Chickens." Inside the booth are several tables arranged in a line across the booth, behind which the booth's operators stand. On the tables are several plastic containers with sample products nested in ice. Sales are packed in plastic bags, and customers are given the option of having product packed in ice. GPP is currently researching the availability of offering insulated bags to customers, either for purchase or in conjunction with a promotional campaign (ie. buy 10 chickens, receive the bag free).

Free Samples

One of the most obvious means of establishing trust with potential customers is have them try your products before they purchase them. Unfortunately, with meats and poultry, this entails cooking your product, and makes free samples much more complicated. From a food safety perspective samples are anything but simple. Foods must be kept above a minimum temperature, be cooked completely, and the operator must have an approved hand washing facility. The city where the farmers market is located may require a temporary restaurant license because product is being cooked; such permits can run $50 to $100 per day. Finally, offering samples may require having a person devoted to this task, meaning either having extra staff, or not having enough staff to keep the booth operating smoothly. Once you have an established customer base offering samples is less critical for attracting new customers than the buzz of eager customers lined up to purchase your products. The issue of whether to offer free samples is one requiring much planning and coordination with both your market management and your local government.

Marketing for Market Management

If you are a market manager you are continually under pressure from customers wanting products, producers wanting to enter your market, and established producers demanding more space within an ever-crowded market. If a new producer (or group of producers) comes to you wanting to enter your market, but requiring facilities your market may not presently have (truck space, electricity, etc.), how do you balance competing concerns, and decide whether it is worth having meat vendors included in your market?

Several years ago as part of a SARE grant GPP and

the management of several farmers markets surveyed customers at various farmers markets to determine the level of support for the sale of local meats and poultry. Customers at three farmers markets were asked questions pertaining to their preferences concerning specific products, and the farmers market in general. Dot Posters were used, a technique developed by two Oregon State University researchers to study the customer base of Oregon's farmers markets (http://www.joe.org/joe/1999october/tt1.html). Dot Posters are a quick, unobtrusive means of surveying customer preferences. A series of posters is set up in a line, each poster containing a single, carefully crafted question, and several specific responses. Customers are handed stickers, and asked to place one on each poster "where it makes the most sense." Only questions that yield a restricted set of short answers will work in this form of survey. Each market conducted a survey on the same day to provide an estimate of total market demand.

Participating customers demonstrated a strong interest in buying pastured poultry (between 48% and 60% in the "Yes" category), and most indicated a preference toward parts over whole birds. Most customers estimated their initial purchase would be one to two birds per month. These statistical results are consistent with actual sales records. The most surprising result was that 16% of the participants thought that having fresh pastured poultry available at the farmers market would make them more frequent market shoppers, and an additional 22% indicated that their shopping might increase.

Dot poster surveys are relatively easy to conduct, and are a good tool for both managers and producers to gauge the feasibility of adding the facilities needed to market fresh meats and poultry.

Farmers markets can be a great venue for direct marketing meats and poultry. Because of the level of investment required, by both the vendor and the market, care and research is needed to ensure a profitable operation. Adequate population of a market is one important consideration, one that the market managers should be able to assist in gauging. Booth layout, product packaging, and product presentation should be carefully considered to meet the demands of both food-safety regulators and the expectations of your customers.

Legal Resources for Direct-Marketing Meats and Poultry

The most difficult aspect of marketing meats and poultry is understanding the myriad of regulations covering this activity. Every state handles this issue differently, even for products processed with USDA Inspection. The best resource for negotiating your state's regulations is your state's Department of Agriculture. Always ask for a copy of the text of the regulations or statutes that pertain to your activity. Bureaucrats are often well-meaning, but occasionally misinterpret statutes. With the actual text in hand you should have greater leverage if faced with resistance towards your activities.

The Legal Guide for Direct Farm Marketing, Neil Hamilton, 1999 Drake University Press. Chapter 12 provides a good overview of the legal issues involving marketing meat, poultry, eggs, and dairy products.

USDA Poultry Processing Regulations: http://www.access.gpo.gov/nara/cfr/waisidx_01/9cfr381_01.html
Rules for Exempted Facilities: section 381.10

State Regulations Listing: http://www.apppa.org see "Resources." A good starting point for research. APPPA, the American Pastured Poultry Producers' Association, also has a listing of members by state who are also great resources for research.

Meat Selling Licenses: regulations concerning the licensing and handling of meat for resale is most often handled by State Departments of Agriculture. Most states require licenses to market meats not processed by the seller.

Marketing Options
Excerpted from "Profitable Pasture: Raising Birds on Pasture"
published by the USDA Sustainable Agriculture Network Spring 2002

The experience of practically every range poultry producer bears this out: marketing your product will take as much time and energy as the actual task of raising and processing your product.

In a survey, 80% of APPPA members cited direct marketing as a top sales method. For most, the best way to reach family, neighbors and others in the community was word of mouth, posting flyers on local bulletin boards, selling products at farmers markets and contacting customers often.

Marketing Tips
Newspaper stories. Mary Berry-Smith doesn't consider herself a marketing genius, but she managed in one year to have her pastured poultry operation featured twice in the Louisville Courier-Journal. Each time, she received a flood of orders that led to people reserving one of her broilers and turkeys well before the 2001 season was complete.

Marketing is all about capitalizing on advantages. The key lies in what some people call "relationship marketing." Berry-Smith worked with a newspaper editor to explain the benefits of the system, and that made the editor more willing to try, and be impressed by, the product. Joel Salatin and his farm were profiled in the national Smithsonian magazine, as well as ABC news.

Farmers who have received ink in newspapers or magazines report that when people read about their product—and the philosophy and practices behind pasture-based poultry systems—their phones, in Chuck Smith's words, "ring off the walls."

Pre-Orders. David Bosle of central Nebraska prints a newsletter every winter for his customer list of close to 300. He includes a self-addressed, stamped envelope to take orders, by month, for the season. "That's a must," he said. "The biggest cost is to get the customer on the belt, and once you've got them there, it's stupid to let them fall off." Bosle takes advantage of the short growing time for chickens and he clusters his flocks around spring, summer and fall holidays, including Memorial Day and Labor Day. With the pre-ordering system, he generally sells his birds prior to growing them.

Samples. Robin Way not only praises the virtues of investing in a colorful, easy-to-spot farm sign, but also recommends giving out free meat. "If they take the trouble to drive down our lane, I'll give the people freebies," she said. "Maybe they'll never show up again, but maybe they'll be one of our best customers."

When they first started raising poultry on pasture, the Ways would bring chickens to auctions and meetings and make donations. All of the meat was accompanied by a farm business card.

Selling Other Products. Tom Delahanty, the New Mexico grower, markets his organic meat under a "Real Chicken" brand that commands varying premium prices--as high as $5 per pound at some upscale groceries in Albuquerque and Santa Fe. Next, he plans to sell organic vegetables he expects will flourish in the manure-rich soil aided by his flocks.

"I've already got the contacts at the farmers markets, grocery stores and restaurants all up and down the valley," he said, "so selling them vegetables the chickens help grow should be easy."

One grower who works with James McNitt at Southern University garners $2.25 a pound for chicken partly because she already has a dedicated list of customers lining up for her organic blueberries. "And people are pushing her to do more," McNitt said.

Molly and Ted Bartlett offer chickens as an extra option for members of their Community Supported Agriculture (CSA) enterprise in northeast Ohio. When joining their farm for the season, customers decide whether to buy a poultry package. "We offer them 10 birds for $90," Molly Bartlett said, "and they can take them all at once, or over the course of a year. It works well, it helps the cash flow, and it provides more variety for our customers."

From Customers Into Converts
By David Schafer Summer 2005

As you and I sit at our humble farmers market booths and influence consumers one shopper at a time, we collectively approach that critical mass where the scales tip and grassfed products accelerate even more rapidly into the mainstream. That's the goal. We know it is destined because our product is so superior in so many ways. What tools do we have to accelerate that process and bring us individually more customers?

Your Countenance

Your healthy optimism is by far the most important. We've had folks come up to us and ask, "I just had to know what you're selling! You two look so radiant!"

That was from a vegetarian who first started buying for her grown children and has since crossed over to glorious carnivorousness and become one of our very good customers.

The point is, you and I are doing something heroic out here on the land. We are on the leading edge, bucking the current, changing the paradigm. We are on a noble quest and it should show on our faces. I'm not talking about snobbery or being high falutin.' Just having a deep conviction in what we're doing.

Your Farm

A brag book is worth its weight in gold. Take happy photos of all your animals. If people are attracted to your booth by the way you look, your farm photos will hook them to stay for a while. As you see which photo they are looking at you may sprinkle in a light dose of educational comments.

"The pigs really love being on deep bedding."

"The chicks move every day in that pen."

"The sheep stay out on clean pasture all their lives. The donkey guards them from predators."

"The cows get a new pasture every day, sometimes twice a day."

You get the picture. Unless you're just really good, don't talk non-stop. That drives most people crazy.

If you can't use photos make bullet sheets for each animal and outline the benefits for people to easily grasp. It could even be a take-home handout. But forget about hawking or indiscriminately handing out fliers. That's a waste of time, materials and effort. Trust that the people who are ready for you will find you. They will.

Your Products

Display your products in styrofoam trays and use "twice ice" underneath your packages. This is my name for the reusable little ketchup sized U-Tek packs we use instead of ice. You can purchase 50 lbs. (1500 singles) for $104.75 from ThermoSafe (800-323-7442) and have a lifetime supply. They also sell the styrofoam trays.

Your packages should be cryovac sealed and displayed attractively. After your shining countenance has attracted her and your photo book and your grassfed educational bullets have gotten her attention she will look into your product trays. You'd be wasting everybody's time to have butcher paper or sloppy packages there.

Handle your cryovac packages carefully and let everybody who handles those packages know that they must not be dropped or banged around (I even tell the customers if I see them handling them carelessly.) A simple rap against a counter top will put a pin prick sized hole in the plastic and the bag will inflate with air. Oxidation will soon follow. Not good for appearance or quality.

Suggest something for trial. Don't bother telling her about discounts for quantity unless she asks. Chicken is the best way to start if it's legal for you to sell it at your market. It's not for us so we just deliver already ordered chickens and take orders for more, leaving one "display" chicken out for all to see. Chickens have an under $10 price tag, a clearly superior flavor, smell and appearance and that photo of them on pasture will stick in her mind.

If not chicken, recommend a pound of ground beef, bacon, brats or something else small. Usually, she will try three or four different items if everything has impressed her to this point and price is not a problem.

Take Orders
Put your brochure in her bag, a receipt if she wants one (we always ask) and point out that your market schedule and phone number are on your brochure. Work towards taking orders for future market dates. After 11 years of selling at the KC River Market, we've reached the point where we average around $2,500 per market day. On this recent pre-Thanksgiving Saturday we reached a new record of $5,500. And that is only in a two-hour delivery window. Over 90% of that comes from taking orders.

Making a Convert out of a Customer
Your customers fall into two categories: Customers and Converts. Your mission is to turn as many customers into converts as possible. "Why? What's the big difference?"

A convert has an emotional investment in what we are doing. They inhale all the data and some will know it better than you and me. They are our sales reps and bring us more people like themselves so we are free to stay home and move fences more often.

A customer buys our products for only one or two reasons. They like the taste. They like to know where their food comes from. They are chemically sensitive.

A convert buys our products for at least three reasons. They buy the whole package, the concept. They share the nobility of the quest and see it as their dollars changing the world, which is exactly what it is. We give them kudos for that in every newsletter and brochure, Christmas card or note.

We are disproportionately advantaged to have this advanced guard of converts out there selling for us. Just as soon as we reach a critical mass of converts in this world our product will be the norm and the paradigm shift will be complete.

Conversion Literature
Before Jo Robinson and Michael Pollan there were precious few authors besides ourselves we could put in front of our public. Now we have some zingers. We give away Jo's *Pasture Perfect* to our best customers if we think they may be convert candidates. Otherwise, we offer it at a greatly discounted price since we bought a case of them (actually two now, or is it three?)

We also display magazines featuring our farm but that is more like a stamp of outside approval than anything else. It doesn't hurt a bit, but doesn't go far toward education and convert making.

I'm quite excited—and this article has been inspired by—two new books that fill critical niches for us grassfed producers and make ideal conversion literature.

Holy Cows & Hog Heaven
Does anyone else besides me have this feeling of what it must've been like to be an Italian artist when Michelangelo was alive? An English contemporary of Shakespeare? I know of no one in any walk of life I admire more than Joel Salatin; no one I think nearly as brilliant and creative. Well, maybe Paul McCartney.

That makes me feel half brilliant because I choose to do some of the same things Joel chooses to do. His latest book pulls no punches on corporate America. I've purchased extra copies and I'm putting it on my market table. I gave one away yesterday to a customer who picked up a $300 order. I told her a little about it and her reciprocation was immediate: she told me about giving our meat away to well chosen friends to try us out.

Holy Cows & Hog Heaven is not only a food buyer's guide to farm friendly food, it is also a guerilla marketing assault weapon for you and your customers, soon-to-be-converts.

Grassfed Gourmet Cookbook
I'll admit, I was skeptical about this book—probably because I didn't have a recipe to contribute. All the more reason for me to own a copy! It also seemed pricey at $25.50 (shipping included). I've since found out the publisher (Eating Fresh 609-466-1700) offers generous volume discounts ($13.77/book for ten) and I guarantee you, I'm going to sell a lot of these books.

Shannon Hayes wrote this book for you and me and our market customers. Like you, me and Joel, Shannon markets directly to her customers, too. This book will go far, I believe, toward shifting the paradigm toward grassfed.

Not only is it the first cookbook dedicated to grassfed meats and destined to be a classic for that alone, but it is a great meat cookbook and tells the whole grassfed story very effectively. The nutritional, environmental and welfare information is all in there, but does not get in the way. Plus, you (and your customers) meet other producers and find out what makes them (us) tick.

I'm imagining some of my longtime customers getting excited about having new recipes specifically designed for grassfed meats; they'll love this book. They will cook our products better and more creatively and probably more often. Plus, through Grassfed Gourmet's presentation of the whole story, I think they will become converts. Maybe not zealots, but conversant in all the pros of grassfed products and more likely to convert others to grassfed because of that.

So far we've only tried a few recipes - all fabulous. I especially appreciate that Shannon is very careful when to recommend dry heat and moist heat and clearly explains the how and why of it. (Do you have an outside-the-oven meat thermometer? I didn't. But now it is a grassfed meat cooking tool I'm sold on.)

The book has sections on beef, bison, veal and venison; lamb and goat; pork; rabbits and poultry; and dairy and desserts. It has 267 pages and is packed with recipes. There are 42 in the red meats chapter alone. Sprinkled throughout are thoughtful sidebars on topics like meat safety, aging, BSE, reheating grassfed meats, marinades, even making and using lard!

Shannon rates the meals for fanciness (from "showcase" to "on a budget") as well as how much preparation it will take and whether it is kid friendly or not. Many of the recipes are from other farmers but it is clear that Shannon has made herself an expert in this area and spent a lot of time in her own kitchen.

Push the Paradigm – Make a Convert Today

How many of your current customers would you put in the "convert" category? I figure about 10-15% of my customers are armed with the knowledge and conviction to influence others about grassfed. What would it do to my yearly sales if I could increase that another 10 or 20%?

Is it unreasonable to think that every customer may someday understand and appreciate all the benefits of our grassfed products? Not at all. The fact that they don't now speaks volumes of the opportunity before us. Put this literature before your good customers.

Consider the power of the gift. Have you noticed that when you give out of the joy of giving that you are rewarded many times over? Well placed product and book gifts will always reward you and the gift recipient.

Our attitude, farm appearance (and display) and appearance of the product attract our customers and bring them back for that second sale. Perfect. Now, to take our marketing to the next level, let's consider the power of information through low-cost, or even free, books to transform those good customers into energetic grassfed converts.

Price Check/ Reality Check
By Merri Morgan Summer 1998

My partner and I raise pastured poultry in West Virginia, and as I write this we are halfway through our third season, selling 1,400 birds to about 85 customers. Periodically we see articles about other pastured poultry producers around the country. We're pleased to read about others' success, but we're always shocked and saddened at the price they sell their chicken for usually about $1.50/lb.

When we established a price for our chicken, we had no idea what other producers were charging. So, we considered what store bought birds were selling for and decided that a vastly superior product was worth a higher price-worth it both to us and to the consumer.

We set our price at $1.85/lb. Is it too high? Are we gouging our customers? We don't think so, and neither do our customers, who often mention how reasonable the price is. One of our first customers was the owner of our local health food store. The frozen organic chicken shipped from California that she sells costs her $2.33/lb. wholesale and retails at $3.10/lb. She says our pastured poultry is so much better tasting than the shipped-in organic that there is simply no comparison. She also understands that ours is actually a healthier bird, and that's why she buys it for herself. Other customers, used to buying chicken at supermarket prices, have also expressed surprise and pleasure that our price is so reasonable.

A March 1998 Consumer Reports chicken article had a sidebar on chicken contamination that also listed the average price nationwide for whole chicken. What Consumer Reports called the "premium brands" (Bell & Evans, Green Pastures, Rocky Junior, and Wellington Farms Free Range) sold for $1.85/lb. These chickens were, by the way, the most contaminated with salmonella and campylobacter. Empire Kosher brand sold for a whopping $2.35/lb. Supermarket brands sold for $1.10/lb. Tyson Holly Farms was the lowest at $0.85/lb.

Pastured poultry producers raise and sell chicken that is incomparably better than any other chicken available, with not even one runner-up in sight in the home stretch. Why then are so many selling it for a price that is 35 cents a pound lower than "premium brand" chicken, chicken that consumers pay more for precisely because they believe it is superior?

Let's not continue to underestimate the value of our product. Raising, processing and marketing pastured poultry is hard work, and anyone who thinks it isn't hasn't done it. At $1.85/lb.; the farmer gets a return for his labor that better reflects both the work involved and the superior quality of his product. At $1.85/lb. the customer is still getting a huge bargain, the best chicken at the same price they would pay in the supermarket for the worst.

Perhaps even $1.85 is too low for pastured poultry. We've been considering that possibility. What's certain is that $1.50/lb. is much too low. Ridiculously low. Insultingly low. My partner calls that kind of gross underevaluation of a superior product "Depression Era thinking." Fine cabinetmakers do not sell their kitchen cabinets at lower cost than big-box store models. Amana refrigerators are more expensive than Sears. Porsches cost a lot more than Fords.

Farmers are all too accustomed to working very hard for very little return. Let's be more realistic about the real value of pastured poultry and charge a price that's fair to both our customers and ourselves.

Developing A Brochure For Your Farm
By Jim McLaughlin Spring 2003

One of the most inexpensive ways to advertise your farm and bring in consumers is through the use of brochures. A good brochure will inform potential customers of the products and services you have available. Current customers will also have something they can provide to their friends to use to direct future customers to you as well. These can be placed on community information boards, which are commonly found in supermarkets. With a little modification they can be used as a mailing to existing and new customers. Have your farm brochure available on pickup day and include them with your customer's order.

A well thought out brochure will include your name, logo, brief philosophy/history, hours you are open, products you have, payment services available, website, e-mail and contact information. Decide if you will use a tri-fold type brochure, or a letter or poster style sheet. Many word-processing programs have templates that can be used to help with the layout.

Professional, professional, professional; your brochure may be the first and last impression you leave on a potential customer. Make sure it looks professional. If your brochure is sloppy customers may equate it to your products the same way. Develop the flow from page to page that is pleasant to the eye and easy to read. When displaying brochures at your point of sale keep them neat and clean.

To begin, write out a brief statement about your farm. This would include information such as why you started farming, how long you have been farming, information on the principles of the operation, and the benefits of your product. Write out your mission or vision statement if you have one. You may want to include other information such as affiliation with consumer groups or producer associations i.e. APPPA, organic groups, etc. Introduce new customers to pastured poultry; many may not know what pastured poultry is so it is always desirable to have a brief introduction to your model. The following is taken from an APPPA brochure "Pastured Poultry is a production system that employs raising chickens directly on pasture. The birds are moved regularly to fresh pasture, which allows the birds to be raised in a cleaner, healthier environment. Pastured Poultry is raised the old fashioned way, on fresh green pasture and wholesome grain. Processing is often done on the farm in a clean healthy, sanitary environment." Many producers also include a narrative of the health aspects and whether they use medications or not. Make your statements as friendly as possible, make your customers feel as if they are part of the family. Consumers have a lot of choices to pick from, even as far as locally raised products are considered. Make the case as to why they should come to your business instead of some else.

Many producers use the brochure to outline their prices. Do you sell per pound or per bird? Do you offer different cuts? What is the average weight per bird? What is the availability of product? Is it fresh or frozen? Does the customer need to pick up on processing day or can you hold it over for them? Do they need to pre order? How about ordering by e-mail? Do you provide additional services such as delivery? You may not need to address all these questions, but try to include the most pertinent information.

Once you have put your thoughts on paper begin to layout the brochure. On the part of the brochure that the customer will see first, you will want your logo or farm name in clear bold type. If you are going to make a color brochure you may want to use "power colors" such as purple or red. These colors catch the eye and cause a reaction that will make the brochure stand out and say, "Pick me up"! Most people will pick up a piece of literature and make up their mind if they are going to read it in between three and five seconds. So you want them to "want" your brochure. Consider, in the case of a tri-fold brochure, the way the customer will open it up. Again, remember you want them to read your brochure, so the opening page is the place to use words that catch their eye. Catchwords today are: *All Natural, Naturally Raised, Hormone Free, Drug Free, Environmentally Friendly,* and *Sustainable*. Set out the point you are trying to make in large letters and follow up with a brief explanation under it in smaller font. If the customer likes what they see in large bold print they will take the time later to read the small print. The opening page should have only a few important statements. As they continue to open the brochure lay it out in such a way as to lead them through your farm.

On the next section list your mission/vision, why you do what you do, a brief history about your operation. Perhaps your short-term or long-term goals, and include some information about the farm and family. Where did your farm name come from? It may be of interest to your customers. Remember to make them feel at home. Don't go into a long dissertation about yourselves or philosophy, this is really about them, providing good healthy food for their families. It is much better to have the customer walking away wanting a little more information than overwhelming them with too much detail. A second informational page may be helpful for the customer to read at a later date. Make yourself available to answer more detailed questions that might arise from your brochure.

Use bullets to emphasize your strong points. As people scan the brochure the bullets will grab them and pull them in. Use underlining to separate thoughts or points. You may want to include a free coupon for new customers to tear off and bring in to redeem a gift.

One of the things I often notice about brochures is the failure to provide contact information. A good place for this is in the middle of the back of a tri-fold brochure because this will be the back of the brochure when folded. Have you ever watched a person pick up a brochure? Most people look at the front and turn it over and look at the back. In those three to five seconds they have already decided if they will keep it or not. So the back is a good place to put your phone number in large easy to read print. Beneath the phone number put your mailing address and other contact information such as a website or e-mail address.

Due to the fact that your brochure shouldn't have a lot of wordiness to it, you may want to include pictures. You know the old saying "a picture is worth a thousand words." Pictures of birds foraging in a field may be great to you as you took them, but what about the person looking at your brochure who isn't familiar with your farm? Make sure the pictures are clear. Give some thought to what the background of the picture is, does it portray a neat clean farm, are the animals portrayed in a clean comfortable setting? Believe it or not many people have a hard time looking at the live version of what they might be having for supper. Pictures that are too "cutesy" of young stock may turn people off when they equate it as dinner on their plate.

If all this planning is to much for you, APPPA now has a Producer Plus membership with one of the benefits being access to a Producer Plus web page. On the site will be several different brochures that you can download and change specific to your farm, then print with very little effort.

Overall, spend some time thinking your brochure through even before you sit down at the computer. With today's color printers and colored papers available at the office discount stores you can print off a few and see if you need to make changes. Brochures are a low cost tool that you can use to promote your business.

We are committed to:

Animal Health and Happiness. The chickens are in the sun with fresh green things to eat.

Organic Production. No synthetic chemicals have been used in the production of the bird's food or at any point during their lives. No antibiotics or hormones have been used. All grain is certified organic by MOSA.

Local Production and Consumption. By buying direct from farmers you keep your dollars in Wisconsin, even in Chippewa County. No big chain profits leading back to large corporations far away. Your dollars really count!

Safe Food. We maintain high cleanliness standards to ensure our eggs and chickens are disease-free. No irradiation to cover up bad food here! If you start with a clean animal that is closely handled and scrutinized, you end up with clean food!

Outside of A Brochure for Chickens and Eggs

2005 Prices

Frozen Pastured Chicken
3-6 lbs each $2.45/lb

Pasture-Raised Brown Eggs
$2.35/dozen

Delivery available in Eau Claire, Chippewa Falls or Eastern Chippewa County for orders of four or more chickens. Call to discuss egg delivery options.

We love to have visitors at the farm, and invite you to come out and visit the animals. We have chickens, dogs, cats, turkeys, sheep, donkeys, horses and all the "wild ones." Please call ahead to make sure we are home. Children welcome!

Thanks for supporting us!

Jody
and the gang at Wild Crescent Farm
715-667-3203

Wild Crescent Farm

Jody Padgham
Pasture-Raised
Eggs and Chickens

2240 310th St
Boyd, WI 54726
715-772-3203
jodyp@cvol.net

APPPA BUSINESS MEMBER PROFILE

Growers Discount Labels
By Karen Wynne Spring 2004

Stu McCarty, Owner

We have been in the printing business since the mid '60s and began to supply labels to growers in the early '90s. Two of us do the design work and specify the material best suited for your application. Dozens of professionals in our factory use their skills and the latest equipment to produce high quality labels in a three to five-day production cycle.

Our office is in upstate New York. The factory is in the Midwest. We have customers from Maine to California and from Oregon to Georgia.

One tool that farmers need is a convenient, one-stop source for their labels. We offer a design and production service for all kinds of labels which can be applied to frozen meat products, eggs, vegetables or any value-added products such as body care products, honey, salsas, cheese, preserves, baked goods, etc. Tell us how the label is to be used and we can recommend the most effective and durable label.

We focus on customer satisfaction and personal service. We relish the challenge of transforming all the information a farmer gives us into a label that will talk to the consumer when the farmer can't be there to tell his story. We can measure our customer's success by the reorders that we receive as products are being sold. Labeling your direct-marketed products is one way the small farmer can succeed by keeping a higher percentage of the consumer's dollar.

One of the requirements of most pastured poultry producers is that the label will stick to a poly bag that will be frozen. Our freezer adhesive has the wide temperature range to work under these conditions. Our pricing for the freezer adhesive is very reasonable compared to other sources we have checked.

We have been organic gardeners going back into the 60s, and began to get involved in local agriculture through the Northeast Organic Farming Association in New York about 15 years ago. We have been editors of their journal and have served on their board of directors.

We have an acre and half of mixed vegetables we grow for a 40-family Community Supported Agriculture project and maintain a flock of laying hens and will soon have a pair of milking Devons. Whatever food we cannot produce ourselves, we buy from local farms. We thoroughly understand the challenges a small farmer faces in producing and marketing their products.

When we talk with our customers, it is very exciting for us to see how local agriculture across the country is thriving. We want to do all we can to help it flourish.

Contact Information: 800-693-1572 growersdiscountlabels@tds.net

Pastured Poultry on the Internet
By Chad Anderson Summer 1999

Why did the chicken cross the information superhighway? To get to the other site!

The internet is a virtual meeting place for people to electronically exchange information, knowledge, and opinions about any subject.

It is also a great means of teaching and learning about pastured poultry. I have been privileged to become acquainted with many people who raise poultry in this manner through my web site and other facets of the internet. In private and public conversations, the exchange of information is always educational, enlightening, and enjoyable.

Though agricultural ideas can sometimes stagnate like farmyard runoff in spring puddles, internet resources provide a means and opportunity for graziers to share, give, and gain knowledge. Since the pastured poultry concept is a dynamic and still-developing system of poultry rearing, there are few experts and many learners. This instant exchange of information through light-speed information transfers gives us a way to learn collectively, cooperatively, and constructively.

The Internet and its Constituent Parts

The INTERNET is the result of millions of linked computer systems around the world, and their information transfer. A few of its components are listed below.

An ISP (Internet Service Provider) is a company which provides internet access to customers. Users may connect to the ISP's server in various ways, including using a dial-up modem, which is the most common method. The user's computer communicates through a regular telephone line to the ISP's computer. E-MAIL is the basis of millions of messages sent between people on a daily basis. Most often limited to text, they are an almost instant means of communicating between individuals. A WEB SITE is an organized group of WEB PAGES built by a person or organization for the primary purposes of informing and/or selling. A web page may include any combination of plain or formatted text, images, sound, or video. HTML (Hypertext Markup Language) is the code used to construct and link pages, which will collectively constitute the "world wide web". A URL (uniform resource location) is a web address given to a specific web site. It may appear as: http://www.something.com, or http://something.com. Web addresses may include a number of forward-slash divisions after the domain name (which is "something" in the above example). These divisions denote different directories within the site. A LISTSERVE or electronic mailing list is an e-mail based means of communication between an individual and a group, whereby an individual sends a comment or question to the listserve, which will send a copy of that message to each member of the list. People may respond to the listserve, and that message will likewise be distributed to the list's membership. A NEWSGROUP is a chronological hierarchy of messages posted by group "subscribers". A MESSAGE THREAD develops as responses to one posted message accumulate after it. These definitions help explain the mechanics of some of the most common means of communication used by people on the internet and will help the inexperienced better understand just how the following resources can be used to gain information.

The internet is a valuable resource for many people, including pastured poultry people, who would like to share and exchange information, experience, and old stories. (And watch out for stray chickens as you speed down the information superhighway!)

Web Pages for Your Farm
By Robert Plamondon Summer 2004

Should you have a web page for your farm? If so, how should you go about getting one?

The reason to have a web page is because it's a fairly cheap and easy way to connect with your customers. For example, I rarely use the phone book anymore. I use www.Google.com to find information about local businesses.

A webpage is especially valuable when your product is hard to order for some reason, such as when it's seasonal or requires advance ordering. You can put the details on your web page, and the customers can figure out what's what, instead of having to call you up and ask a series of more or less random questions before they understand what you are offering.

What Goes Into a Web Page?
The web page should have the same kind of information you'd put into a flyer or brochure. This ought to include:
• What you sell and why it's good.
• How to buy your products. This should include production schedule, order forms, and anything the customer needs to place an order.
• How to contact you: mailing address, phone number, fax, email.
• A very clear description of what part of the country you're in, to keep people from far away from bothering you by accident.
• Directions to your farm, farm stand, farmers market as appropriate.
• A brief, feel-good description of your family and your farm.
• Color photos that include as many of these elements as possible:
> • Bright green grass
> • Happy poultry or livestock, preferably babies
> • Brilliant blue sky with white fluffy clouds
> • Children. Extra credit for overalls. Children feeding baby animals would be ideal.
> • You. Extra credit for overalls.
> • Nostalgia elements: an old wooden barn, old tractor, etc.

How Do You Make a Web Page?
You can make a web page the way you do anything on a family farm: make the kids do it! It's a skill that comes easily to many high-school and even middle-school kids, who tend to have some relevant computer skills already.

Or you can learn the skills yourself. Simple Web pages are easy to make. The libraries are bulging with books on this topic, written for beginners. In short, though, the process consists of simply saving existing word-processing documents as web pages (Microsoft Word can do this), or using a web-page editor to create a more customized web page.

I recommend that you look for some kind of beginner's guide to web page design, and use whatever editor it recommends, especially if it's free. Many such books come with a CD that includes everything you need. (I'd be more specific, but it's been a while since I was a beginner, and all the products have changed since then.)

Do I Need A Domain Name?
Yes, you need a domain name! It's so inexpensive that there's no reason not to go for it. For example, I registered http://www.nortoncreekfarm.com. To do this, you either have your web-hosting service register your domain for you, or you go to the web site of a "Web Registrar" and do it yourself. It's not difficult at all. I use http://www.godaddy.com. A domain name should cost under $10 a year.

That just gives you the name. You also need a web hosting service to actually make your page available to the teeming millions on the Internet, and to make email addressed to your new domain available to you.

How Do You Put Your Page on the Web?
It seems as if there are millions of web-hosting services. A lot of people use a hosting service provided by their Internet Service Provider, but you can buy web hosting from innumerable services.

There is ferocious, cutthroat competition in the web hosting business. I noticed that godaddy.com (the Web Registrar I mentioned above) is offering Web hosting for $3.95 a month. There are innumerable Web hosting services that will give you more features than you will ever use for $7.95 per month or less if you prepay a year in advance, or $9.95 per month if you don't. I found the reviews on http://www.top-10webhosting.com useful, though I haven't tried any of the services listed there yet.

How Do You Make People Notice Your Site?
If you're selling to local customers, people will notice your site largely because you will (or ought to) put your site's address (or URL) on everything you give your customers: business cards, flyers, egg cartons, broiler tags, refrigerator magnets, pencils. Everything. One of the main uses for your web page is so existing customers can find simple information like your hours of operation or how to place an order.

To pull in new business, people will need to be able to find you through search engines and online directories.

The most important search engine is Google. People sometimes find our farm by typing search strings into Google such as "Corvallis free-range eggs." This can lead to new customers and doesn't cost anything. To add your site to the Google database, go to their "Add your URL" page at http://www.google.com/addurl.html and fill in the two-line form. That will add your Web page to their database in the fullness of time (it may take several weeks).

There are a great many people who will list your site in their directory for a fee, or submit your URL to a million search engines for a fee. My advice is to save your money.

Maintain Your Site
Getting your site up on the web is interesting and exciting. Maintaining it is dull. But it works a lot better if it has new content from time to time. If you can have periodic updates, people who drop by for the first time will see information that's more timely, and others will have a reason to come back. This doesn't have to be fancy at all. New photos and updated calendars that don't stretch too far back into the past are really all it takes. A little bit of chatty news about the farm helps, too.

That's about it. My own farm web site at http://www.nortoncreekfarm.com is about as simple as they come; just a few paragraphs of text, a few links, and a copy of our farm newsletter. You can do this, too!

Measuring the Interest for Marketing: Pastured Poultry at Farmers Markets
By Aaron Silverman Spring 2002

Direct marketing grassfed meats is becoming increasingly attractive for diversified family farms. Marketing fresh meat and poultry at farmers markets presents different challenges from marketing produce or flowers. Health regulations, product handling by both vendor and customer, and facility limitations of the market require careful consideration by potential meat vendors.

Greener Pastures Poultry®, a collaboration between four family farms in Western Oregon, established a cooperatively operated licensed processing facility. Under a unified label, Greener Pastures seeks to market premium pasture-raised poultry to restaurants and farmers markets.

With a grant from the Sustainable Agriculture Research and Education Program (SARE), Greener Pastures Poultry sought to develop a model for gauging the feasibility of using farmers markets as a major outlet for marketing pastured poultry, as well as other fresh meats. Dot Poster surveys were used to conduct customer surveys at three farmers markets, with the goal of measuring the initial sales potential at the farmers markets.

The Dot Posters proved to be an easy, effective way to establish some general baseline sales data at farmers markets. The results of the surveys, combined with sales data from a farmers market Greener Pastures marketed at in 2001, suggest that there is substantial demand for fresh pasture-raised poultry at farmers markets, and the addition of fresh meat vendors would be a valuable addition to the markets.

Project Methods
The project sought to measure the demand for a specific product at farmers markets, as well as the impact the marketing of the product might have on the farmers market as a whole, through the use of Dot Posters. Based on a project by two Oregon State University researchers to study the customer base of Oregon's farmers markets, the Dot Posters are a quick, unobtrusive means of surveying customer preferences. Each poster contains a single, carefully crafted question, and several specific responses. Customers are handed four stickers, and asked to place one on each poster "where it makes the most sense." Questions must be broad enough to allow for a short list of answers (i.e., yes/no/maybe), while focused on revealing useful information. An overall population count must be conducted simultaneously, especially if an accurate statistical result is desired. Many markets take periodic counts regularly, and all the markets in this project provided attendee estimates for that day.

The Dot Posters were set up in a linear fashion, with a table containing display information about Greener Pastures Poultry at the head. Two to three volunteers asked people to participate and handed out stickers, while another volunteer answered questions from participants.

The Questions Asked for this Project:
¤ Question 1: Greener Pastures chickens are family-farm raised fryers grown on grass pasture and a custom, natural grain mixture. Unlike conventional and "free range" birds, after three weeks of age Greener Pastures chickens live in portable pens and ALWAYS have access to fresh pasture and sunshine. Greener Pastures Poultry is considering offering fresh, pastured poultry at this market. The current price is $— per pound for whole chickens.

Would you consider purchasing Greener Pastures poultry at this farmers market?
 Yes Probably No

¤ Question 2: What Greener Pastures Poultry products would you be most interested in purchasing at this market?
 Whole birds only
 Parts only (quartered, boneless breast, legs and thighs)
 Mostly whole, some parts
 Mostly parts, some whole

¤ Question 3: How many chickens would you consider purchasing from Greener Pastures Poultry each month if it was available at this market?
 1-2 3-4 5 or more

¤ Question 4: Would the availability of fresh pastured poultry at this market make you a more fre-

quent shopper of this farmers market?
   Yes   No   Maybe
   No, but I am interested in pastured poultry

Results

Greener Pastures conducted surveys at three farmers markets during the 2001 season. The results from each market were surprisingly consistent. (figure 1) The response rate for each farmer's market with a population count of 3,000-4,000 was 9%; the weekday market surveyed had a participation rate of 30%, but only a population count of 1,500. Lower participation rates at the busier markets may have been a function of understaffing the survey booth.

Participating customers overwhelmingly were interested in pastured poultry, and most indicated a preference towards parts over whole birds. Most customers thought their initial purchase would be one or two birds per month. The results suggest that initial sales of over 200 items is possible at each market.

The most surprising result was that between 38% and 40% of the participants thought that having fresh pastured poultry available at the farmer's market would cause them to shop the market more frequently.

Greener Pastures Poultry was a regular vendor at a farmers market not surveyed during this project. The market had an average customer population of about 4,000 over the course of the season, June through October. Number of items sold increased over the course of the season, eventually surpassing the 200 items by the end of the season (see figure 2). Most customers purchased two to four items each month. The ration of whole bird sales to parts sales are reversed from the survey results largely based on supply. Over 90% of supply was sold at every market.

	Figure 1: Dot Poster Questions for GPP Market Survey			
	Est Market Pop	**Survey Resp Rate**		
Corvalis Weds 8/22	1500	30.0%		
Corvalis Sat 7/21	3500	9.0%		
Portland Fri 8/10	3800	9.0%		
Question 1	**Yes**	**Probably**	**No**	**Total Y&P**
Corvalis Weds 8/22	52.9%	29.3%	17.8%	82.2%
Corvalis Sat 7/21	47.6	30.3	22.1	77.9
Portland Fri 8/10	59.6	29.2	11.2	88.8
Question 2	**Whole Only**	**Mostly Whole**	**Parts Only**	**Mostly Parts**
Corvalis Weds 8/22	10.6%	17.9%	36.2%	35.1%
Corvalis Sat 7/21	8.9	20.5	42.2	28.4
Portland Fri 8/10	10.9	31.9	22.9	34.3
Question 3	**1-2**	**3-4**	**>4**	
Corvalis Weds 8/22	81.2%	16.4%	2.4%	
Corvalis Sat 7/21	82.1	16.0	1.9	
Portland Fri 8/10	75.0	24.6	0.4	
Question 4	**Yes**	**Maybe**	**No**	**Total Y & M**
Corvalis Weds 8/22	16.5	21.5	62.0	38.0
Corvalis Sat 7/21	17.0	22.9	60.1	39.9
Portland Fri 8/10	15.1	23.6	61.3	38.7

Conclusion

The Dot Poster method proved to be a very adaptable tool for conducting an interactive, unobtrusive survey. While it has demonstrated its value for looking at general customer preferences, this was the first time Dot Posters were used at focusing on a specific product.

Careful attention must be paid to the crafting of both the questions and answers for each Dot Poster. Too many choices made the data from Question 2 mostly irrelevant. "Yes, No, Maybe" types of answers are easier and quicker to answer, perhaps giving a better picture of how people might really act.

The Dot Posters clearly indicate that a significant demand exists at these farmers markets for fresh, clean, local poultry—and therefore most likely other meats as well. A defined initial sales potential is critical for producers to determine the feasibility of marketing through farmers markets.

The fact that a sizable portion of those surveyed indicated that they would be more likely to shop the market if this type of product was available, is important for a prospective market making decisions about growth, or adding the facilities needed for marketing fresh meats. Dot Posters are a fairly easy way for a market to gauge the needs of its customers, and determine how the market can better serve them, and the farmers.

Figure 2: Greener Pasture's Sales at Eugene Farmer's Market 2001			
Date	Whole	Parts	Total Items Sold
June 30	70	46	116
July 28	66	41	107
Aug 18	80	49	129
Sept 22	73	62	135
Oct 20	149	93	242
TOTAL	60.1%	39.9%	729

Give Them What They Want!
By Kip Glass Summer 2004

A new customer calls and asks if you sell pastured chicken; you answer, "yes we do."

Then they proceed with more questions: Do you sell the chicken cut up? Do you sell just breast meat? Do you sell legs? Wings? We sure would like some halves for grilling. And, of course, you give the same answer, "No ma'am, we just sell whole chicken."

Have you ever wondered if it would be worth your while to sell the chicken cut up, or in individual pieces, or parts? One thing to remember: you are raising the chicken for the consumer, not for yourself. Have you ever looked in the grocery store at the chicken case? If you have, you will notice there are more packages of different chicken cuts and pieces than there are whole chicken offerings. That is because the consumer wants choice, and they will pay for that choice.

For the last two years we have offered chicken cut in half, cut up into all the pieces, bone-in split breast, leg/thigh combo, wings, and stock bags consisting of backs, necks and broken wings and legs. Our costs, on average, with extra packaging, labeling, and labor are approximately 75 cents extra per bird. But, at our current price structure on a 4.25 lb. bird, we make an additional profit of $2.50 on each bird selling by the piece than by selling it whole at $2.25 lb.

Breast meat is a hot item, so we charge a premium for it. That higher pricing also balances the fact that the other cuts don't sell as well. Our current price on bone-in split breast with skin is $5.95 to $6.05 per lb., depending on which marketing avenue we are selling at. I can hear you say, "no way will I get that price." But listen to this- on the day that I'm writing this I sold over 32 pounds of breast meat at our local farmers market. This was the highest price per pound of all the choices I was selling.

Out of a 4.25 lb. chicken I will have a little over 1.2 lbs. of breast meat on this individual chicken (see chart.) At $6.05 per lb., I will bring in $7.26 on that breast cut alone. This covers my costs of raising that bird to the sellable point, plus some profit left over. I then have total profit on the other cuts: legs/thighs, wings, stock meat, liver, hearts, etc.

We have even been more creative and developed a market for the heads and feet and extra stock pieces by grinding them up and selling them for $1.25 per lb. for dog food.

Wings are another hot item: we consistently sell them for $2.85 per lb. At ½ lb. per bird, that is excellent money. My wife and I have been working with the genetics of breeding a chicken that has 6 wings to capture that extra profit per bird; we accomplished that goal, but we can't catch the darn thing. You ought to see that bird fly!

Offering a whole chicken cut up is a big hit with the younger crowd that doesn't know how to cut up a chicken, or the older, arthritic customers that can't do it anymore. We charge an extra 25 cents per lb. for a cut up bird.

Just remember: the customer is always right and you have to meet their needs as best you can. Believe me, they will pay extra for what they want.

Chicken-Cornish Cross		Chicken Parts %		
Total lbs. of edible	2.7	3.55	4.1	4.2
Wing 13%	0.35	0.45	0.55	0.5
Leg/Thigh 32%	0.85	1.15	1.3	1.35
Breast 30%	0.8	1.1	1.2	1.25
Neck/Back 25%	0.7	0.85	1.05	1.1
Head/Feet - extra lb. for dog food	0.20	0.25	0.30	0.30

Almost Organic?
By Jody Padgham Summer 2004

At more than a few public events I have attended this year (including APPPA's own presentations at the PASA conference) I have been in the frustrating position of having to "bite my tongue" over misstatements surrounding the use of the word "organic." I would like to take this opportunity to offer clarity on when this term can--and should--be used by producers of any product, and specifically poultry, in the U.S.

This is an issue dear to my heart, as I work part-time for an organization that educates people about organic production (the Midwest Organic and Sustainable Education Service, MOSES- www.mosesorganic.org).

You may be aware that organic labeling is now governed by the USDA through the National Organic Program (NOP). I have never been a great fan of getting the government involved in private lives, but the NOP was a rule that was demanded by CONSUMERS because of their desire for clarity on just what they were buying when a product was called "organic." Farmers, consumers, researchers and others worked hard for 10 years to come up with a rule that would work. The public comments after the draft rule was released were of phenomenal volume, and the rule was changed to reflect the desires of the public. This rule may not be fantastic, but it is something customers wanted and was put together by a very broad coalition of folks to the best of their ability.

So, now we have rules written down that say what it means to call something "organic" all over the U.S. I agree with those who worry that "the big boys" are trying hard to compromise the definition. I applaud those who say we should revolt and "go beyond organic," but at the same time I wish that the energy put into creating alternative systems was put toward defending a not-too-bad system that has not been truly tried yet. As a small producer that strictly follows organic practices and markets direct, I am happy to be able to explain my methods of production straight to my customers without having to worry about a government translation of my activities. Rule, labeling and translation concerns apply most to those producers that market large-scale or wholesale. These are the folks that really need the National Organic Program.

Regardless of what you feel about the government getting involved in how you farm, I'd like everyone to set that aside a moment and think about your fellow farmers. There are a significant number of APPPA members that are raising larger numbers of organic chickens, following the NOP, spending the money to become organically certified and using the USDA organic label when selling their birds. Our past board president, and winner of the 2004 "Real Chicken Award," Tom Delehanty is one, Peggy and Richard Sechrist, former APPPA board members, my neighbors Deb and Mike Hansen here in WI, friends Shelia and Ron Hamilton in Alberta, Canada, and many others- follow the rules of the NOP and sell certified organic birds. These folks aren't Tyson or Gold 'n Plump- they are small family farmers, trying to make a decent living and do good things for their families, like you and I.

So what has me all in a tizzy? Hearing people make statements that they are "almost organic" or "basically organic." I have heard variations of this statement more times than I care to count, and hope that I never hear it again. Let me help you understand why I feel this statement is such a bad idea.

First of all, why do people feel compelled to say this kind of thing? The market for organic products has been growing more than 20% per year for the last several years—organic is the largest growing agricultural opportunity. Consumers have been studied, and by and far they buy organic because they feel organic products are healthier for them. They want a product that is grown without chemicals. With the prevalence of genetically modified organisms (GMOs) in corn and soybeans, organic is one of the only ways those who care can be assured their food is GMO free. Organic producers have worked had to communicate with consumers, and have succeeded in fetching strong prices that reflect increased production costs and higher perceived value.

So, ok, you say- my birds are healthy. I'll give you that- grass-raised birds are the best there is, and you can get a lot of marketing mileage on the taste alone. But, can you say they are raised without chemicals, "organic" or "almost organic?" Most likely not—if you buy your feed at the local mill or feed store. Unless you have evidence otherwise, whoever grew that corn and those soybeans sprayed LOTS of chemicals on those grains and beans. The corn is most likely genetically modified, the soybeans roundup ready GMO. Chemicals were used to control insects and rodents during storage and additives may have been added to preserve freshness. If medicated, feeds contain low levels of antibiotics as a "preventative" measure.

I am not one to judge what other people choose to do. If you choose to buy feed that has these things in it, that is ok with me. Lots of people do. There are people that will argue me into the ground that their products raised with chemicals are lots better for people, animals, the environment, or whatever. What I have a problem with are those that feed this feed and say their birds are "more or less organic." If you have been unaware of the difference between organic and "feed store" feed, I hope this helps clear things up. If you support the use of chemicals in producing your food, then don't try to please your customers by telling them something that isn't true.

Those who negotiate with their neighbors for homegrown feed, or raise their own without chemicals have a stronger leg to stand on in this argument I am making, but unless they strictly follow organic practices (and become certified if above the $5,000 per year small producer exemption) they legally can not use the term "Organic Chicken." Now, I expect that many of you will disagree with this.

This is when we need to think of our friends the Delehantys, Sechrists, Hansens and Hamiltons again. These folks work very hard, and pay a lot extra for feed, health care and processing, to be able to say they have raised their poultry according to the abovementioned national rule. I feel that we need to honor their commitment by NOT using the same terminology unless we are AT LEAST walking the same mile they are. I know many APPPA members will disagree with me here. You are welcome to write in and voice your opinions to stir a debate, if you feel the need.

Organic certification is not for everyone. Many of us raise poultry the same way we have been doing it for years, with corn raised by Joe down the road and healthy pastures that have never been sprayed. If you are selling birds to the neighbors and relatives, you can basically ignore everything I have said here today. If, however, you have a display at a farmers market, you are taking ads out in the local paper, you have a web presence or you are wanting to beef up your brochure, think twice about saying you raise birds that are "almost organic" unless you are ready to explain exactly what part of organic you don't believe in.

Many people resort to the term "natural" in lieu of "organic." Due to the lack of true definition there is no line drawn that lets us know what the term "natural" refers to. I know plenty of people that use chemically raised feed store feed, including chick starter laden with antibiotics, and call their product "natural" because the birds are out on grass rather than in cages. Although it seems kind of goofy to me to call this "natural," it is technically just fine, as no one has defined what that term "natural" really means. Organic production, for better or worse, has the line firmly drawn. Antibiotics and GMOs are out, outdoor access and non-chemical feed are in. Let's support our fellow farmers, and the consumers behind them, and keep within the lines. Let's commit to being organic or not, but never to being "almost organic."

Defining Organic Production
Just what does organic production of poultry under the national rule entail? There are several points in the production of organic poultry.

1. All birds must be fed, from the second day of life, 100% certified organic feed. Certified organic feed is grain that has been raised on land without prohibited substances (a big term basically meaning no synthetic chemicals, GMOs, treated seeds, sewage sludge.... For more on how to grow organic crops, go to the ATTRA website at www.attra.ncat.org) for three years.

2. All additives to the feed, such as mineral mixes and supplements, must be "certifiable." This is why we love the Fertrell Company, APPPA business members, as they do the organic feed homework for us! (www.fertrell.com)

3. No antibiotics may be used. Most conventional medicines are not allowed. Natural remedies are promoted- check out the new book: *Remedies for Health Problems for the Organic Layer Flock* by APPPA member Karma Gloss, available free on the web

at http://www.kingbirdfarm.com/Layerhealthcompendium.pdf.

4. Poultry must have access to the outdoors when seasonally and age appropriate. No sweat for us, right? A very big deal for the 'big guys" like Tyson and Gold 'n Plump.

5. Vaccines are allowed, as long as they are not GMO. Not many pastured poultry producers use vaccines.

6. Processing must be at a certified organic processing plant. This is a huge issue for many producers. It is hard enough to find a plant that will process your birds, much less at a reasonable cost. Now to be certified organic..... Luckily, it is not monumentally difficult for a processor to become organically certified- there is little that they do that is not ok. The main issues are the separation of organic product from non-organic. Some producers have worked through this by having their processor certified only for their birds, which are run first thing in the morning on clean equipment. In this case the producer will often pay the cost of having the plant certified for organic for the several days that are needed. Check with your local plant to see if this is a possibility.

Those of us who home-process get caught up by this point. Since I home process and direct market, although I follow all of the other organic rules they cannot be certified, or technically labeled as "organic."

7. You may still buy chicks from whomever you choose. The chicks do not need to be "organic," but need to be raised organically from the moment you get them at one or two days old.

8. You must keep very comprehensive records on anything associated with the birds--labels from feed sacks, notes on pasture movement, remedies used, sales documentation. See the ATTRA website for more information.

9. If you sell less than $5,000 each year in organic product, you need to follow the rules, including keeping excellent records, but you need not be certified.

10. Those selling more than $5,000 per year in organic product must be certified. To become certified, you must first apply to a USDA accredited Certification Agency. You may find these agencies by visiting www.ams.usda.gov/nop/CertifyingAgents/Accredited.html or calling the USDA at (202) 720-3252. I recommend that you ask any neighbors that may be into organics (They often have signs out front that they are organic) who they are certified with. All certification agencies certify to the National Rule--they have different fee structures and office policies that may make some more attractive to you than others.

The certification agency will send you an application (it may cost money- $50 is common). That application will become your "Farm Plan"–another requirement of the NOP. Once you submit the application, it will be reviewed by the agency, and if it is complete, an organic inspector will be assigned. The inspector will come out and look at your farm and poultry operation. They will look at all your records, go to the fields and look at any handling areas. They will submit a written report to the agency. The agency will offer a determination as to whether you qualify for organic certification or not.

This all sounds rather harsh, but in my experience is a rather interesting blend of sterile rules and real people that come together to a very human, functional and admirable process.

11. Once certified you may use the USDA "Organic Seal" and label your product as "USDA Certified Organic."

12. You will pay an annual fee for certification. The average small poultry operation will be $300-500 per year, varying with whether you need crop fields certified. You generally pay costs associated with the inspection, and some certification agencies require a small percent of the annual organic sales.

Most states have a cost share program in place for organic certification. Contact your local Department of Agriculture to see if there are funds available to you.

13. You re-apply for organic certification each year, with an annual application and inspection.

APPPA PRODUCER PROFILE

Elmwood Stock Farms
By Karen Wynne Summer 2005

Mac Stone attributes the solution to the biggest bottleneck in his operation to APPPA networking at the annual meeting with Acres-USA in Indianapolis. He had been hauling water in buckets, and that was definitely the limiting factor in getting things accomplished, when he heard two other producers talking about using spaghetti tubing to distribute water to their birds. This tubing, which is about the diameter of a pencil, runs to the float valve in a water bucket and allows the water to flow constantly from the water source to two waterers in each pen. "It really took us to the next level in our poultry production," says Mac.

Mac does not have a lot of spare time to be hauling water. In addition to working with his wife, brother- and father-in-law to run the largest certified organic farm in Kentucky, Mac spends five days a week working for the Kentucky Department of Agriculture as division director of Value-Added Plant Production. AND he's an APPPA board member!

In addition to 350 layers, 3,000 broilers and 125 turkeys per year, all certified organic, Elmwood Stock Farms produces certified organic beef, lamb, almost 60 acres of organic and commercial produce, and 16 acres of tobacco including one organic acre. Each member of the family is responsible for a different part of the operation; Mac and his wife Ann Bell manage the poultry, sheep, and marketing. Farm-wide decisions are made together; Wednesday mornings are reserved for a meeting over breakfast. While Mac works off the farm during the week, Ann spends five days each week at farmers markets, selling the majority of their produce, eggs and meat there. Efficiency in their operation is essential to get everything done in the time that they are on the farm, and developing more efficient systems is an ongoing challenge.

Elmwood Stock Farm's Cornish Crosses are day-ranged in eight batches of 300, which is a good number for their brooder and field housing to carry. The first batch of chicks shows up in March and the last is processed in early November. The birds are housed in hoophouse pens that are surrounded by one length of electric netting; the hoophouse is moved daily while the netting is moved every two to three days.

In an attempt to reduce the time spent moving the houses and fencing, this year Mac will try out a new system. He plans to use two lengths of netting to create a corridor of pasture; two shorter lengths of netting running perpendicular to this corridor will create smaller areas of pasture. This method is similar to rotational grazing used with sheep and cattle. As the birds progress down the corridor, only the houses and shorter lengths of netting will be moved. Only once the birds have traveled down the length of the corridor of netting will the whole system have to be moved.

The layers, a mix of Barred Rock, Rhode Island Red, Production Red, and Black Star breeds, are housed in two mobile houses, each on a wheeled wagon frame two to three feet above ground. These houses are moved with a tractor every one to two weeks and are surrounded by electric netting. Water flows on demand using automatic bell waterers with a large reservoir. Range feeders allow feed and oyster shell to be maintained weekly when the houses and netting are moved to fresh pasture. This limits daily labor to egg gathering.

APPPA PRODUCER PROFILE   *Elmwood*

Two years ago, Mac and Ann tried some slower-growing rainbow broilers on a trial basis. Mac says that the customers loved them – the birds are not as fat, have more dark meat, and their flavor is superior. However, it was a challenge to keep the rainbows on pasture for an additional four weeks and keep the predators away. It was also complicated to integrate their schedule with that of their Cornish Crosses, so for now they are sticking with one schedule and one type of bird.

Mac and Ann also raise 125 turkeys each year: 100 commercial breed and 25 Blue Slate and Royal Palm heritage turkeys. They will try some Bourbon Reds this year. (The turkeys are named for Bourbon County, Kentucky; the county line is only three miles from the farm.) The turkeys are all sold on special order and customers pick them up at the farmers market. It's a good thing to have at an early winter market, Mac says. "It was snowing sideways at the farmers market the day before Thanksgiving, and all the other farmers were huddled by their trucks, not selling anything. Meanwhile we had a line of turkey customers at our stand."

The pasture is rotated between the sheep and poultry. Chickens and turkeys are kept in separate pastures to prevent blackhead disease, a disease that is relatively common in chickens but deadly to turkeys. Pastures are improved by overseeding white clover in the spring and are either mowed or hayed to maintain a shorter forage for the birds.

Growing certified organic poultry requires certified organic feed. Elmwood Stock Farm gets its feed from an organic egg farm located two hours north. The local feed mill sends a truck and they store the feed in hopper bottom feed tanks. To adjust for different nutritional needs, they add roasted organic soybeans to the broiler feed while turkeys get extra organic soybean meal.

Mac says that generally organic certification "isn't too tough once you're in the system." It does require a good recordkeeping system, which is a good idea for any farm. Each year the certification is updated with a new application and fee and the farm is visited by an organic inspector. And higher-priced feed results in a higher-priced end product.

Kentucky poultry producers have been working with a mobile processing unit for a number of years, and now a new small USDA-inspected plant has opened in Kentucky. It's located 2 ½ hours from the farm. They take 300 birds to process every three weeks; that batch can be sold fresh for up to four to five days. The fresh birds are always held at least 48-hours before freezing, which seems to help with tenderness and set the flavor. Ann and Mac can then sell the frozen birds for the next two weeks.

After reaching 11 used chest freezers to store their meat, they finally graduated to a walk-in freezer last year. Mac says that this resolves his constant concern about whether the freezers are working properly and makes organizing the freezer space much easier.

Five farmers markets are the mainstay of their sales. The seasonal markets begin in mid-April and last through Thanksgiving, and one indoor market lasts through the winter. To haul their product to market they use a covered cargo trailer complete with three freezers and two small apartment-sized refrigerators. Maintaining a clean environment is very important in ensuring a reputation as a professional and safe operation. Customers appreciate details such as having separate meat and produce scales.

In addition to five farmers markets, restaurants account for about a quarter of Elmwood Stock Farm's direct sales and the farm is beginning to dabble in online sales. (Check out their website at www.elmwoodstock-farm.com.) Broadening their customer base helps maintain a good market for their higher-priced organic products.

Each year is a new experiment and a different challenge for the increasingly diversified farm. As the website says, the land has supported six generations of Kentucky family farmers. The family is working hard to ensure that Elmwood Stock Farm is ready for the seventh.

Mega-Farms Make Your Marketing Easier
Andy Stevens Fall 1998

Modern-day agriculture is quickly evolving into a mixture of small farms and "mega-farms." The mega-livestock farms have received much publicity during the past few years--and most of it has been negative. Combine this attention with the food poisoning scares and you have created a dynamite market for food that is produced on small farms.

More and more people are wanting to buy their food from people they know. There is a desire to buy food closer to home. This is a marvelous marketing opportunity for you.

Consumers are spurred to local food sources by their increased concern over health and chemical contamination of food. This concern results in a dramatic growth in direct marketing of food by farmers. But this new marketing approach means that we farm differently than our ancestors. More marketing savvy is needed now.

This marketing niche markets itself with a mother earth approach. Do you know where that chicken on your plate is coming from? Do you know what hormones are in that beef? If consumers answer no and are alarmed, you have your foot in the door. And traditional farmers have the door shut in their face.

You have an advantage over these bigger, traditional farms because you can control the price of your product by selling directly to the consumer. The traditional farm has little, or no control over the prices of its commodities: it takes what commercial or co-op buyers offer.

The big farms are taking some tough blasts these days. The war against the proliferation of "mega livestock factories" is being waged in many states. And the opponents of these big farms are not confined to "environmental groups." Neighbors, both farmers and non-farmers, join in the battle.

For instance, neighbors of a proposed mega-hog operation near Springfield, Illinois are concerned about the volume of manure and possible environmental hazards that the farm could bring. The Prairie View Farm will produce 41,600 sows and more than 800,000 piglets each year. According to some studies one sow can generate about two times the manure of a human.

Owners of the proposed farm said that all waste will be used as fertilizer on crop and pasture land. The fertilizer will be stored in concrete pits under the confinement buildings and extend the full size of each building. This has brought up the issue on the legality of the use of pits.

Critics argue that Prairie View's handling of waste skirts state requirements related to large-scale livestock operations. Currently there is no legislation governing the use of pits, only earthen lagoons. In addition, construction guidelines or inspection procedures do not exist.

Whether true or not, mega-hog farms are perceived as an environmental disaster waiting to happen. And the opponents of the big farms say in some cases it already has happened. They say the danger comes from the hog lagoons which store the immense amount of manure produced on these giant farms. The smell is so bad that residents living close to them have filed suit because of the stench. The hog lagoons also tend to rupture, spilling out over neighboring land, and into the water supply.

Illinois experienced its first lagoon rupture in April 1996, when 25,000 gallons of hog waste discharged from a Heartland Pork facility in Edgar County, killing fish in four miles of stream. In 1995, 22 million gallons of hog manure burst through a lagoon wall in North Carolina. The excrement swept across crop fields and a rural highway before spilling into the head waters of the New River. Fish were killed 20 miles downstream. Opponents say these kinds of manure spills have become fairly common in states with a lot of factory farms.

The whole point in bringing up this mega-farm controversy is that consumers are exposed to this negative information about farms and food daily. Many consumers are concerned about food safety and the environmental costs of the low-priced food in the supermarket. The mega- farms' bloody noses and black eyes could spill over to smaller farms. That's a real danger. This certainly is not the time for small farmers to jump on the let's-pour-gasoline-on-the-blaze bandwagon to punish the mega-farms. We could also be burned by that flame.

"The fact is," writes Joel Salatin in his book *Pastured Poultry Profit$*, "that many consumers want to exit conventional food channels, some of necessity, some of conviction, some of mistrust, and some simply because somewhere they've tasted clean food and found it memorable. Pastured poultry offers an alternative to all of these consumers."

Salatin says that the whole notion of confinement housing is flawed. He points out that the inherent fecal contamination in such buildings causes all sorts of health problems. "Feathers, eyes, beaks, nostrils--nothing is exempt for a layer of fecal dust and its pathogen-laden microorganisms," reasons Salatin.

"The birds breathe in the fecal dust, which contains a high percentage of ammonia, and this causes lesions of the respiratory linings, the fragile mucous membranes."

For these reasons, small farmers have an opportunity to market clean, humanely-produced livestock products to customers who are searching for alternate food sources. The animals with the most potential for this market are the ones that have the most vertical integration. This includes poultry for meat and eggs, hogs, beef and dairy.

Integration in the food system brings with it a demand for uniformity of products coming from our farms and ranches. Today a large percentage of poultry, eggs, beef and pork is produced more for uniform weight and shape than taste and nutrition. Isn't it time for you to participate in a dramatic growth in direct marketing of food by farmers? You can reap the profits, too.

APPPA BUSINESS MEMBER PROFILE

Pennsylvania Association for Sustainable Agriculture (PASA)
Winter 2004

PASA is a nonprofit membership-based organization with over 3,000 members, many of whom are active in shaping programs and policy. There are six full-time employees and a number of part-time workers. PASA education, policy, and market-based programs reach thousands more people in Pennsylvania and the mid-Atlantic region. PASA is unique in that it serves both consumer and farmers, and shows the inextricable link between the two. The PASA Mission is: Promoting profitable farms which produces healthy food for all people while respecting the natural environment.

Serving the Interests of Pasture-Based Farmers for Over 15 Years.

Although our educational programming, marketing services and policy work are limited in focus to the state of Pennsylvania, 20% of our members live out of state. This can be attributed, in part, to the fact that members in bordering states can easily take advantage of our programming.

PASA was started in 1992 by a group of people interested in forming an association of like-minded farmers and folks with an eye on the sustainability of the farm and farmer in Pennsylvania. We are an organization working to improve the economic and social prosperity of Pennsylvania, and work with the farmers that grow our food, the consumers that eat the food, and those concerned with the ecological well being of our environment and natural resources.

PASA creates networks and markets to strengthen the ties between concerned consumers and family farmers. PASA is building statewide channels that link farmers with farmers, farmers with consumers, and consumers with markets. As our organization has grown, we've had some real successes with a variety of educational programs, both on and off the farm, that are shaping new partnerships that enhance the lives and livelihoods of producers and consumers. Our "products" are actually services, primarily in the form of educational programming, market creation and growth, and policy work. Last year we featured a pastured poultry track at our conference, which was organized in cooperation with APPPA. In addition to conference workshops and farmer-to-farmer field days, staff works hard to create opportunities for marketing pastured poultry. We have established farmers markets and farm-to-chef networks which support the endeavors of pastured poultry farmers who are direct marketing to consumers.

The annual "Farming for the Future" conference is PASA's signature event and our main vehicle for community building. Widely regarded as the best of its kind in the East, this diverse event brings together an audience of over 1,400 farmers, processors, consumers, students, environmentalists, and business and community leaders annually. The sheer numbers and diversity of business and organizations that are associated with the conference are notable, through sponsoring, exhibiting and presenting. Special features of the conference include; youth & teen programming, a babysitting program, a Silent Auction, the Sustainable Trade Show and Marketplace, and conference meals featuring sustainably, organically, and regionally raised foods.

PASA Membership can be a meaningful experience for poultry producers and comes with a myriad of benefits, including 6 issues of our 24-page newsletter PASSAGES, discounts to most PASA educational events, including field days, the conference, and special speaker engagements, a directory of all PASA members, and access to regional member list-serves. Membership materials are available at www.pasafarming.org or by calling 814-349-9856.

As we look at like-minded organizations throughout the country and keep and eye on the national newsfront on food, farming, and food safety issues, we are encouraged to see more and more people actively seeking and building a safe, reliable and healthy food supply. It has been our pleasure in this past two years to begin to forge a stronger relationship with our friends at APPPA, and it was a great pleasure and privilege for PASA to host APPPA's annual meeting at our Farming for the Future Conference in 2004.

Contact Information: PASA PO Box 419, Millheim, PA 16854 814-349-9856 www.pasafarming.org

Why Grassfed is Best
By Jo Robinson   Spring 2000

Much of the excitement about grassfarming in recent years has focused on animal welfare, the environment and the economic well being of small family farms. Now the spotlight is shifting to its nutritional benefits for consumers. Compared with conventional products, meat and eggs from grassfed animals are lower in fat and calories but higher in key vitamins such as beta-carotene and Vitamin E. What's more, they are a rich source of "good" fats such as omega-3 fatty acids and conjugated linoleic acid (CLA)- nutrients which may reduce the risk of cardiovascular disease, cancer, diabetes, obesity and other debilitating conditions.

A Bounty of Good Fat

Although gassfed meat is low in "bad" fat (such as saturated fat), it has from two to six times more of a type of good fat called "omega-3 fatty acids." Meat and eggs from pastured poultry are good sources of omega-3s as well. An egg from a free-range hen can have up to 20 times more of this good fat than eggs from caged hens. Many people are surprised at the notion that some kinds of fat can be good for your health. They've gotten the idea that all fat is bad, and they should eat as little as possible. omega-3 fatty acids are not only good for you, they play a vital role in every cell and system in your body. For example, of all the fats, they are the most "heart friendly." People who have ample amounts of omega-3s in their diet are less likely to have high blood pressure or an irregular heartbeat. Remarkably, they are 50% less likely to have a serious heart attack. These good fats are essential to the function of the brain as well. People with a diet rich in omega-3s have a reduced risk of a number of mental disorders, including depression, schizophrenia, attention deficit disorder and Alzheimer's disease. omega-3s offer protection against cancer as well. In humans, omega-3s have slowed or even reversed the extreme weight loss that often accompanies advanced cancer and they have hastened the recovery from cancer surgery. Omega-3s are most abundant in seafood and certain nuts and seeds such as flaxseeds and walnuts, but they are also found in grassfed animal products. The reason that grassfed ruminants and pastured pigs and poultry are a good source of omega-3s is that these essential fats are formed in the green leaves (specifically the chloroplasts) of plants. Sixty percent of the fat content of grass is a type of omega-3 fatty acid called alpha-linoleic, or LNA. When cattle are taken off grass and shipped to a feed lot to be fattened on grain, they loose their valuable store of LNA. Each day that an animal spends in the feed lot its supply of omega-3s is diminished.

Most commercial poultry are raised in large indoor facilities housing 10,000 or more birds and fed a grain diet laced with antibiotics and appetite stimulants. This artificial diet makes their eggs and meat unnaturally low in omega-3 fatty acids, vitamin E, vitamin A and beta-carotene. The only difference between some types of "organic" chicken and regular chicken is that the birds have been fed organic grain and have not been dosed with drugs or chemicals. Unless they have also been allowed to forage for greens, they will be deficient in omega-3s and other key nutrients. For maximum nutritional benefits, chickens and turkeys need to be "pastured" poultry. The universal practice of raising animals in confinement is one of the reasons our modern diet is so low in omega-3 fatty acids. Only 40% of Americans consume a sufficient supply of these nutrients. Twenty percent have levels so low that they cannot be detected. Eating grassfed meat, eggs, poultry and dairy products is one way to help prevent this widespread deficiency.

Nutritional Analysis of Pastured Poultry Products
By Barb Gorski Winter 2000

A collaborative effort among four pastured poultry producers in Pennsylvania, Double G Farm, Forks Farm, Lone Pine Farm, and Stricker Family Farm, resulted in a successful grant application for nutritional testing of pastured chickens and eggs. Partial funding for the work reported here was provided by a grant from the USDA Sustainable Agriculture Research and Education (SARE) Program. American Westech, Inc. in Harrisburg, PA provided analytical testing and consulting for the project. Our goal was to obtain scientific evidence of the nutritional soundness of our product. This information can be used to promote our products--we always thought our birds were better, but we wanted proof.

Our project looked at three areas of nutritional analysis: standard nutrients, omega-3 and pmega-6 fatty acids, and conjugated linoleic acid (CLA). A table of the standard nutrients and fatty acids is included with this article. You will notice that CLA is not mentioned in the table because there was no CLA found in the chickens or the eggs. We then learned that CLA is produced by ruminants, which is why you hear so much good news about grassfed cattle, goats, etc. but there is little CLA in chickens.

I think the nutritional comparison table speaks for itself, but I will mention some of the more important findings. The pastured whole chickens were found to have significantly higher levels of Vitamin A than the standard; the same was not true of the skinless breast meat, as Vitamin A is a fat-soluble vitamin that will be found mainly in the skin. The Vitamin A in the pastured eggs was also greater than in the standard. Pastured whole chickens and eggs were also found to have less saturated and mono-unsaturated fat than the standard.

Speaking of fat, we all know that saturated and monounsaturated are the bad fats and polyunsaturated is the relatively good fat, but there's more to the story. Polyunsaturated fat is made up of omega-3 and omega-6 fatty acids. Americans tend to eat too much of the omega-6 vitamins and not enough of the omega-3 vitamins. In the book, *The Omega Diet* by Jo Robinson and Dr. Artemis Simopoulos, it is recommended that the ratio of omega-6 to omega-3 fatty acids should be four to one, yet the average American diet is in the range between 14 to 1 and 20 to 1. The pastured whole chickens and eggs had omega-6

Nutrient	Whole Chicken, meat and skin, raw amount per 100 g			Skinless Chicken Breast, raw amount per 100g			Eggs, Whole, raw amount per 100g		
	pastured poultry	standard reference	variance %	pastured poultry	standard reference	variance %	pastured poultry	standard reference	variance %
Calories	177.65	215.00	-17.37	108.74	110.00	-1.15	134.46	149.00	-9.76
Calories from fat	106.29	135.54	-21.58	15.30	11.16	37.10	79.74	90.18	-11.58
Total fat-gram	11.81	15.06	-21.58	1.70	1.24	37.10	8.86	10.02	-11.58
Saturated fat- g	3.04	4.31	-29.47	0.49	0.33	48.48	2.69	3.10	-13.23
Monounsaturated fat-g	4.34	6.24	-30.45	0.55	0.30	83.33	3.44	3.81	-9.71
Polyunsaturated fat-g	3.82	3.23	18.27	0.58	0.28	107.14	2.29	1.36	68.38
Omega-3 fatty acids-g	0.36	0.18	100.00	0.06	0.03	100.00	0.27	0.07	285.71
Omega-6 fatty acids-g	3.40	2.96	14.86	0.49	0.21	133.33	1.96	1.29	51.94
Ratio Omega 3: Omega 6	9:1	16:1		8:1	7:1		7:1	18:1	
Cholesterol- mg	68.00	75.00	-9.33	50.00	58.00	-13.79	280.00	425.00	-34.12
Sodium-mg	47.00	70.00	-32.86	34.00	65.00	-47.69	130.00	126.00	3.17
Total carbohydrate	0.00	0.00	0.00	0.00	0.00	0.00	1.39	1.22	13.93
Protein- gram	17.84	18.60	-4.09	23.36	23.09	1.17	12.29	12.49	-1.60
Vitamin A-IU	210.00	140.00	50.0	0.00	21.00	-100.00	1100.00	635.00	73.23
Vitamin C-mg	0.74	1.60	-53.75	1.04	1.20	-13.33	0.00	0.00	0.00
Calcium-mg	7.00	11.00	-36.36	4.00	11.00	-63.64	47.00	49.00	-4.08

to omega-3 ratios considerably better than the standard. I should add here that all of the chickens tested had been fed Joel Salatin's ration, which includes fishmeal. Fishmeal, seaweed and flax seeds are some of the highest sources of omega-3.

Five birds from our farm were tested for various forms of bacteria. The results of these left us both pleased and surprised. The bacterial tests for listeria, salmonella, and campylobacter, all very harmful, showed zero levels, which we were very pleased to see. Our surprise however, came with higher levels of E. coli. This left us perplexed and looking for answers as to why this occurred. We carefully hand process every bird, we don't rip intestines and we don't have fecal matter floating in our chill tank. After conversations with American Westech, two reasons may explain these higher levels. One, commercial birds are subjected to chlorine baths as well as tested immediately after slaughter, whereas our birds were tested two days after processing. Secondly, our chill tank water may not have gotten the birds' body temperature down to 40° F quickly enough to inhibit the bacteria growth. E. coli levels can double in 20 minutes at the bird's body temperature. At a temperature reading of 50° F, these same E. coli levels double in approximately forty minutes. You can see the advantage that rapid testing may have and the importance of cooling down the birds to at least 40° F as soon as possible. So, our first purchase of the new season is going to be a thermometer for our chill tank.

We are very thankful for all of these test results because we were able to learn a lot and we will use them to educate our customers. I hope that you have found this article to be valuable as well as a possible marketing tool.

How to Sell Grassfed Science
By Shannon Hayes Fall 2002

This article appeared in the November, 2002 issue of GRAZE, a nationally focused publication written "by graziers, for graziers". The article is reprinted here by permission. Graze, PO Box 48, Belleville, WI 53508.

I was seven and out playing in the front field one afternoon when a fancy car pulled in our driveway. A man got out, wearing a dark suit with his hair parted neatly and slicked down to his head--an anomaly in these parts. My Mom came out to see what he wanted.
"Hello. Is your husband home?"
"No, he's out back in the fields," she replied.
The man leaned in close to my mom's face. He popped his eyes open wide--so wide that I wondered if they were touching the lenses of his glasses. "And what if he doesn't make it back? What will happen then?" He leaned down to me. "And what is your name?"
"Shannon."
"What will Shannon's future be like? What will your future be like?" At this point, he launched into his sales pitch for a suspicious sounding life insurance policy. My mom hated him. I hated him. He never made the sale.

Every product needs a sales person, whether it is a fly-by-night life insurance plan, or wholesome food grown in harmony with Mother Nature. Many of us who sell grassfed meat and dairy products are excited about the health benefits offered by the high levels of conjugated linoleic acids and omega-3 fatty acids in our products. This new information adds to the science we already knew--that our meat is lower in fat and calories; that it is not full of antibiotics and hormones; that grazing our animals aids in carbon sequestration, increases biodiversity and restores soil health.

But how does an honest grass farmer avoid sounding like a slick insurance salesman firing off too many complex details and too many reasons to buy? When we talk to potential customers face to face, we rarely have more than 30 seconds to give then a convincing reason to buy our products. Thirty seconds is not enough time to teach the necessary biochemistry and environmental biology required to fully understand the benefits of our meats.

If improperly communicated with too many details, this type of information may cause the customer to start fantasizing about the express line at Shop and Save. Properly communicated, this information can change occasional drop-in buyers into committed customers who want to buy all of their food from your farm.

Marketing is about linking what you have to sell with what your customer wants. If he or she is interested in health issues, then you can spend your 30 seconds talking about CLAs. But if your customer is interested in gourmet foods or food safety, talk about how great your meat tastes, or the security they'll feel in knowing the source of their food. Don't waste the short amount of face-to-face time you have talking about subjects that do not interest your customer.

Still, over the long run, the science is a potentially powerful tool in marketing your grassfed products. While you may only have 30 seconds to speak to someone about your products, you still have plenty of time to educate them. Customer education—teaching the biochemistry, environmental biology, social justice and economic principles that set your products apart from anything that can be found in the grocery store or through a glitzy mail order catalog—is an important part of the marketing process. You need to spend some time on this.

The first step is educating yourself. Read and listen. Pay attention to what's going on. Articles on grassfed farming and products that appear in newspapers, magazines, television and radio help you to learn what the mainstream is thinking, as well as the latest developments within the industry. It is also good to know what kind of information your customers might be discovering.

There is so much information flying around about grassfed livestock and small scale farming that it is difficult to keep up with it all. There are a few shortcuts. Some of the major newspapers, like the New York Times, have free online subscription services with "news trackers," which are search functions that use key words to send you specific articles that may be of interest. My news trackers are set up to send any articles that contain the keywords "meat" or "livestock." List serves such as graze-l are also helpful. Frequently, when a relevant story comes out, short synopses can be read over your e-mail. Another great shortcut lies in industry publications and newsletters from sustainable agriculture organizations.

Now that you have educated yourself, your job is to help your customers read and listen. Collect your customers' contact information and send out a farm newsletter once or twice a year. Keep an e-mail distribution list and forward any articles that might be of interest to them. Maintain a lending library of essential literature, such as Jo Robinson's *Why Grassfed is Best,* Sally Fallon's *Nourishing Traditions,* and Eric Schlosser's *Fast Food Nation.* Generously share them with customers. When you meet people out in public, be able to refer them to the latest writing, rather than using your 30 seconds to explain years of research. At your point of sale, post articles that have appeared in the mainstream press on a bulletin board, give away photocopies, or keep a scrapbook of news clippings that people can peruse while they visit your farm, roadstand or market stall. Again, you only have a few seconds to talk face to face, so let some other reputable sources do the bulk of the education for you.

Sales require persuasion, and effective persuasion is more about what you do than what you say. If you are going to do an effective job of informing your customers about the health and environmental benefits of your meats, you have to practice what you preach. Maintain your own health. Be conscientious with your grazing practices. Your newly informed customers will notice, and they will become fiercely loyal.

Probably most of your new customers will come to you with little specific knowledge of CLAs, Omega-3 fatty acids and many of the other scientific findings that excite us. It's your job to inform and educate over a period of time.

I recently spoke with one of our customers, someone who originally came to the farm in search of gourmet food. She said, though, that she kept learning more from us about health implications, social benefits and environmental issues with each successive visit. She said that, bit by bit, through brief conversations with our family and through articles and books we'd shared with her, she and her husband completely changed their diets. What was once an occasional visit for a gourmet turkey or chicken grew into a serious commitment to allowing us to provide most of the meat they eat.

Ultimately, marketing grassfed products with their numerous selling points is as challenging as selling a scam life insurance policy. The few seconds we get to speak to our customers is not enough time to fully educate them about our products. However, by engaging in the continual process of keeping them informed, we can be educating and marketing quietly, effectively and with ease.

CHAPTER TWELVE
Record-Keeping and Insurance

Many people say they just want to farm, and don't like the part about keeping records. However, if you wish to be successful in any size commercial farm business you must keep at least minimal records so that you can tell if you are balancing costs and expenses with room for profit. Our authors in this chapter offer insight on record-keeping systems that have worked for them.

Setting up insurance for your pastured poultry operation can be a real challenge. We offer several articles with cautions and suggestions for protecting your farm and business assets.

Tracking and Using Information About Your Farm
By Diane Kaufmann Winter 2000

How do farmers collect information about their farm? How do they keep information? How do they figure out what to do with it once they have it and make it useful for decision-making? What helps farmers do their own research?

In an information-rich society farmers have access to multiple tools to help answer those questions in addition to their own unique talents and gifts. I don't know whether to admire or hate people who can tell me that their grain feed intake was 30% less with pasture and rate of gain was 100 percent more! How do they do that?!

Whether you're number-challenged or your mind dances delightfully with numbers, the key to understanding your farm is observation. Every farmer I spoke to about how they collect and use information on their farm mentioned observation first. Writing those observations in a notebook was an essential next step for these farmers in developing usable records and accurate information that led to good farming decisions.

Tom Delehanty, an organic pastured poultry producer in Socorro, New Mexico, used to observe and track every detail of his poultry operation in notebooks until everything was working the way he wanted it. Now, keen observation tells him if he's still on track or if something needs to be tweaked. Laura Rogers, from Woodbine, Kentucky, custom hatches eggs for her poultry operation and for area farmers. In the beginning of learning the craft of hatching, a notebook of her observations helped her track successes and failures. That discipline and attention to detail helped develop her experience level to the point where she can quickly spot problems before they occur.

While observational skills appear to be the primary tool in the information toolbox, farmers have access to many other tools as well. Nita Walton grazes beef and poultry in the Columbia Gorge area of Oregon. When she decided a few years ago to begin actively managing her pastures for intensive grazing, she turned to satellite maps of her property and her years of observation and experience of living on that land to help her develop paddocks that complemented the geography of the land under her stewardship. When pastured poultry was added to the operation, a columnar pad and calculator came out of the toolbox. Each batch of chicks is tracked from the hatchery to the freezer. Information like death loss, cause of death, feed costs, weight gain of the chicks, which customers bought chicks from that batch, and even what the weather patterns were like during the production period of that batch is easily accessible to Nita from one sheet of paper. She is developing invaluable knowledge that is indigenous to her land. A limitation she notes of her seasonal grazing operations is that it takes longer to see impacts and trends than if she were able to produce year-round, an advantage that Tom Delehanty can enjoy in the more moderate climate of New Mexico.

Dan Bennett of Ottawa, Kansas, pulls a computer out of his toolbox. While he's still building the basic physical and computer systems that will give him access to the production information he wants to see about his farm, he's currently using Quickbooks, an off-the-shelf financial software program on his laptop computer to help him develop enterprise analyses of his beef and poultry production systems. But he quickly points out that financial analysis is only one part of the equation in analyzing his farming practices and what he ultimately produces. "Does my labor pool like it?" and "Does my customer like it?" are driving questions in Dan's mind as he continues to develop his family farming operation.

If observation is the primary tool in the information toolbox, the key to unlocking the toolbox is time. Joel Salatin, also a poultry and beef grazier in Swoope, Virginia, believes that the daily "have-to-do" chore time on a farm shouldn't exceed four hours a day. When the "have-to-do" chores consume more than four hours, the result is farmer burnout and the loss of creative, thinking time. It's the time spent in observing and letting those observations meander around in the mind that eventually leads to creative problem solving, new ideas, and the development of indigenous knowledge about a farm.

It's perhaps that indigenous knowledge, an intimate acquaintance with a physical area, that can make farmer-research so valuable. While rudimentary observational skills and enough time may eventually ingrain a farmer with an intuitive knowledge of the land and how to best work with its natural cycles, a deliberate and thoughtful use of simple tools in an "information toolbox" will shorten that time frame and yield an abundant harvest of meaningful information.

Successful Poultry Production Begins With Record-Keeping
By Aaron Silverman Summer 2000

By now the growing season is in full swing—chicks have arrived, batches are growing and processed, and the days are packed with endless lists of tasks. We are in the portion of the farming cycle that consists mainly of action, rather than contemplation and planning.

We raise about 2,000 pastured broilers each year, processed every other week. These are all sold with an advance $2 per bird deposit. In addition, we raise five-acres of produce, herbs, and cut flowers for restaurants and a small subscription program. All this is done with one and one half farmers, one full time on-farm employee, and two or three part time helpers. Efficiency is essential, as time and energy are at a premium.

We've all heard that it's wise, even necessary, to operate with a plan. Any plan is merely a road map, a set of landmarks to help keep things on track during the times of pure action. Rarely do things go strictly according to plan. Things go wrong, we surpass our expectations, and new variables pop up; but a plan is still a helpful tool to have.

The problem is always in coming up with a plan in the first place. Pulling numbers out of the air can be a starting place, but the best plan for your farm is occurring right now. By spending a few minutes recording what you are currently doing in your production system, you establish a rational base for future planning.

This year (2000)
¤ How many birds were sold?
¤ Average weight?
¤ How much feed was fed?
¤ How many birds were lost?
¤ To what?
¤ How many birds were culled during processing?

Next year (2001)
¤ How many chicks to order?
¤ How much feed to purchase?
¤ How many birds to sell?

Records established from this year's production serve as guidelines for deciding next year's. For instance, by looking at loss and culling records, we may see that we should order 20% more chicks than we plan to sell.

We divide our information gathering into two distinct realms: Production Records and Customer Records. Production Records consist mostly of the type of questions above, and consist of relatively simple data. Customer Records, on the other hand, entail much more complicated information: customer personal information, how much individuals purchase, when they purchase, deposits paid, future orders, etc. We track customer records directly on the computer; production data is collected on simple sheets in the field and barn then entered into the computer later.

The Computer

For us the computer is as essential a tool on our farm as the cordless drill or duct tape. The exact type of computer and the programs used are not as important as being efficient in using the software you have. We use three types of programs to track all information on our farm: a spreadsheet, an accounting program, and a relational database. Single programs exist that perform all the functions of these three, but the more functions a program performs, the more complex the program. Base your choice of software on your personal comfort as the user as much as the abilities of the software.

The Spreadsheet: Microsoft Excel

Spreadsheets are the computer version of the old green ledger book. It is a simple, yet powerful way of recording and analyzing data. Data is entered in columns and rows, and calculations are easily and automatically performed. Excel is used to track loss records, feed consumption, and processed weights. The enclosed graph is an example of the usefulness of tracking data in Excel. We have always noticed a shift in weight averages throughout the season, and wondered if there were some pattern to them. By charting out weight averages with Excel, we immediately noticed a common trend in weights, regardless of breed produced. We can now have a better handle on our cash flow, and direct customers preferring larger birds to those batches that normally have heavier birds.

The Accounting Program: Quickbooks Pro

Quicken, Quickbooks and Quickbooks Pro, all written by Intuit, are common accounting programs available for both the Mac and PC. Although they each have varying levels of complexity, they are designed to track financial data: income, expenses, accounts expendable, accounts receivable, etc. Quickbooks provides the background for all our data collection: how much money came in versus how much money went out. Both income and expenditures can be tracked with great detail if desired. Beyond questions of gross profitability, Quickbooks helps keep track of customer information, especially deposits paid by customers, and the deposit balance. At the end of the season, we may find a customer with outstanding deposits due to our error—great customer goodwill in getting a call notifying them of a 10 bird deposit towards next year's order.

The Relational Database: Filemaker

Filemaker, by Aldus, and its PC counterpart, Access, by Microsoft, are complex programs that allow a vast amount of related information to be managed. A single customer has contact information (address, phone, etc.), as well as their order for the season. Each processing batch has a group of customers that make up the orders for that date.

The relational database allows us to create a report for each processing batch that contains all the customers and their information, or a report for an individual customer. It is the foundation of all our Customer Records, and allows us to keep this storehouse of data organized and accessible. The downside to relational database program is the large time investment required to set the program up. However, the ease in which data can be managed makes this investment well worth it.

Production Records

This is the easy part to record, but the hardest to actually do. In the constant rush from task to task it is tough to take the couple seconds needed to mark down data on the various charts used. The key to keeping field records is finding a format that is EASY. If it requires thinking, it probably won't get done.

Feed records are kept on a clipboard by our feed. A page of paper split into columns is all that is needed: feed in, feed out to brooder, field, etc. We record all feed usage in buckets (5-gallons= 30 lbs.) when the feed leaves the barn. At the end of the season the feed log is entered into the spreadsheet. General feed consumption ratios derive from this data. More importantly, the pattern of feed usage is clearly visible, allowing for more accurate estimating of next season's feed requirements -a ton of feed in May may go much farther than a ton in August.

A small notebook is kept in our brooder house. We record temperatures, activities, and observations of the chicks here, as well as field observations and losses, and transfer of birds from brooder to field, or among field pens. Notation is simple: losses are described in the lined portion of the page, and noted at the top right corner allowing for quick counting. For example: (5) + 190 = 190 in for Batch 5; (3) -2, (2) -1 = lost 2 in Batch 3 and 1 in Batch 2.

We also keep a clipboard for recording processing information. Again, a blank sheet of paper with columns to record the processing date, number of birds gathered for processing, number of birds culled for wings, legs, weight, etc... number given to labor, and number going into our freezer.

Customer Records

All our customer records are kept in Filemaker, in four basic layouts. One layout is for entering their personal information (address, phone, source, etc.), as well as summaries of their current and past orders. This is very useful when paring down your mailing list. We generally keep people on the list for two years if they do not place orders; this year we received orders from four customers who hadn't ordered in two seasons, and several more that skipped last year.

The customer order form allows us to track an individual's order, their deposits, and all other information related to their order. We give out free birds as incentive for folks to send their orders in early, and these are tracked here, as well as interest in other products.

From this format, two other reports are easily created. The first is the general Sales Report, which tracks target sales orders to date, and room left in each batch. This allows us to immediately ascertain whether we can continue to take orders for a certain date, or if those orders should be shifted to later batches.

The second report is the Customer Pickup form. This has proved invaluable for tracking sales and payments. This report is basically a listing of all the customers with orders for that batch. Their order, deposit, free birds, and notes are automatically filled in by the program. Blanks for quantity purchased, weight, total price, and payment method are filled in by the person conducting the sale. The data from this form is entered into Quickbooks to track income, and Excel to track weights. Future orders placed and deposits taken are noted on this form as well, helping to ensure that they will be remembered. Little is as embarrassing as a customer showing up for chickens they've paid deposit on, and having no record and no chickens for them.

The exact format of your record-keeping, the software you may use (if you use any), is not as important as simply keeping records. Everything can be recorded simply on forms you work up by hand. The computer's main function is assisting in the analysis and organization of the information kept. Record-keeping is a habit worth establishing, and even data kept for part of the season will be of great assistance this winter when trying to determine the success of your operation, and planning for next year's chickens.

Sample Reports Used by Aaron Silverman

Batch 4 report 7/29						
Silverman, Kelly		ph 935-7954				
# ordered	# bought	pounds	gross price	less deposit		Balance paid
10	10	51.23		20.00		

Batch	1	2	3	4	5	6	7
Chicks Ordered	150	180	180	180	180	180	180
Target sales	120	153	140	153	153	153	153
Orders to date	97	167	135	120	84	134	85
Room left	23	-14	5	33	69	19	68
Total sales to date: 822							

Average Chicken Weights						
	Batch 1	Batch 2	Batch 3	Batch 4	total	
1997	6/19	7/03	7/17	8/14		
Avg lbs	3.38	3.81	4.43	4.49	4.03	
# sold	63	50	53	31	197	
notes	90% Kosher King cockerels, rest various cornish x's; 21% protein feed; brooded 2 wks					
1998	6/18	7/02	7/16	7/30		
Avg lbs	4.76	4.85	4.67	4.54	4.71	
# sold	95	103	104	70	372	
notes	8:5 HY-Y strt run: Kosher KIng cockerels; 21% protein ration; Nutribalancer/ no bonemeal; 8 wks, 2 brooder, 6 pasture					

Record-Keeping
By Dan Bennett Winter 2003

I've heard it said that you can't manage something you're not keeping track of. In other words, record-keeping is a vital link to the success of any business. Good decisions are always based on good information! Farming is no exception to this rule. Those of us in alternative agriculture who have the added complexity of direct marketing, combined with our many production and financial challenges must particularly focus a great deal of attention on record-keeping.

When the subject of record-keeping comes up, many people think of only the accounting and bookkeeping functions of a business. I would argue that every major phase of our businesses has a record-keeping need. In this article, I will attempt to cover the topic of record-keeping across all the major functions of a farm business including marketing, production, and finance. I'll also talk a little about the common problem every business, small and large, experiences after we begin to collect information about the different areas of our businesses: the difficult job of integration--putting data together into useful pieces of information. In addition, I will give examples of the record-keeping techniques and tools I use on my farm to help bring a little practical application to this potentially boring topic.

Marketing
The critical information needed by every good marketer is about customers. The more information we can collect and manage about our customers, the more successful we can be. At the Bennett Ranch, we collect this information in two places. The first customer database we keep is in Microsoft Outlook. This database allows us to store vital demographic information about our customers—name, address, phone, email etc. Computer tools like Outlook or another popular one called ACT, are generally categorized in software lingo as "Contact Managers." With a good Contact Manager, you can keep track of any conceivable detail about your customers you would like. These tools even allow you to electronically file all customer related correspondence for easy retrieval. The second place we store customer information is in Quickbooks, which is our financial system. Here we keep track of all customer buying information–most notably, what they buy, how much they buy, and how well they pay. The great thing about Quickbooks and either Outlook or ACT is that they integrate together to keep your customer databases in sync.

Production Record-Keeping
At the Bennett Ranch, we primarily collect our production information manually using a table-type form or chart and then enter the data into a spreadsheet program. We use Microsoft Excel as the computer tool to collect and interpret the information. Our primary production information collected is listed in the chart below:

Eggs	Broilers	Turkeys	Beef
Daily Egg Count	Daily Death Loss by Batch	Daily Death Loss	Weights
Daily Death Loss	Weekly Orders	Orders	Slaughter Weights and Count
Weekly Orders	Slaughter Weights and Count	Slaughter Weights and Count	

With this information we manage the flow of product to customers, estimate future production capability, and determine production efficiencies. In our operation, production numbers don't integrate electronically with our marketing and finance functions. They obviously relate in many practical ways, which means that we integrate this information manually and then make decisions accordingly. Most of our decisions made with our production record keeping are essential to the operation and are not something we do only if we have time.

For example, we must compare our production information with our orders to know how much additional marketing we need to do before processing. We must know if we have enough product this week to meet our customer commitments. We must also know slaughter weights in order to create invoices.

Production information can also be used after the fact to look for trends in production techniques and analyze the effect of long term improvements. Either way, production information is critical to the successful operation of every farm and a way must be found to collect it.

Financial Record-Keeping

At the Bennett Ranch, we utilize a software product created by Intuit called Quickbooks. Many of you have heard of and perhaps used its sister product, Quicken. Quickbooks is the business version of this popular personal financial management software. Quickbooks is a fairly comprehensive tool for small businesses and works well for a small farm or ranch. Like Quicken, Quickbooks has the capability to automate your checkbook and tremendously eases the effort to keep it balanced. Its capability goes way beyond this, however. With it, you can create invoices and manage your accounts receivables. You can keep track of all your bills and better manage your cash flow. Other highlights include sales tax payments, full financial reporting, and sales tracking by product. Quickbooks even has complete payroll capability. We use most of the capability of Quickbooks, with the exception of payroll.

There are obviously many benefits to taking the time to gather financial information. First and foremost, it helps to determine the difference between making money and losing money. Additionally, we've realized from this effort that our accountant has a very easy time in preparing tax returns. We simply email our Quickbooks file to her and she does the rest.

Like any computer program, Quickbooks takes time to understand and begin to derive benefit from it. With some time and effort, most of us can master this tool. There are other software options available for you to choose from. We chose QuickBooks because it appears to be the most popular small business accounting tool available. This has afforded us the benefit of finding help when we need it and working with a company that is committed to continually improving the product. We have used QuickBooks since '98 and are currently using the 2003 version.

Once all this record-keeping is complete and accurate, then the hard work starts–interpreting the information. This many times requires the manual integration of information prior to attempting to make wise decisions. The interpretation phase is the tough part because it requires time and energy. Often times, we do the work associated with collecting the information then run out of time and energy to interpret the information properly. This is the key to good decision making. We must have the discipline to complete the job – I'm talking to myself as much as I'm talking to you!

Many times, a team approach is helpful in the interpretation of the business information. A family member or business partner can be very helpful in this process. We have found it very effective to combine a team approach with an established time for interpreting our business information and creating a plan for the coming year. In January each year we set aside a few days to brainstorm and plan for the new year. This forces us to prepare the necessary year end information collected from last season to begin the plan for the new year. This planning effort also helps us to establish specific action items for data interpretation and decision making that need to be done prior to the new season starting in the spring.

That's kind of the way it is with record-keeping. None of us really like to do it but the benefit you derive from the effort far outweighs the effort expended. We must remember that once the job of record-keeping is done, the job of more effectively managing our businesses begins. Good luck and much 2004 success!

APPPA PRODUCER PROFILE

Dan Bennett

By Aaron Silverman Spring 2002

Dan Bennett farms 210 acres of pasture in Ottawa, Kansas, 30 miles from Lawrence, and 50 miles from Kansas City and Topeka. The farm is entering its fifth full season with plans of producing 15,000 broilers, 2,000 layers, 250 turkeys, along with grass fed pigs and cows. The whole family is involved in the farm, and the farm is their main source of income. Bennett Ranch emphasizes value-added products, and direct-marketing to restaurants, stores, and individuals.

Broiler Production System
Dan raises 600 broilers per week during a 27-week season March through October. Typical harvest age is eight weeks, and the Bennetts limit feed for the last two weeks to manage for desired weights. The average death loss grew last year from around 10% the three previous years, to a little over 15% in 2001. Dan attributes some of this increase to expanding production faster than management skills. He notes that half of the losses were in the brooder, so he plans on replacing the bedding between each batch in an effort to bring brooder loss back to the 3% he expects. Half the field losses were attributable to weather; the other half were due to Dan's experimentation with feathenet/ day ranging approaches. Dan feels that the weather in Kansas is just too variable for these approaches to be successful in the manner he used.

The Bennetts use 12 x 10 ft. PVC pens designed by Brower. Stocking density is 80 to 100 birds per pen. Pens are moved daily, and use Plasson gravity waterers and large Brower PVC feeders. Dan plans on transitioning the water system to cup waterers by GQF, with one gravity tank per six pens attached to a pasture water line. After his unsuccessful experimentation with day-ranging, Dan is planning on producing exclusively in the Brower-designed pens, 30 for the 2002 season. After three years of grazing chickens on native prairie, the Bennett pasture is in transition to a cool season fescue pasture with no irrigation.

Feed
The Bennetts use the "standard" Fertrell broiler ration, composed of locally produced grains and mixed at the local coop. The feed costs him about $250/ton, delivered.

Brooder Management
Nine brooders are used, each approximately 10 x 10 ft., and containing 300 chicks. Electric heat lamps and hovers are used. Chicks are kept in the brooder for three weeks.

Processing
Bennett Ranch has all their birds processed at a local State licensed processor. Cost of processing is an average of $1.35 per bird; actual cost depends on whether birds are whole, parted, and packed. Very little product is lost due to parting, and everything that is processed is eventually sold. Some birds are frozen for winter sales.

Labor
Most labor on the farm is provided by the family. Dan and his wife Jenny manage the farm full-time, along with their daughter Jamen, sons Max and John, and nephew Justin. Part-time help is hired for collecting and processing eggs. Dan did a labor analysis last year which determined that the broiler operation used 1,570 labor hours per year. Spread over 30 weeks of production, this averages 52 labor hours per week.

APPPA PRODUCER PROFILE — *Bennett*

This labor figure covered the following:
- picking chicks up at post office
- brooder preparation
- brooder management
- transfer of chicks to pasture
- field management
- gathering chickens for processing
- processing: final packaging & marketing
- delivery
- building projects
- repair and maintenance

Feathernet and Hens at Bennett Ranch

Marketing
About 70% of Dan's sales are through direct marketing to restaurants and grocery stores. Of this volume, a little less than half goes through a distributor to a large grocery chain. Plastic totes with lids are used to deliver whole chickens to restaurants. Some of his customers will then cut large quantities of birds to sell as pieces.

The other 30% of sales are through on-farm sales. Most customers are generated through word of mouth. Bennett Ranch has a web site, www.bennettranch.com, and a newsletter published twice a year. For 2002, the Bennetts are trying a more aggressive buying club strategy to target customers from farther distances who resist driving to the farm.

Record-Keeping
Dan uses a computer program called Quickbooks and spreadsheets to track his farm's performance. Spreadsheets are used to track losses throughout the life cycle by week, weekly customer orders, frozen product inventory, and daily egg production. Sales information for products and customers are tracked through Quickbooks, as well as the other financial bookkeeping. A database is used to compile customer information for mailings.

The Bottom Line
Bennett Ranch is projecting approximately $200,000 in gross sales for 2002. The Bennetts have a mortgage on their land, which is both their largest asset and primary liability. Labor is almost entirely supplied by the family, but monthly land payments are substantial. Dan's goal is for the whole farm to show a 25% profit margin.

Unfair Advantages
The most substantial advantage Dan has is the availability of a processing facility in close proximity. Another advantage in marketing is his location, close to both the college town of Lawrence, and metropolitan Kansas City. Dan's background in marketing gives him an advantage in forming relationships with distributors, store purchasers, and restaurant chefs. Being in Kansas, feed costs are much less than other parts of the country.

Ed note: In 2004 Dan Bennett and family sold their ranch to another family, who has continued with many of the same systems and markets.

Profitability- Will it Make Money?
By Anne Fanatico and David Redhage, Winter 2004

Summarized from the ATTRA Publication "Growing Your Range Poultry Business: An Entrepreneur's Toolbox" By Anne Fanatico and David Redhage. To obtain a free copy of the entire 63 page publication, go to www.ncat/.attra.org/ or call ATTRA at 800-346-9140

To project whether your enterprise will be profitable you should complete an income statement. An income statement lists income and expenses for a given time period, usually a year. The income statement lists all business receipts (cash and non-cash payments received from the sale of goods and services or other sources) and expenses (operating expenses and depreciation) related to the year's production. Expenses are then subtracted from receipts, and the amount remaining is net farm income. Net farm income represents the return to the operator's labor and management time, unpaid family labor and equity capital.

Estimating Expenses
There are two types of expenses:
¤ Capital costs: relatively major purchases that are made infrequently. Examples include land, processing equipment, a building for processing, poultry housing etc.
¤ Operating costs: recurring expenses that are a regular part of the production cycle. Examples include feed, chicks, utilities and interest and principle payment on debt for capital costs.

High capital purchases expose you to greater financial risk, while high operating costs can put you at a competitive disadvantage.

Budgets are used to estimate what your income statement will look like in the future. Attached here is a sample budget with some numbers filled in. The numbers were generated by a survey done of several producers. You can use them as guidelines but will need to fill in your own number to get an accurate picture of your own operation. As you fill in the budget there are several things to think about:

Estimating Costs:
¤ Feed prices affect profitability a lot since feed is the major cost in pastured poultry production. Organic feed is especially expensive.
¤ Mobile housing has a higher annual depreciation cost per bird because it does not last as long as permanent housing.
¤ Marketing costs depend on the market. A common rule of thumb is that 3% to 4% of total income will be spent on marketing costs.

Estimating Income
Most direct marketers charge $1.50-$2.50/lb for whole dressed broilers. Larger marketers have more complex pricing and organic birds sell for more.

Sensitivity Analysis
It is important to ask "what if" questions when making a budget. Proponents of unique enterprises tend to be too optimistic about potential income. You need to ask: What if prices are 25% below my estimates? What if I have a weather problem? What if it takes twice as much labor as I think it will? What if a lot of chicks die? Since you know the future is uncertain, you may want to examine different possible price and yield scenarios and see how your strategies perform.

A sensitivity analysis is used to determine how changes in various assumptions change the costs of production and processing, which in turn will affect the profitability of an enterprise. Make allowances for worst-case scenarios. After your sensitivity analysis is complete, you may want to make adjustments to your budget to reflect the effects of such changes. Important production costs and determinants of profitability are:
¤ Feed cost: feed consumption is directly related to the quality of the rearing environment, including housing, insulation and time spent on range in cold weather.
¤ Finishing age and the price per bird: if you plan to raise a slower growing bird (such as a Label Rouge type bird) it will cost more to get it to the same live weight as a fast-growing Cornish Cross. Although many direct markets will not sustain the higher price needed, other markets may.
¤ Operation size and cost efficiencies: profit generally increases with scale since larger units can spread out overhead costs. Input costs such as feed and chicks,

transport, processing, packaging and marketing also tend to be higher for small producers because of the small quantities involved. Producers try to lower costs by buying in bulk and using family labor. However, small units may not be profitable unless the poultry is processed on farm and direct marketed. Smaller units require less capital investment, but housing and labor costs per bird are generally higher than in the case of larger units where economies of scale may be significant. Larger scale production is possible through capital investment, automation, and collaboration between producers, particularly in the area of packing, processing and slaughtering.

Break-even Analysis
A break-even analysis determines the profits possible at various levels of output. Break-even calculations will show the level of production where the enterprise will cover operating costs for different output prices, wages and costs of raw product.

How Much Profitability Do You Need?
Even if you have figured out how to enter all your numbers into a budget, it may not mean that your operation is profitable. Profitability measures include net returns per bird, net returns to labor and management and dollars earned per hour of labor. A 17-25% margin is often needed to cover fixed costs. The level of profitability you need is related to whether or not you include your labor as an expense. If you do, then 0% profitability could be acceptable—the enterprise breaks even and you have made a job for yourself. The level of profitability is also related to your standard of living. Some families draw a lot more on an enterprise for their living expenses than other families would and therefore require a higher profit.

Financial Reality—Can You Afford To Do It?
One of the most critical items for a small business is having enough cash to meet needs throughout the year. Even a profitable enterprise can be sunk by cash-flow problems. You need to know how much cash will be needed for day-to-day expenses (operating costs, family living needs and debt payments) and where the cash will come from (customer sales, loans, membership equity, etc.) If sufficient cash is not available, cash-flow analysis will tell you the amount of debt you can afford.

A cash-flow analysis is a summary of the amount of cash that flows into and out of the business over a given period, generally one year. Monthly or quarterly cash-flow statements showing the timing of cash-flow are especially critical in an agricultural business, which is seasonal in nature. Winter bills must be balanced out with summer revenue.

By completing a cash-flow statement you can determine the amount of capital needed to finance the business, as well as the repayment ability of the business if money is borrowed.

Pastured Poultry Production with On-farm processing SAMPLE BUDGET

Income	# birds	Lbs/bird	$ per lb	Total
Sell Birds	999	4.5	2.00	8991.00
Expenses				
			sample	
Fixed				
1. Brooder facility			$320.00	
2. Processing structure			320.00	
3. Processing equip			157.86	
4. Pens			160.00	
5. Brooder equip			27.86	
6. Watering system			25.00	
7. Feeding equip			25.00	
8. Misc			20.00	
Total Fixed Expenses:				1055.72
Variable:				
9. chicks			684.00	
10. bedding			150.00	
11. feed			2520.00	
12. Utilities			20.00	
13. Bags & labels			79.92	
14. Marketing			400.00	
15. labor (production)			1584.00	
16. Labor (processing)			1152.00	
17. Liability Insurance			250.00	
18. Pasture rent/acre			30.00	
19. Miscellaneous			400.00	
Total Variable Exp				7269.92
Total Expenses				8325.64
Inc-Exp (Net Income)				665.36
	Cost per bird (breakeven)			$8.33
	Net income per bird			$0.67

You should develop a current cash-flow statement and a series of projected cash-flow statements for the first and second years, and for a future average year. Note that you need to know your monthly payments on a loan for this exercise.

Most businesses do not turn a profit for the first few years. A new enterprise with a payoff in five years may look good strictly from a profitability standpoint but may not pay the bills between now and then.

Obtaining Financing

You may need to borrow money to start or expand your range poultry enterprise. Lenders want to understand your business idea and see what evidence you can offer to show that there is a market and that it is likely to turn a profit. They also need to know the amount you want to borrow and if you can repay the money. Financial institutions look at the "5 C's" (character, cash flow, collateral, conditions and capital). The owner should be putting up 30-50% of the capital needed. It can be hard to identify sources of financing for nontraditional enterprises. Besides borrowing, you can pursue investors, business angels or venture capital.

What Will I Do Next Year?
By Kip Glass Winter 2004

Our last chickens were processed four weeks ago, turkeys were done the week before Thanksgiving. If you had a productive year, you're thankful that things went so well. Now it's time to rest.

But not so fast! As the year went along, you probably at some point said to yourself, "next year I'll do this differently." Perhaps customers told you they would love it if you had smaller birds, or maybe even larger birds to offer. You stumbled over a water pipe once too often, or you decided you really should sit down with a calculator and figure out how much feed each batch is wolfing down.

The point is, now is the time to plan. While all these ideas and suggestions are fresh in your head, get them down on paper, crunch the numbers, and see what can work better for you.

What areas am I looking at for next year? Most are related to labor, weather, marketing, areas for cost cutting, possible new enterprises, and capital expenditures.

Each producer/farmer's situation is going to be different, each person's way of doing things is different, and thus their costs will be different. Let my suggestions be just that, suggestions. Let this spur your own thought processes into high gear, and use or adjust my ideas to your situation.

Labor
Let's start with labor, which in my situation is the most difficult or most important to address. We have no kids, I work a full time job, and my wife (who is a small person) can't do any serious manual labor. I am always looking for ways to make things easier and work for themselves, so called "shortcuts," without sacrificing productivity.

An example of what we did for the 2004 season to eliminate a lot of labor was to mount a four-ton bulk feed bin on a wagon chassis to roll along near the field pens so we could fill buckets as needed to feed the chickens. Before this, every day, we loaded buckets up in the truck and hauled them out to the field. This bin saved us many hours of labor and work over the year, and was well worth the $900 cost of buying the used bin and securing it to a used four-wheel wagon chassis.

Ask yourself what areas of your production currently require an extra few minutes of labor. Is there something that you totally despise doing that could be made more enjoyable? Think of ways to make that part of your production easier. Or better yet, go on the APPPA pro plus email list and ask questions of all the producers that participate. I have been amazed at all the wonderful suggestions that people have come up with that I would never have thought of.

Weather
You should probably think about weather, even if you haven't experienced any ill effects from it. An example: in the past few years we have had no severe weather problems growing our turkeys out to the

week before Thanksgiving. In fact, we've had milder than normal weather during this time period the last couple of years. I was out one chilly morning this year and realized that my water system was not adequate if we had an extended cold period that kept the waterers frozen. I am going to change that part of the system for next year. What about extreme heat? Or extreme drought? How are you going to handle these potential problems in your situation?

Marketing

Marketing improvements can cover lots of things. Changes can range from ideas your customers have told you, to areas that you have seen that you could increase your income or even better ways to manage your customer lists.

We have all heard of "value-adding." My article "Give Them What They Want" (page 187) on cutting up chicken and selling the individual parts is one suggestion. Start small next year, offer a value added product, and keep detailed records of the costs and labor associated with that enterprise. See if it fits your operation. How about different markets- can you go to a farmers market? One thing we did this year was offer smaller sized chicken to our customers. Most of the smaller families or singles want a smaller portion. We charged extra for the smaller birds because the fixed costs are the same whether it is a small or larger bird. But, we gave those people what they wanted, which made them dedicated customers.

Marketing is an art all to itself. While many of us work hard to improve our skills at producing a good product, I feel we often let one of the most important parts of the equation, marketing, go by the wayside. I recommend that you read books, and talk to successful people in the marketing arena. If you are direct marketing your product you need to learn all you can to be efficient and successful in marketing.

Cost Cutting

This is one area, in my opinion, that you should be careful with. You have to remember that you are offering a unique and special product. If you start cutting the wrong corners, you can start sacrificing the quality or uniqueness of your product. I have learned that people will pay extra, a lot extra, for that unique offering you have. Exploit the extra care, special feed, or extra costs that you have in that product. These benefits will be recognized by your discretionary consumer and they will not balk at paying extra.

One example I have is a new poultry product we were offering at the farmers market this year. A competitor at the market offered a similar product. Similar, but not the same. They didn't raise theirs on grass, and they fed a far less superior "commercial feed". Yes, their costs of production were less but we were selling ours at two and a half times over their price. We out sold them three to one. We explained on a sheet why our product was better and people gratefully paid the extra price.

There are legitimate areas to cut costs, but closely research what or why you are doing the cost cutting. You can lose more than your saving if you cut in the wrong places.

New Enterprises and Capital Expenditures

This area could take up a whole article in itself. The guidelines I follow for starting a new venture are as follows:
- Does my current knowledge give me the expertise to do this?
- Does it follow or complement the other enterprises I'm currently doing?
- What regulations will I need to follow?
- What risk is there, and what new problems will be created?
- What will the market acceptance be?
- Start small and keep good records to see if it is worth pursuing.

You can elaborate more on each of these different questions, even add your own, but you get the idea. You don't want to jump blindly into any new enterprise without careful thought and planning.

As far as capital expenditures, I ask myself, if the new purchase can be used in multiple applications. In other words, can I get multiple use out of the item or will it sit most of the time because of only one application need? Will it generate income and reduce labor costs? Again, I think through the purchase carefully, and put a pencil to any new capital investment.

Hopefully, this discussion will get your own gears turning. And again don't wait to plan for next year. DO IT NOW !

Creating a Business Plan
Fall 2000

If new goals are your destination and the resource base is your means of getting there, a business plan serves as a kind of road map. A business plan sets objectives and priorities, providing a format for regular review and course corrections. Useful business plans contain concrete programs to achieve specific, measurable objectives, assign tasks to appropriate people, and set milestones and deadlines for tracking implementation.

Begin by developing a mission statement, critical factors, market analysis and break-even analysis. This kind of plan won't tell you how to run your business, but it can indicate whether an enterprise is worth pursuing. Try the following:

¤ Write a mission statement that addresses why your business should exist, who your customers will be and how the business will benefit them.

¤ Determine what factors are critical for the enterprise to survive and whether those requirements can be met. Adequate parking, hours or seasons of operation and location of market outlets are all examples of critical factors.

¤ Conduct a simple market analysis. Define what characteristics make someone a potential customer and think about where those customers are shopping now, Estimate how many customers you may have and how many you will need. Simply observing traffic flows and the types of products people buy at farmers markets or specialty stores, and attending farm tours or farm-related community events can provide needed information about who your customers might be and ways to target them. "Find out how the market works," advises Herman Beck-Chenoweth, who direct-markets poultry and vegetables in Ohio. "Research a farmers market to learn what sells and for how much. You don't want to take 40 dozen eggs hoping to sell them at $2 a dozen when someone else is already selling eggs for 50 cents a dozen."

¤ Analyze basic break-even scenarios. Project sales volumes and prices, and complete a preliminary production plan to figure out the costs of producing the goods. Knowing the costs of production will tell you whether prevailing market prices will cover those costs. Many direct marketers set their prices too low. Prices should be based on what the market will pay to ensure a reasonable return over the costs of production.

¤ Assess how many units of sales are needed to cover costs. Be realistic: Add up costs for rent, advertising and other overhead, figure out how much money you'll make for every unit you sell after its specific costs of production and calculate how many units you need to break even. Estimating profitability under best, expected and worst-case scenarios for yield or sales, costs and prices can provide a better feel for the risks. While higher-risk activities tend to generate the highest profits, you will have to decide how much risk you are willing to accept.

Once you digest this information, the potential viability of the enterprise should be apparent. If it seems worth pursuing, the creation of a full-fledged business plan is warranted.

Conducting Market Research

Failure to judge the true demand for a product is a common cause of failure in many business ventures. To improve your odds, thoroughly research your ideas. Market research includes ferreting out potential business, competition and consumer trends. Good research also entails finding out as much information as possible about your planned products or services.

Gather information on demographics, consumption, and current and future trends from libraries, government agencies, chambers of commerce, universities and trade publications.

Pinpoint trends that would most likely affect your enterprise, such as customer preference toward specialty shops, existence of local direct marketing associations, attendance and sales figures for farmers markets, popularity of farm tours for school and senior groups, and so on. Local and regional sustainable farming and direct marketing associations are also good sources of advice.

Collecting data yourself can help fill the gaps. You may want to do the following:

¤ Talk to other farmers. Ask them what kinds of buyers they attract, what kinds of service they offer and how they promote their products. Most small-scale farmers are happy to offer such information. Visit market outlets at different times to see what they have to offer.

⌘ Evaluate marketing methods and consider new approaches that put a new twist on an existing product. Not only might you produce homemade jam, but you also could offer it in cases. Hook up with community centers or jam-making groups, or offer to teach the old art of canning.

⌘ Design surveys to find out about customers' buying habits and preferences, and whether there is a need that you can fill. Personal interviews are time-consuming but will yield valuable information. Combined with samples or other promotional materials, surveying doubles as advertising. Be careful when you interpret the responses. What people say about how they spend their money is often very different from what they actually do. You want to get a realistic idea of whether people will in fact spend money on your product.

⌘ Talk to store owners to assess your potential to sell your product. Compare stores to determine which ones best meet your strategies and needs.

Investigate as many marketing options as possible and identify several that look promising. The more ways and places you have to sell your product, the better your chances of success.

Using the results of your market research, you can target the customers or businesses you want to attract and pinpoint your strategy. Estimate the number of customers in your target market and how often they buy similar products. Your target market may already be satisfied by the competition, and you will need to rethink your strategy.

Promotion and customer relations must be part of your marketing plan. A common rule of thumb for promotional expenses is three percent of projected sales.

Some ideas:
⌘ Network, then rely on word of mouth.
⌘ Make attractive, eye-catching signs for your displays, to direct traffic, to advertise your stand, etc.
⌘ Offer promotional items and don't be shy about passing them out to interested visitors.
⌘ Advertise in local or state guides to organic foods. Contact your county extension agent or selected state Departments of Agriculture for suggestions.

⌘ Offer school and other group tours of your farm or facilities. Contact schools to encourage visits and tours.
⌘ Conduct cooking demonstrations.
⌘ Offer samples (if health laws allow), at farmers markets and stores.

Legal Considerations for Direct Marketers

Marketing activities are guided by a wide variety of laws and regulations at federal, state, county and city levels. Some regulations vary by type of enterprise and location, while others are more general. Legal considerations include the type of business ownership (sole proprietorship; partnership, etc.), zoning ordinances, small business licenses, building codes and permits; weights and measures, federal and state business tax issues, sanitation permits and inspections, food processors' permits and others.

If you plan to employ workers, you must meet more requirements, such as acquiring an employer tax identification from the IRS and getting state workmen's compensation insurance. Environmental laws also are becoming increasingly important to farmers.

Adequate insurance coverage is essential. Every operator should have liability insurance for your product and your premises, employer's liability insurance to protect you if employees are injured, and damage insurance to protect against loss to buildings, merchandise and other property. General comprehensive farm liability insurance often does not cover on-farm marketing operations such as agri-tourism businesses. Check with your local insurance agent about liability and loss insurance specifically designed for direct market farmers,

Executing the Plan

The best-laid plans go to waste without good management. Track actual spending and sales, then compare the results against the plan projections—a technique called variance analysis. Once you have the variance, follow up with course corrections, new plans, revisions and more follow-up.

Holistic Management begins with the assumption that every plan is "wrong"—a safe bet when you consider future weather, capricious markets and other unforeseeable factors. Managers engage in a repeating cycle of planning, monitoring and re-planning that adjusts the course of the business as circumstances change.

How Much Insurance is Enough? And Other Questions About Your Least Favorite Expenses

By Lynn Byczynski   Spring 2000

One of the business matters that farmers must think about, like it or not, is insurance. There's not much enjoyment to be had in visualizing the possible disasters that could befall the farm, and nobody gets excited about spending money on insurance, but few would deny that insurance is a necessary evil for direct market farms.

But working your way through the insurance maze is not easy. You need insurance, but what kind? How much? And how much should it cost?

There's no simple answer that will work for every farm, because every farm is different in the amount of 'risk it faces and the amount of assets it has to protect. But it helps to know what's available before deciding whether you've got the right stuff, and. enough of it. If your business has grown or your marketing has changed since you bought your insurance, now might be a good time to reconsider whether your insurance has you covered.

There are four types of insurance that farmers need to think about (besides the personal insurance issues of health, disability and life insurance): farm liability, product liability, employee coverage, and vehicle insurance. Your best resource for figuring all this out is your insurance agent. Although he or she is in the business of selling insurance, if you've got a good one, you can trust the advice you'll receive.

The most important thing to know about talking to an insurance agent is that you have to be completely honest about every aspect of your business and make the agent understand exactly what it is you do. Neil Hamilton, director of the Agricultural Law Center at Drake University and author of *The Legal Guide for Direct Farm Marketing*, advises growers not to understate any aspect of your operation in the hope of saving money on the insurance premium.

"If you don't disclose the full nature of your business there is a greater likelihood that the insurance you buy will be inadequate," Hamilton says. "Then, if something happens and you ask the insurer to cover you (which is why you bought insurance in the first place) you may find out your policy does not cover the situation. Then you are in the worst possible situation - you have paid good money for an insurance policy which was not what you needed and now you have a problem for which you are uninsured."

Before you go to see an agent and explain your business, though, it helps to know some of the basics about direct marketing risks and policies.

How Serious a Risk?

"Some farmers don't buy insurance because they don't expect to get sued. Their operations may be small, they may not have people out to the farm, or they may feel they know their customers and don't worry about them suing. That's the optimistic view, and there are two things you need to know before you decide to adopt it.

The most important thing to consider is that someone who is injured on your farm or by your products may be forced to sue you by his or her own insurance company. They may like you, even love you, but they have signed an insurance contract that allows their company to seek repayment from you if they get injured on your farm. This is known as "subrogation." Neil Hamilton describes how it works in *The Legal Guide for Direct Farm Marketing*: *Consider a situation like the one discussed in Chapter Nine, where a CSA member was injured on the farm. If Jimmy break his leg on the CSA his family will go to their insurer who will pay the medical expenses based on the insurance contract for first person coverage. But the insurer will also ask, "How did Jimmy break his leg and where did it happen?" Under the "subrogation" clause in the policy, the company has a right to seek recovery from someone else if they are responsible for what happened to Jimmy. If the company believes such recovery is possible, they could sue the owner of the CSA to recover from the owners insurance (or sue the owner personally if there is no insurance). Under the subrogation clause the company can ask their insured to be a "use plaintiff" so the suit will be in their name. Insurance companies usually don't bring suits in their own names because it might prejudice the jury. The insured is obligated to cooperate with the subrogation and to help with the case, such as by testifying. If the insured party refuses to sign or cooperate, because the third party being sued is a friend, the company can refuse to pay the coverage or seek repayment from the insured. It is important to understand for this reason you cannot depend on the fact you deal with your friends, to assume they won't sue you if something goes wrong. In most cases they will not be making this decision, the insurance company will, and the insurance company is not interested in friendships.*

That's probably enough to scare you into calling an insurance agent, but if you're a gambler, you might also want to know the frequency of lawsuits against direct marketers. Charlie Touchette, director of the North American Farmers Direct Marketing Association (NAFDMA), studied direct marketing insurance for several years while creating a policy specifically for direct marketers, and he says there's just no industry data on direct marketing claims.

"The whole thing of direct marketing is relatively new--even 20 years is new when you're talking about insurance statistics," he said.

Touchette looked at the claims history in the cut-your-own Christmas tree business-surely one of the most potentially dangerous direct marketing ventures-and found "they were far less significant and fewer of them than the insurance companies imagined," he said. The
NAFDMA also surveyed its members for anecdotal information about insurance claims and found only a handful. The biggest was "a badly twisted ankle at a pick-your-own apple orchard that turned into a $40,000 settlement," Touchette said.

He has also managed the liability insurance coverage for the farmers markets in Massachusetts. In 12 years, with 50 to 70 markets covered each year; there have been only three successful claims. All three involved wind accidents—signs or canopies blowing over and hitting customers; all three customers went to the hospital and the claims were settled for $12,000, $26,000 and $32,000.

In other words, the statistical risk of an accident is probably small, but accidents do happen and farmers and markets do get sued. Touchette says that for about a $300 annual premium, farmers don't have to worry about accidents or defending themselves against a lawsuit, but adds that it's a personal decision. "Sometimes it's just for peace of mind; it's hard to want to spend $300 but what kind of productive energy is lost worrying about it if you don't have it?" he said. "Of course, there are those people who don't worry about it, and if they're not losing even mental time, okay."

Farm Liability Policy

If you've decided you'd better have insurance, the first policy to consider is your liability policy. Many growers, when they first sell produce, assume that their homeowner's insurance will cover them both on the farm and at a farmers' market. That may or may not be true. Your homeowner's policy will cover an accident on the farm to a guest or visitor, but once that guest is paying you for your products, the relationship changes. For example, if you let a friend pick a bouquet on your farm, injuries would be covered by your homeowner's policy. If you charge that friend $20 to pick flowers, it might not be covered. Some companies won't quibble about small commercial transactions, but if you're making more than a few hundred dollars in farm sales, you'd better check to find out whether that business is covered In some cases you can just add excess liability coverage, called an umbrella policy, for your business activities. If you're currently buying only a homeowner's policy, read it carefully for mention of commercial activity, particularly the exclusions, and have a talk with your agent.

Once you start farming in earnest, you need a farm liability policy, which will cover all activities related to farming in addition to the usual liabilities of owning property Whether your direct marketing activities are included in the company's definition of farming activities will vary, particularly if you're buying from a company that does most of its business with traditional farmers. Again, read the exclusions to find out if roadside markets, off-farm farmers markets and pick-your-own (PYO) operations are covered. Generally, PYOs will require additional coverage because the exposure, or potential for someone to be injured, is greater when there are more people visiting the farm.

Farm liability policies may contain two types of coverage: personal liability and medical payments to others. At my farm, for example, our insurance company would pay up to $1,000 to any person who was injured on our farm if we had not been negligent. If the injured person decided to sue us, alleging negligence, we would be covered up to $500,000 and the insurance company would handle the defense.

How Much Coverage?
This brings up the point of how much coverage you need to purchase. The old insurance maxim is "cover your assets." In theory, if someone was injured seriously because of your negligence (in the eyes of the court), the damage award could take everything you own and even attach your future earnings. In the Northeast and on the West Coast, where the price of real estate is high, and on farms with a lot of buildings and equipment, many direct marketers insure for $1 million. Farmers of more modest means might decide to go with $100,000 or $300,000 coverage.

The difference between $300,000 coverage and $500,000 coverage is relatively small—$25 a year on the premium, in our case. It would cost more to increase the medical payment for non-negligence accidents from $1,000 to $5,000 than it would to increase liability coverage to a half-million because the risk of a small injury is greater than the risk of a big, lawsuit-producing one. My insurance agent tells me it's unlikely that a court would force a farmer to sell the farm, but cash assets would be an easy target for the opposing attorney. And there have been cases where defendants' homes, while not taken away from them, have been put in a trust that reverts to the injured person upon the death of the owner.

Covering Employees
Your employees should be covered for injuries one of two ways: either through your state's workers' compensation program or through your liability insurance. Here's the situation on whether you need to buy workers' comp insurance, according to Neil Hamilton's book:
× Agricultural workers must have workers' comp in 12 states: Arizona, California, Colorado, Connecticut, Hawaii, Idaho, Massachusetts, Montana, New Hampshire, New Jersey, Ohio and Oregon.
× Employers can choose to buy workers' comp, but don't have to, for agricultural workers' in 13 states: Alabama, Arkansas, Indiana, Kansas, Kentucky, Mississippi, Nebraska, Nevada, New Mexico, North Dakota, Rhode Island, South Carolina and Tennessee.

If your state isn't on one of the two lists above, call your state labor office to find out what the current situation is for seasonal farm workers.

Hamilton advises farmers that they should view workers' compensation insurance as a benefit both to their employees and themselves. Obviously, the employee benefits because in the case of an injury there is a standard recourse to compensation. The employer benefits because if the farm does have workers' comp, an injured employee is limited to workers' comp as the sole source of recovery. That means the employee can't sue you seeking greater damages or huge punitive damage claims.

However, workers' comp insurance is expensive compared to the cost of adding employees to your farm liability policy. Our insurance agent estimated that we would pay $400-500 a year to participate in the workers' compensation program in Kansas. Adding four employees to our liability policy costs an extra $80 a year.

The negative side to covering employees through the liability policy is that if the employee were injured he or she would have to sue and your insurance company might decide to defend the suit. You would, under your contract, be obliged to help with the defense. If you care about your employees, you would want to see them compensated for injuries, not forced to face you in court to fight about it.

Products Liability
Your general farm liability policy may or may not cover an accident in which your farm products made someone sick. Check to find out. If you're selling fresh produce only, you're probably covered. If you're doing any value-added products you may need to purchase separate products liability coverage.

Some stores won't buy from you unless you have a products liability policy. Some insurance companies won't even insure for farm-made products, like jams, salad dressings, baked goods and so forth, so you may need to shop around to find coverage.

Vehicles
It goes without saying that you need to tell your car insurance company if you have employees driving your vehicles. In any vehicle accident, the vehicle insurance is considered the primary insurance policy that handles claims first. If someone other than the person named on your policy is driving at the time of the accident you could be in trouble.

Lynn is the Editor and Publisher of Growing for Market. This article appeared in Volume 8, Number 11, November 1999 and is reprinted with permission. For more information on Growing for Market, call 800-307-8949.

Chickens Galore LLC, or How to Protect Your Business though Legal Structuring

By Jody Padgham Summer 2003

Recently we had communication about liability insurance and options pastured poultry producers have for insuring product they may process and sell from the farm. After polling several APPPA Board Members, it became clear that one of the best options for now is for a producer to create a legal structure for their poultry business and insure that business as a separate entity from the farm.

Most of us may not know it, but by reporting taxes we do more or less have a "business structure." If you haven't filed for any other option, your business is most likely considered by your state government to be a "Sole Proprietorship." This means that you are the owner, you have complete control, and that all of the assets (of your farm, household and life) are lumped together. Any profits or losses of your business are reported on your personal income tax return. The advantage of this kind of business is that it is very simple--and often requires no additional paperwork, though in some states requires a simple registration. The disadvantage is that any debts incurred by your poultry business also put personal assets at risk.

Another disadvantage of a Sole Proprietorship for pastured poultry producers is that it can be very hard to find product liability insurance.

One way to get around this problem is to legally organize your poultry business in one of the ways that will "limit your liability." There are several ways of doing this--including forming an S- or C-Corporation, a Cooperative or a Limited Liability Company. Larger operations may consider one of the first three types of incorporation, but the Limited Liability Company is the option that many smaller producers are using to protect their farm assets and get liability insurance.

All this talk of liability can be pretty confusing. When talking of a Limited Liability Company (LLC) we mean a business structure that will separate the assets of your home and farm from your poultry business. Product Liability Insurance is that needed to cover your assets in the event that a customer becomes ill or has other problems with your products. Farm Liability Insurance protects your farm if a visitor gets hurt while on the farm. Many of us are able to get Farm Liability Insurance as a rider on our normal farm insurance policies, though some of us that process on-farm are even having problems finding this kind of coverage. It is very difficult to find Product Liability Insurance for home processed poultry, even though states allow home processing and the product is generally extremely high quality and safe.

I admit to feeling mixed about writing about how to situate yourself so that you can get more insurance. I, like most of you I imagine, am none too fond of the insurance world (apologies to any insurance reps out there). But, unfortunately, many farmers markets require product liability, and though I am not aware of any, in this world of libel we have today, I wouldn't be surprised to hear of a suit against a pastured poultry producer for a product problem some day. Seeing how hot coffee at a fast food restaurant can bring a customer thousands of dollars (or was it millions?), don't think that you'll get off easy if someone decides that the grit in an under-cleaned gizzard was responsible for breaking their expensive dental bridge.

So, after that diversion- just what is a LLC and how do you go about getting one? The LLC is one of the newest forms of business structuring. It was first used in the late 1970s, and is seen as a more-or-less blending of the positive aspects of partnerships with the limited liability of a corporation. Though laws will differ from state to state, (and before your move forward you should contact your state small business office or department of regulation,) an LLC is generally made up of two or more individuals that sign an operating agreement to do business together. For those of us on family farms, this can be family members, but can also include outsiders that are interested in the business or want to contribute financially.

By organizing as an LLC, in most states all "owners, managers and agents" are protected by state law from personal liability for debts and other obligations. LLC laws are very general, and dictate only that those forming the business come to an agreement on how to manage the business, and how it can be dissolved. Any profit division, democratic issues, etc., are left to the discretion of the business owners, as in any partnership. However, an LLC does carry another benefit of a corporation, in that profits made through the business are taxed only at the individual level, and are not noted at the LLC business level. This sounds confusing now, but believe me that it is a good thing that your accountant can better explain to you.

The first step to forming an LLC is to contact your state authorities to find out the particular laws governing LLCs in your state. For those with access to the web, the Small Business Administration has a wonderful website with links to every state licensing bureau at http://www.sba.gov/hotlist/license.html. Those without web access can look in the phone book for their local Small Business Administration office or call the national SBA for a local contact. (800-827-5722.)

Many states will require that you fill out articles of organization with your Secretary of State or Department of Financial Institutions. These will ask who is involved, where they live, the name of the business, the business address, an agent, and not a whole lot more. You will have to pay a fee to file (I think it is $50 in Wisconsin). After that, in some states you will have to file an application with the Secretary of State or other office to reserve a limited liability company name. The words "Limited Liability Company" or "LLC" may be required. Some states will have a search service to help you find out what names are already registered.

Following the above, the members of the LLC will have to prepare an "Operating Agreement." For those who have seen a set of bylaws, this agreement is often somewhat similar. The details of the operating agreement are decided by the members. Generally included will be details on distribution of ownership, monetary investments, voting procedures, management responsibilities and dissolution procedures. It is useful to have a lawyer advise the creation of the operating agreement to ensure the LLC will be structured properly. If you don't already work with a lawyer, your banker may have a suggestion. Also, a list of agricultural lawyers can be accessed by contacting the American Agricultural Law Association at the University of Arkansas College of Law (phone 501-575-7389) or www.aglaw-assn.org. The operating agreement is kept as an internal document (much like bylaws).

Annual reports must then by filed by the LLC with the Secretary of State or other state office. Taxes are claimed on the individual tax returns of the members, much like in a partnership. The LLC itself does not file tax forms. In most states managing members of the LLC are treated as self-employed individuals, and must pay self-employment taxes.

When forming an LLC for an on-farm poultry business, you will be separating out the assets of the poultry business (the birds, equipment, some housing etc) from those of the other parts of your farm and life (home, cars, perhaps dairy barn or other buildings). The liability of the owners will be equal to their investment in the company. You may structure it that the LLC owns the birds but pays rent to the farm to use brooding facilities or winter hen housing, for example. It is important to be clear on what is a part of the LLC and what is the farm or home. Costs and income will need to be clearly separated.

Once the LLC is set up, you can explore the possibilities for insurance. Insurance companies have tended to look more favorably on farm businesses that have separated out the poultry into an LLC.

Those wishing to operate "under the radar screen" may not want to formalize as an LLC, as the registration will be filed and accessible in state offices.

For more information on the legal issues of direct marketing, I highly recommend the 1999 book *The Legal Guide for Direct Farm Marketing* by Neil D. Hamilton of the Drake University Law School. The small format 224 page spiral bound book sells for around $20 and may be ordered by calling 515-271-2065.

On Coyotes and Lawyers
Using Legal Entities to Replace or Supplement Farm Insurance
By Richard Gaskin Spring 2004

As I am new to the chicken farming business, I get a lot of my advice from the online pastured poultry forums. Dealing with threats like coyotes is new to me. Recently the discussion turned to insurance, entities and lawsuit protection, and I realized I might be able to suggest some strategies for dealing with a different and less ethical coyote.

Asset Protection
Asset protection is a worst case form of risk management. The objective is to keep as much personal/family wealth intact in the case of a successful lawsuit. Traditionally, under common-law, a person could not be found liable unless he knew, or reasonably should have known, that his actions or omissions were likely to be harmful. This is no longer the case; a person may be held liable for not knowing what no one could have known. Further, liability is apportioned not by blame but by the depth of your pockets. I believe most of us take responsibility for our actions, but this is no longer the function of our legal system.

No asset protection is totally secure. What we are trying to create is a series of barriers. Think in terms of your electric fence or your rifle and still the coyote gets a few of your chickens. But not all. In this case the varmint looking for an easy meal is an attorney. He's probably paid by contingency so he's looking for big payoffs that are relatively easy.

Unless you are very wealthy, asset protection is a do-it-yourself project. There are few attorneys or financial planners that are very good at asset protections and a lot that are not. There are also a lot of hucksters and product sellers. Buyers beware.

For this discussion we will quickly cover four basic levels of protection:
1. Lower your profile
2. Procedures and documentation
3. Insurance
4. Structure (Entities and Other)

Lower Your Profile
Roosters notify the coyote of a free meal at a great distance. Many people, who should be smarter than roosters, make themselves lawsuit bait with fancy lifestyles and owning all their wealth in their own name. Your lifestyle I leave to your judgment. An advantage of some of the structures we will discuss is the ability to retain control while titling them in another name.

Who are you doing business with? Fire any customers who are difficult, who feel somehow you owe them your efforts at a lower price or who have an entitlement mentality. The best practice management advice I ever received is to fire 10% of your customers each year. Your profits will increase and your life will be better.

Procedures and Documentation
We know our farms are safer than the industrial models and that our chicken is healthier. Now...convince a jury or a bureaucrat.

Your word on what you told a customer concerning safe product storage or safe behavior on a farm will not do you nearly as much good as dated photos of warning signs, a safety standard operating procedure (SOP), and a letter of farm conduct agreement signed by the customer. A safety video with a log of viewers would be a nice touch.

As an added benefit perhaps we can find ways to run safer farms. Confession--I have a finger recovering right now from a run-in with a wood chipper. Steel trumps flesh. Even ecological farms can be dangerous if the operator gets careless. And the insurance companies know it.

Insurance
Investors often take a small but highly leveraged position in opposition to the way the majority of their investments are placed. This is known as hedging, and the investor hopes to lose that money, but if the market turns against him it will prevent great losses.

This is the role of insurance in society. The best thing that can happen is to never need your insurance. That doesn't make it wasted money.

In my mind the purpose of insurance in the asset protection hierarchy is to:
1. Provide for a legal defense
2. Pay for legitimate and reasonable damages
3. Position yourself as a reasonable and prudent businessperson.

Unfortunately, insurance also acts as lawyer bait.

Many poultry producers report that their agent has told them that insurance is unavailable if:
1. They sell at a farmers market
2. They process on farm
3. They allow customer pickup

Without knowing the details, I suspect they are carrying a farm coverage form attached to their homeowners' policies and the above processes require commercial lines of insurance, which their agent doesn't carry and doesn't understand.

Do not, by the way, omit informing your agent of the above activities. If you have an accident, they will void coverage back to the beginning and return your premium. You may as well have gone naked.

When I think of how outrageous my insurance bill is I try to imagine a little girl getting mangled on my farm. Nothing I own can make that up. Fault isn't the issue. At least the insurance can give her a decent life. There is a moral argument in support of insurance. On the other hand, a million or two liability won't cover many of today's legal judgments, and many of us cannot afford all the coverage we need. Let's structure our assets to preserve some of them.

Structure(s)

I'll first run through a laundry list of entities and other tools with a few brief comments on best use. Then I'll briefly comment on putting it all together.

Corporations: Best for holding business or revenue producing assets. Owners of the corporation are not personally liable for its debts. Ordinarily this protects personal assets from business liabilities. If your actions had anything to do with the hazard, the injured may try to sue you personally, particularly if you have many personal assets. The corporation does protect non-corporate assets from the actions of employees.

For asset protection C-Corporations and S-Corporations are about the same, however C-Corporations are far more flexible in how they are set up. For tax benefits, if there is a profitable segment of your business, I recommend this being the cornerstone of your plan. Do not use to hold real estate. Unlimited life. Get rid of any advisor who warns you about double taxation. It's not an issue and they don't have the imagination to advise you.

Corporations are legal persons, able to contract, borrow money etc. Be sure to sign all documents as a representative of the corporation. Avoid personal guarantees to the bank if the corporation borrows money—it sets a precedent that you personally guarantee the actions of the corporation.

Limited Liability Companies: Easy to set up, are not as well tested for asset protection, but should be OK. LLC's are nice for re-titling property as many states do not even require the names or members or managers. Their main advantage is lower corporate formalities. This does not mean you don't have to maintain a separation between you personally and your LLC.

Limited Life Partnerships: Useless. No asset protection. No tax advantages. Limited life. Most partnerships are created by accident—if you are in business with someone else and haven't set up anything else you are a partnership. This includes husband and wife, although the IRS doesn't require a separate return.

Trusts: Only irrevocable trusts have any asset protection value–not the living trusts promoted in the various media. Trusts are useful ways to maintain use of an asset for your life while placing it out of judgments' reach. The life is governed by the state, but usually limited.

Family Limited Partnerships: A personal favorite. Simple, inexpensive, good estate tool, and good asset protection. If a creditor successfully attacks a FLP, at best he gets a charging order. This entitles him to a share of distributions, if any, and the right to pay a share of the partnership tax, whether or not he received a distribution. The charging order becomes a tar baby. Coyotes know this and generally leave FLP's alone.

Unfortunately, there must be a general partner who is personally liable for all partnership debts. The general partner can be a corporation, partnership or trust. Good for holding real estate, mortgages, equipment etc. Under laws of the 49 common law states, an active participant in a business cannot be a limited partner.

Place any entity that might attract litigation in a separate entity.

Pensions: Assets in most qualified pensions (not IRAs) are untouchable by most creditors. Something to consider before taking a lump sum distribution. Profitable C-Corporations can establish qualified pensions.

Homestead: Your homestead is your home and the immediate property around it. Most states have some protection against creditors for a homestead. In Florida it is unlimited, making a home a good place to store wealth.

Mortgage: Creditors have little interest in highly mortgaged property. Your Limited Partnership can hold mortgages on assets.

Life Insurance and Annuities: Generally protected from creditors. Some policies can be used like a bank while keeping your funds out of reach.

Leases: As Allen Nation of the *Stockman Grassfarmer* states—"the goal is to control the land, not necessarily to own it." Leasing land and equipment makes good financial sense and is protected from lawsuits. Already own enough land? Set up your own leasing LLC and transfer land and assets into it.

Occasionally I hear it said that corporations and other entities no longer protect assets. I would agree that 90% of the closely held businesses in America will not protect the assets of the owners. This is not the fault of the entity structure—the problem is:
¤ The structure was improperly set up or is the wrong structure
¤ The owner did not comply with the corporate formalities such as minutes and board meetings
¤ The owner did not maintain a wall of separation between his entities and his person—such as commingling funds.

Do not set up an asset protection structure if you will not maintain it. Maintaining it can be a pain in the vent. But, paying a lawyer to incorporate your farm and then forgetting about it is a waste of money and will give you false sense of security.

Register all property in the entities' name. Sign papers as an entity official. Lease the use of equipment between entities. Do not enter into sweetheart transactions. Keep proper accounts and financial statements. Hold required meetings and votes. You must demonstrate that your entity is not a sham. Each entity must have its own bank account. Each entity also deals with insurance independently.

As farmers we have two particular problems. First, we tend to see no separation between our businesses and our persons. And second, we are usually personally involved with anything that may be considered a hazard. This means that both our entity and we personally are open for lawsuit.

Because we are personally open for lawsuit, if we own the shares in our name they are available to settle any judgment against us. The whole structure will collapse.

The solution, of course, is to have few or no shares in the corporation in your name. If you know anyone uninvolved in the business you trust to title the shares in (and it's not a casual decision) then place them in a Family Limited Partnership. If a creditor successfully attacks the Family Limited Partnership they still will not have control of the corporate shares.

Do not automatically title assets in both spouses' names. It is convenient for probate but you can have the same result with a revocable trust.

Designing Your Plan
Your asset protection plan must be set up for your own circumstances. You must take into consideration how much you have to protect, your tolerance for risk, your tolerance for corporate formalities and tradeoff of expense vs. protection desired. Start simple and add as necessary.

The process I use for a full-blown protection plan is:
1. Analyze all your operations for the amount of risk and the assets you wish to protect. If you work on the farm, you personally are a risk area. If you are selling chickens directly, especially with on-farm processing that is a risk area. If customers visit your farm, that is a risk area. Segment risk areas from other assets and operations.
2. Choose a protective entity for your non-risk assets. I generally use a C-Corp or Family Limited Partnership. An LLC is probably OK but of less use for tax

planning.

3. Place each risk activity (i.e. chicken operation) into a separate entity. LLC's are good for this. You are trying to isolate damage if any entity is successfully attacked. You personally are an entity already.

4. Try to remove any assets from the threat activity. Perhaps place your land in a trust and any equipment in an LLC that will lease it back to the risk activity. Title your personal assets in a noninvolved spouses name or place large mortgages on them.

5. Meticulously maintain the formalities of the structure.

So, for example, we have:

If it seems like a lot of trouble, it is.

It is also what is necessary to give the level of protection that most people think they get from a single corporation around all their assets.

C-Corporation	Family Limited Partnership	LLC	You
Contains:	Contains:	Contains:	Contains:
Profitable farm or side business	Land Equipment C- Corp Stock	Chicken operation Rents pasture and Equipment from FLP	Own little of interest

You may decide that a lower level of protection is acceptable. Asset protection planning is usually a side benefit to good tax and estate planning in an integrated system where the costs are offset by tax savings.

Worst case planning is a good thing to do and easy to put off, however it is a bad place to live mentally. Business isn't much fun if you think of your customers as potential problems rather than friends. We hear about large lawsuit settlements, but they are unusual. Make your plans, implement and maintain them, and relax and enjoy life.

Rest Assured, You Can Find Insurance
By Jane Eckert Winter 2004

As published in the Vegetable Grower News – August 2004

In speaking to farmers across the country, and now working with the Kansas Department of Commerce on its statewide agritourism initiative, I have found one common challenge facing producers: liability insurance.

Farmers are having difficulty finding insurance carriers that will write liability policies. In fact, most of you are having problems even getting quotes or worse, even getting insurance agents to return your calls!

It seems that this issue of liability has become a bigger problem for some folks than selecting products, creating new activities or growing an agritourism enterprise.

So it's time to crack the code on liability insurance!

I spoke with Steve Hall of Anderson, Hall & Marsh & company in St. Louis, Mo. Anderson, Hall & Marsh & Company is a commercial insurance "broker" that contractually represents about 18 insurance companies and is licensed to place insurance in 42 states. The company has served as broker for my family's farm for more than 26 years.

Here is how Steve answered the questions most frequently asked by farmers.

What's the problem when farmers get into agritourism?
⋈ When farmers go from traditional agriculture to agritourism, adding activities such as petting corrals, wagon rides, corn mazes, playgrounds, etc., they become entertainment enterprises. Once you allow

people to enter your premises on a commercial basis all of the rules change and the farmer has to recognize that an entire new way of doing business is necessary.

¤ When farmers invite the public onto their property for these activities, they take on a different kind of risk

Why can't farmers get liability insurance through their local insurance agents?

¤ Insurance agents in smaller communities typically represent smaller local insurance carriers and only a few, if any, large, well-known insurance companies.

¤ Large insurance companies believe that farmers as entertainment operators are a big risk because they won't practice the proper safety standards needed to minimize losses nor do they have the expertise and skills required to handle public liability exposures.

¤ Since these companies have been hit hard recently, they are "cleaning-up their books" by getting rid of what they consider to be high-risk policies, such as agritourism liability.

¤ Also, there is no way for these large companies to develop a standard policy that fits all agritourism operations, since each one is different from the next. These companies prefer to cover similar types of commercial businesses and ones with which they are comfortable and understand.

What kind of insurance companies will cover farmers for agritourism liability?

¤ Farmers need to find an insurance company that will take on the unusual risks of agritourism. In the jargon of the insurance world, these carriers are called "excess and surplus lines companies."

¤ Companies providing "excess and surplus lines coverage" are not the large, well-known, household name companies (Hartford, Travelers etc...) that typically want to avoid this type of risk. Rather, the excess and surplus community specializes in difficult placements. Often these carriers are actually subsidiary companies of the big well-known insurance groups.

Can my local agent find this kind of coverage for me?

¤ Usually not. Your local insurance agent typically works only with one or two companies.

¤ You must find an insurance "broker" who represents multiple insurance carriers, some of whom will write "excess and surplus lines coverage." Brokers typically can write policies in many states with a large number of companies.

How do I find an insurance broker?

¤ Ask your local agent. Your local agent understands the insurance system and knows how to access the insurance brokers.

¤ In many states you can call the state department of insurance, explain your situation and they can give you a list of people to call. You can also call the Independent Agents Association and they can refer you to someone in your area.

How high will the premium be?

¤ I can't give any figures, since they would be different for each farm.

¤ The premiums will be higher than policies for traditional agriculture, but they will be based on the types of activities you offer.

¤ In the E & S business there will almost always be a minimum premium. That is, the company will not issue a policy for less than "$x."

Good news if you live in Maryland, Pennsylvania, North Carolina, Ohio, Virginia or West Virginia. The Farm Bureau insurance carrier in these states is now looking to expand their product offerings to include agritourism liability policies. Check with your local Farm Bureau agent if you reside in these states to give you a quote.

CHAPTER THIRTEEN
The Good, the Bad and the Ugly: things to keep you thinking

Those interested in pastured poultry have very active minds—there are many topic streams we have followed in the APPPA GRIT! that won't fit into the categories we have set out in this book. Some will get you thinking, some will make you think twice, and some will basically entertain you. In this chapter you will find all of the above.

What If
By Kip Glass Winter 2003

So, you want to raise pastured poultry? A little to put in the freezer for your own consumption, sell a few to pay for yours, and what the heck, why not raise a few more to make a little profit. And who knows, if it goes well you might make a small production out of it and make a lot more profit.

After all it can't be that hard, letting the birds have a wonderful life in the fresh air and sunshine, running around on that green grass. Spending maybe an hour a day moving pens, feeding and watering, and watching them play and catch bugs. Surely the processing can't be that bad?

When we first got started with our pastured poultry enterprise, we thought we would cover all possible problems that would ever happen (Remember Murphy's Law).

Let's start with brooding; what could possibly go wrong there? Are the chicks warm enough? Why are we losing so many? Is it coccidiosis? How can we improve this part of the growing process? Fortunately we never had any catastrophic brooder losses. This year, neighbors of ours had a major outbreak of coccidiosis, and lost over 120 out of 200 chicks in the first three weeks. Their problem turned out to be poor brooder management. Caked manure, dirty waterers, damp bedding under the waterers, and running out of feed often, causing the chicks to search for food in that bedding breeding-ground. Cleanliness in the brooder is a must. Proper stocking rates, plenty of feeder and waterer space and feed in front of the birds at all times are also critical. Recommended reading is Robert Plamondon's *Success with Baby Chicks*.

Now it's time to move them out to pasture. Is the weather going to be good for the next three days? How low are the night time temperatures? Poultry producer Jim Protiva in south-central Missouri turned me on to a trick of using the wide plastic wrap-type material that they wrap product on pallets in the shipping industry. Wrap the field pens in it, closing up the large open areas that can allow in blowing rain and cold air that is so common early in the season. Allowing an extra few days of protection for the newly pastured birds can go a long way in their survival and lead to better weight gain.

The orientation of your field pens is very important because it isn't going to be 70° F and sunny all the time. This is going to vary with your pen design and the prevailing winds of your area. In our area of SW Missouri we face the open end of our pens to the east.

One spring, with over 800 chickens in field pens, the rains came and didn't stop. The field the birds were in had very little slope. With the ground totally saturated, I went out to check the birds at midnight after the fourth round of heavy rain had moved through in eight hours. It had rained so hard the water couldn't run away fast enough. Birds from three and a half weeks of age to seven weeks of age were wading in three to four inches of water. Of course, the three-and-a-half-week-old birds were over belly deep, soaked, and hypothermic from the cold water.

There was nothing to do but take them to safety before they died. Waking up my reluctant (very helpful) spouse, we hauled chicks in a wheelbarrow, two crates at a time, to the truck waiting on the roadway as we could not drive through the soaked field without getting stuck. We loaded over 400 of the 800 birds and took them to the safety and dryness of the brooding area. We then hauled straw to the remaining larger birds to elevate them out of the soup. Fortunately, we only lost four out of the 800 and were able to get the others moved back outside two days later, after everything dried up somewhat. Now our spring birds are put in fields that have a significant slope to allow the spring rains to flow away. (Lesson learned)

Heat: how well will the birds handle it? We have found that the white material commercial tents are made of is the coolest covering for our field pens. Again, we learned this through trial and error. Do your water reservoirs have a large enough capacity to handle your largest birds, stocked at the highest stocking densities at the hottest temperatures? Do you have contingencies to quickly replace or repair broken waterers after one breaks or malfunctions, draining your reservoirs dry during the hottest time of day? Can you get your reservoirs filled quickly? Do you check your chickens several times a day during the high heat? Birds die quickly without water in 100° F heat. These problems have happened to everyone and they will happen to you.

Now, you've successfully gotten your broilers to processing day. They are fat and happy, you have several orders to fill, people are counting on you to sell them a healthy product.

Do you have dependable help that will be by your side no matter what? What if the heating element on the scalder quits? Can you replace it? Do you have an extra one on hand? One time we had over 150 chickens to eviscerate and due to three helpers not showing up my wife and I had to eviscerate them ourselves with customers coming in a few hours. STRESS!!!!

I feel that processing is the biggest stumbling block to everyone in this industry. Invest in good equipment to make the toughest part of this enterprise a lot easier.

The birds are processed, customers are coming..... What! You haven't acquired any customers yet? I've seen a lot of people in this industry raise birds with the hope that someone will buy them. Develop your market first, it relieves a lot of stress, and keeps you from buying another freezer. As your demand increases, so should your production numbers. Remember, it is only eight weeks from chick to processing day; your customers will wait and be that much more appreciative.

Basically this article was written to get you to thinking, "WHAT IF?"

Go through every phase of the production process. What if this happens, or that happens? Have a plan for backup with everything. Sometime failures are time critical, plan for it. Don't think you won't have problems; it's inevitable. Just learn from your mistakes, read all you can get your hands on. Why not help and learn from someone already in the industry. Don't be afraid to do things differently, you don't have to exactly follow what another person does. Remember there is always a better way, search it out, but always think of the, "What If" in every thing you do.

APPPA PRODUCER PROFILE

Kip and Jackie Glass
By Aaron Silverman Spring 2003

Kip Glass and his wife Jackie live on 9 acres in southwest Missouri, 8 miles west of Springfield. They produced 2,900 broilers and 130 turkeys in 2002. Kip is a full-time UPS driver of 24 years, and his time available for farming is limited. Production for this year is planned at 2,000 broilers and 160 turkeys. 2002 chicks arrived in mid-March, and the last broilers were processed the last week of October. Turkeys are processed the week before Thanksgiving.

Broiler Production System
Kip was awarded a Missouri Sustainable Agriculture Grant in 2001 in which he compared moveable pens and day-range system. At the conclusion of the study he elected to use moveable pens due to even manure coverage on the pasture, better weight gains, and quicker turn-around in restocking the field pens after harvest. Kip now uses pens 10 ft. wide by 12 ft. long by 3 ft. tall. The houses are constructed with a top and bottom frame and uprights made of 2 x 4" lumber. The back of the pen is covered in white vinyl tent fabric supported by seven pvc pipe hoops. The door in the front is a pvc frame wrapped with three ft. wide poultry wire. The pen is wrapped with the poultry wire to keep out both aerial and ground predators. A six inch gutter runs the width of the pen in front of the door, eliminating the need to remove feeders when moving the pen. Two bell waterers provide water to each pen, supplied by a 15-gallon drum laying on its side with tubing running to the bells under the pen frame. The drums support the largest chickens at the highest stocking rates during the hottest days and are filled once a day when the pens are moved and serviced.

Stocking densities in 2002 were mostly 85 to 90 birds per pen but will probably be reduced to 75 to 80 for 2003. Kip raised pullets only in 2002 in an effort to make sizes uniform for batch harvesting. Kip felt the trial worked well, the pullets avoiding the bullying behavior cockerels develop in moveable pens, and dressed out at 4.25 lb. average at eight weeks. The detraction was an average feed consumption of over 18 lb. per bird. He plans on going back to raising straight run chicks this year, forecasting an average feed consumption of 13 lb. to 14 lb. per bird, harvesting birds at seven and eight weeks.

Brooding is done in a 40 x 30 ft. greenhouse structure. Kip began brooding in the greenhouse last year, and says he loves it. His primary challenge was that temperatures were unacceptably hot, even with adequate ventilation and two layers of shade cloth on the greenhouse, an 80% black shadecloth and a 50% white shadecloth. This year he plans on covering the structure with a white poly tarp during the heat of the year, blocking the sun while keeping the house bright. Kip says it is obvious the bright environment makes it healthier for the chicks. Brooding is usually for three weeks with the usual variations due to weather, delaying in cool damp conditions or moving out earlier in hot weather.

Kip's field mortalities have been between 2% and 3%. His brooder loses were higher, around 5%. Kip hasn't identified anything specific causing this level of losses, and plans to implement suggestions Dan Bennett outlined in his "Brooder Nightmare" article (on page 15.)

Feed
Kip uses single feed ration, a 19% protein ration developed by Fertrell. His feed costs are relatively cheap compared to much of the country, averaging 11 cents per pound in 2002. He uses roasted soybeans and has feed delivered in bulk, typically in three-ton batches.

Labor and Processing
Kip and Jackie together manage the brooder and service the field pens. Kip estimates they can move and

APPPA PRODUCER PROFILE *Glass*

service 14 pens in the morning for one and a half hours, and half an hour in the brooder.

Processing is done on the farm, with help brought in for evisceration. Kip handles the killing, scalding, and plucking, averaging 70 to 80 birds per hour. Two eviscerators average 25 birds per person per hour; Jackie handles quality control before the birds are chilled in a 250-gallon bulk milk cooler. Kip went from tabletop evisceration to eviscerating in shackles supported above a stainless table last year. He says that after the first 160 birds they processed, they couldn't think of going back to tabletop evisceration. "It is so much easier, less stressful on your hands, and much more sanitary," he says. Offal is composted and used in the garden and fields.

Marketing

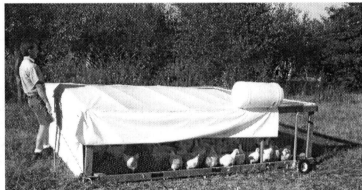

All birds are sold from the farm. Customers bring their cooler on processing days and pick up their pre-ordered birds. "I feel we have an advantage somewhat in the way we got started marketing the birds. Jackie is a certified natural health professional, so we had a pretty large list of contacts to start off with. Then, it grew by word of mouth," Kip states.

The Glasses began raising pastured poultry in 2000, selling 300 that year, and 2,500 in 2001. They have developed a customer base of over 300 families. Kip feels people should be patient and let word-of-mouth grow the business to the level they feel comfortable – "don't rush it, they will come." They are currently selling their broilers for $2.10 per pound. "Anyone raising pastured poultry should not be giving their labor away, and should not be under $2 per lb." He says he raises a quality product, and that should be reflected in the price.

Kip experimented raising some Heritage breeds of turkeys last year, and is offering this as an option to customers this year. Although the Heritage turkeys are considerably more expensive ($3.40/lb versus $2.30/lb for the broad-breasted whites), customers ordered 50% more of the Heritage breeds than the white.

2002 Kip Glass cost of production			
	Cost Goal	Percent of total	
Chicks	0.55	13.0	
Feed	1.95	46	.13 per lb. times 15 lb.
Bedding	0.02	0.08	
Processing equip.	0.25	6	
Building & Pen cost	0.25	6	
Fuel	0.01	1	
Water	0.01	0.04	
Labor	1.20	28	
Misc. (bags, gloves,)	0.01	0.04	

Record-Keeping

Kip keeps meticulous records, using Excel spreadsheets for production records, and Access for a customer database. He keeps track of all areas of expenses and income so he knows where adjustments can be made. "You can't run a business blindly and run it efficiently," he says.

The Bottom Line

The main challenges Kip and Jackie face are keeping up with demand. With only two of them and Kip's full-time job, life's balances have to be managed. One great advantage they feel is their close proximity to a large, diverse metropolitan area.

"Never quit thinking outside the box—always challenge yourself to do or offer something better. Listen to your customers."

Zero Mortality: A Compelling Design Condition
By Andy Lee Spring 2000

Usually, designing a new farm enterprise has a lot to do with how much money or how much time we have available, or the size of the marketplace in which we hope to sell our products. Rarely do we look at systems design from the perspective of the animals that will live there.

Pastured poultry systems traditionally are based on inexpensive housing and low start-up costs. The hoped-for outcome of raising poultry with fresh air, sunshine and green grass is a healthy, tasty, clean food that delights our customers and keeps them willing to pay us a premium price for our products. Simply put, our task is to not disappoint our customers.

However, when we broaden our scope to include quality of life issues in our management schemes, we must examine the circumstances in which we force our poultry to live. When we look through their eyes, we might be dismayed at what we see.

For the moment, let's consider the poultry industry as a continuum, like you would see a pendulum swing. At the far right, we have the integrated poultry industry, where thousands of broilers, turkeys and layers are raised in huge, expensive barns in a climate controlled environment. On the far left of the pendulum swing, as diametrically opposed as any two paradigms can ever be, we have broilers, layers and turkeys living in very inexpensive pasture pens which provide only minimal shelter from cold winds and rain or blistering sun.

Which of these paradigms is best for the poultry is not the question. The real question demanding to be asked is, "Can't we do better?"

At the far right of our poultry continuum, a new barn to house up to 50,000 broilers will cost as much as $256,000 to build and equip. This figure is taken from a recent prospectus given to an acquaintance who is considering building such a broiler barn on her farm. That 1/4-million dollar investment pays for zoning and building permits, site engineering, building design, land clearing and grading, building construction, automatic ventilating and heating systems, feed bins, automatic feeders and water fountains, and training for the new owner-operator. After all of that the owner-operator can expect to raise six batches of broilers per year, with 50,000 broilers per batch. Working full time, raising 300,000 broilers per year, the owner-operator can expect to earn a pre-tax, net income of about $20,000.

It is my contention that if any of these farmers had that $256,000 available in cash, they would not be so likely to invest it in a poultry barn for an expected annual return of only .07 percent. They would more likely invest it in a mutual fund that would return at least $25,000 pre-tax income without having to lift one finger or do one day of work. What we are beginning to see instead in the industry is offshore production and migrant labor filling the gap. If the commercial broiler industry ever collapses it will not be because of consumer complaints or from government interference. It will be simply because the rate of return has crept so low that intelligent money is no longer attracted to the investment.

But, what about zero mortality and the view at poultry eye level? After the farmer gambles her land and commits all her working capital and assets to this commercial broiler house are the poultry actually raised in what we would consider humane conditions? To pay for such an expensive facility requires that every single square foot must be used to maximize productivity. Broilers are crowded into 3/4 ft. sq. per bird. There is no fresh air or sunshine. In fact, some broilers never see the light of day until they are loaded onto semi-trucks and carried to the processing facility.

Obviously, in this big poultry scheme, "mortality" is only a category because it costs money to lose birds. No thought is given to the comfort of the poultry, or inhumane conditions or handling. As long as the birds grow to a reasonable size on a minimal amount of feed within an acceptable period the grower and the integrator company are satisfied.

Charles Wampler Jr. is the son of the man who started the integrated poultry business in central Virginia some 50 years ago. Wampler Jr. has been quoted as saying that when he was young it took 16 weeks and 20 pounds of feed to grow an acceptable broiler. Today, it takes seven pounds of feed and seven weeks to get the job done. To agri-business, this looks like progress. To those of us in the know, it's just so much chicken manure.

But, those of us who have chosen to do something about it may have swung the pendulum too far in the other extreme. I am not in any way defending the confinement poultry industry. However, in our haste to distance ourselves from the integrators, we should not ignore the very real benefits of raising poultry-in confinement. For one thing, the confinement raised broilers or turkeys are inside an insulated building. They rarely get too hot, or too cold, and they are never subjected to scorching sun or cold rain, as is all too often the case in chicken tractors and pastured pens.

In our erstwhile pursuit of better designs for outdoor poultry production we've either overlooked the mortality rate, or chosen to ignore it. In how many businesses is it possible to make a profit when you kill off 1 out of every 10 products you grow?

In my years of growing broilers in chicken tractors, I have lost poultry to cold rain, cold nights, predators, and heat stress. The question I have failed to ask myself until just recently is, "why put the poultry out there at all if they are destined to suffer and even die?" I've always assumed that because I was growing them "on pasture" and not "in confinement" then my model was better than "their" model. How many of you have felt this same way?

The reality is, we pasture poultry producers lose far too many broiler chicks and turkey poults because we put them in circumstances where they simply can't survive. For years, we've railed against genetics as the primary cause of mortalities. What we've chosen to ignore is that the broiler strains we are using are pretty darned tough after all, if they can survive the way we treat them. If you over-heat a chick one day, then drench it with cold water and chill it to the bone the next day, and then it dies, can we blame that on genetics? Is this model any more humane than "their" model?

When a broiler chicken dies it may be sudden, as in the case of being crushed during a pasture pen move. Or the death may take weeks, and finally occur at the end of a long period of stress. Only a few chicks may actually die from heat stress under a baking August sun, or hypothermia during a cold rainy night in May. What we overlook is that ALL the chicks suffered the same stress, but only the weakest few actually succumbed. The survivors will not perform up to their potential. And we blame that on "genetics."

I've heard some growers say that these modern day broiler chicks and turkey poults are pre-disposed to die for no apparent reason. In the case of Hubbard's Hi-Y males, that might actually be true. The Hi-Y may be the one strain of Cornish Cross that has the apparent genetic potential of suicide by over eating.

However the other Cornish Cross broiler strains such as Cobb, Arbor Acres, Ross, Peterson, Hubbard Classic and Hubbard White Mountain will gain weight quickly, efficiently and without leg problems. But, only if they are given an adequate diet and plenty of clean water, warm and dry conditions, and protection from crushing and predators.

After years of budgeting 10% losses in my flocks I have changed my primary design directive. I no longer make futile attempts against all odds to make the poultry go on pasture straight from the brooder. Taking young chicks from a 90° F brooder that is protected from wind and rain and putting them on cold, wet grass in 50° F weather just to get bragging rights to "grassfed" is foolish.

The jury is still out on this one, but it looks to me like the grass has very little value to the broilers anyway, other than a minimal amount of vitamins. Certainly they can't gain weight on grass, even in the lush spring growth, and especially not in the dry summer. On our farm we are typically using 16 lbs. of feed (more than two times as much as "their" model) to get a four-pound dressed weight bird, and it takes us a week longer to do it.

So, why do we want to double our feed bill and kill off a bunch of chicks just to be able to say they were grown on pasture? Surely, we can take the time to harden them off from the brooder, and give them their last two to four weeks on grass, which should be enough to produce a healthy bird with a free-range flavor. Asking for more than that just causes more feed consumption and unnecessary mortalities.

What I am advocating is to take the very best aspects of confinement models and merge them with the very best aspects of pasture-based systems. The resulting model will be one that provides the poultry, and the grower, the best possible environmental and biological security, the highest possible quality of life, the lowest investment, and will optimize (versus maximize) the pasture-based economic profile.

Using the internet as a design modification tool we can instantly exchange information and experiences through poultry chat rooms and list-serves. We can compare on-farm research results and debate the merits of a hundred different ways of doing the multitude of tasks associated with a successful, profitable and satisfying pastured poultry business. Instead of going through an entire season to test our suppositions, we can query an e-list and get instant feed back from one or several growers who have already tried whatever it is we're thinking about.

There are several ways I think we can come closer to zero mortality. One is through better chicks. Not by new genetics necessarily, but by maintaining on-farm (or regional) breeder flocks. When we order broiler chicks from any of the national hatcheries we can never be sure which of the several different Cornish Cross strains they are sending us. What worked well on the last batch may not work at all in future batches.

Several of us are raising broiler chicks from hens raised on pasture, either from our own breeder flocks or from Tim Shell in Virginia. We don't have research results to back this up yet but our collective experience has been that the Pastured Peeps show better health, more aggressive foraging, and faster gains. Last year I grew out 300 peeps from Tim Shell's breeders, and 700 peeps from my own breeder flock. We averaged a four-pound dressed weight in seven weeks with very low mortality.

Not every area of the country has a regional pasture-based breeder flock, so you may need to start your own on-farm hatchery. It's not hard or expensive to do this. Low cost incubators are readily available, and keeping a breeder flock is no more difficult than keeping a layer flock. An on-farm hatchery gives you chicks that are not stressed in transport. Of the 300-plus we've already hatched here in April we lost only two chicks. Taking them directly from the incubator's hatching tray to the brooder means we don't have to rely on airlines that won't carry them except during certain temperature ranges and post offices that leave them sitting on a loading dock during a cold, windy, rainy night.

Better feed can also have a positive effect on broiler and turkey performance. I've always considered that the ideal small farm system would be one in which the poultry manure is used to fertilize the fields where the poultry feed is grown. It takes about three acres of grain field to produce the feed for 1,000 broilers. Interestingly, the manure from 1,000 broilers is enough to fertilize about three acres of land.

Better shelter is another major key to survival in pasture based poultry systems. I am not defending the confinement broiler houses with their environmental controls. These facilities are simply too expensive and don't offer the access to the outdoors and green growing pasture that we want.

Likewise, I am unable to defend the typical 10 x 12 ft. pasture pen or the 4 x 8 ft. chicken tractor. These pens are too expensive and labor intensive for the small number of birds they hold. More importantly, they simply do not provide the protection from cold wind and rain or blistering sun that these birds need.

Some of us are working with the day-range system that uses mini-barns on skids combined with electrified poultry netting to shelter the broilers, layers or turkeys and to protect them from predators. This system works remarkably well, given that it is also about 50% less expensive and 50% less labor intensive than other pastured poultry models.

There are some drawbacks, of course. One is that the system requires more thought from the human manager. Rather than just routinely moving the pasture pen each morning, the manager needs to be able to read the grass and understand the poultry behavior to tell when it is time to move them to a new site. Another drawback is the airborne predators. The electric poultry netting will stop all other predators, but we don't yet have a good way of locking out hawks and owls.

We are doing several trials this year to determine if the lower start up and labor costs combined with increased weight gain and lower stress-related mortality will substantially offset any losses to airborne predators.

As always, I encourage everyone to think like a wing walker. Don't let go of one hold before you have the next one firmly in hand. Pasture pens and chicken tractors have brought us this far with enough good results to keep us going. Let's continue to use them, hopefully more wisely than we may have in the past, until such a time as a new and demonstrably superior system presents itself.

APPPA PRODUCER PROFILE

Karla Tschoepe
By Jody Padgham Summer 2004

Images of west central Colorado turn to mountains, skiing (Aspen, Vail…), movie stars and money. Probably not chickens. But Karla Tschoepe is happy that she found her little piece of paradise on the dry western slope of the Rockies, where she has been growing and selling grassfed chickens for over 35 years. "We're at 7000 feet, with a 180 degree view of mountains in every direction. It's not the easiest place to raise anything, but I wouldn't trade with anyone" she quips on a recent afternoon. "We've never had a lot of money, our equipment is kind of beat up, but we sure eat well."

Farming on 200 acres plus an additional 400 she manages for the neighbors, at one point Karla raised 2,000 broilers per year, but now she's scaled back to around 700 plus about 150 turkeys and several head of beef. "I've gotten too old for all that work," she says. "The demand is always growing, though, so I had to convince some of my customers to grow their own, and now a few of them are really doing a great business themselves." Karla was only able to talk her customers into growing their own birds by agreeing to become the local processing plant. "I bought my Pickwick Jr. Plucker before I even got my first chickens," she says. "I knew that I wouldn't last long raising chickens without a good system for processing." Karla continues to put that plucker to good use, having built a non-inspected processing plant with two rooms (one for plucking, one for evisceration and cooling) where she happily processes her own and the neighbors' birds. Karla, with two helpers, has refined the system so that they can process 20 birds per hour.

Karla has a unique customer base among the wealthy and aging "Hippie" population that took over the area just after Karla bought her Wildwood Ranch in 1969. "Money isn't really an issue with these folks," she tells me. "They appreciate good food, and the fact that I get as much organic feed as possible is important to them." Karla's marketing has been entirely word of mouth. "When I started with organic, it wasn't a popular thing," she says. "But now, 35 years later, the neighbors who thought I was a kook think organic is a pretty good way to go."

Farming in dry land is no easy thing. Crops must be irrigated, and water is always an issue. This year is especially hard, as the west experiences a severe drought. It is hard to grow grain in the area—corn is almost impossible, and soy protein must be trucked in. "The best we can do this year is non-GMO (genetically modified) from Nebraska," Karla says. "I sure would like to find a feed recipe that will grow good broilers without soy protein."

35 years ago, when grocery store chicken was going for 30 cents a pound, Karla told the folks interested in her family chicken production that they could get a bird for 75 cents a pound. They didn't blink an eye, and now she is still selling to some of those same customers for $2.50 per pound. As I talk to Karla, it is obvious that customer service is her forte. "I know, or ask, what my customers want," she tells me. "If they want a five pound bird, I wait till I have some that size and then I call them and tell them their chickens are ready." She sells whole fresh broilers and turkeys, processed with giblets, and will freeze the birds for an additional charge. "It is illegal for us to cut them up, so we don't do that." She adds. To this point she puts the giblets in the chicken cavity, but she's thinking of putting the organs in a separate bag. "I get these calls, 'why didn't you take the guts out'?" she says. "Some people don't know what a gizzard is. If I separate them out, maybe it will help them figure it out."

APPPA PRODUCER PROFILE   Tschoepe

A few things really stand out with Karla: her willingness to do what needs to be done, and the energy she has available to help other folks out. "We had to get feed brought in from Nebraska or further," she says, "so I talked one of my neighbors into starting up a feed mill with organic feed. He's doing a great business now, and it's a real service to the area." They are very happy with poultry pre-mix and other products from Helfter Feed (APPPA Business member from Osco, Illinois).

Karla sees the biggest challenge in pastured poultry right now to be improving the genetics of a pastured bird. "It used to be better, but the last 10 years have been terrible for leg problems and deaths." Karla figures they have been losing about 20% of their birds each year due to health problems. "We've tried new chicks, we've tried new feed, we've tried new management." She claims. "I think it's just that the genetics are going downhill." She has been working on a neighbor that has done a lot of breeding of chickens for fly-tying feathers, hoping he'll help devise a grass-based broiler chick for their region. "I think we need to go back to local hatcheries." She says. "We will loose shipping privileges before too long, and then where will we be? We all need our own hatcheries, working with genetics that are right for our region. Not every chicken thrives at 7,000 feet." Discussing the fact that special chicks would probably cost more, Karla notes that with the losses she has been experiencing, at even double the price she pays now, she'd still be doing ok in the long run with a bird that would thrive in her region on grass. She suggests that APPPA could help facilitate the growth of regional hatcheries and pastured genetics.

Karla has three "rickety wooden houses" as shelter for her birds. She runs them in permanent fenced pastures, which are planted to alfalfa (renewed as needed). Six-foot fences keep out predators, and the dogs discourage coons and skunks. The dirt floor houses are shut up tight at night and opened before dawn. Water is distributed from a 50-gallon drum perched up on legs, with hose running to a PVC pipe cut lengthwise, sealed on the ends and set up with a float valve of the type found in air conditioners. Karla adds OxyBlast from Helfter to the water (a hydrogen peroxide additive), and thinks it makes a difference in chick thriftyness. She broods chicks in the same houses, using alfalfa hay leaves for bedding. "I go and rake around all the hay piles to get the fine leaves," Karla tells me. "The chicks really love to nibble on those leaves."

Karla is a great fan of Joel Salatin and Charles Walters (Acres USA), so it is no surprise that our conversation eventually turns to philosophy. "I'm feeling really heartened," she says "people these days are really interested in good food again." Karla feels that the world, and people's relationship to food, really changed after World War II. "Before the war, most people lived on farms and knew what it took to raise food." She says. "Then the young men went off to war, and a lot of the folks that stayed home came to the cities to work in factories. All of a sudden everyone had money, and they bought food rather than grew it. Houses were built, and everyone had a job. The '50s were fabulous." She continues: "The children of the '50s (baby boomers) were given everything, they didn't have to go hungry or work for their food. They became the generation of the '60s, when the drug culture took over. People just lost interest in good food." Now, Karla is seeing a mellowing of the '60s generation, and a renewed appreciation for good food and those who grow it. "People want vital food again." She notes. "There is a new way of thinking."

It seems like a good thing that Karla, and HER "new way of thinking" have set up shop in the western foothills of the Rocky Mountains not far from Aspen, Colorado. A lot of good people are thriving on the good food that she has provided for so many years.

The Biological Integrity of Pastured Poultry
By Richard Gilbert Spring 1998

In our eagerness to help beginners with the nitty-gritty how-to of this exciting pastured-poultry enterprise, I fear we may risk glossing over the bedrock principles of biology, agronomy, and animal husbandry that make it powerful and sustainable. Our tips on processing and marketing may crowd out a proper emphasis on the biological underpinnings of our methods.

Are beginners getting the most vital message from experienced practitioners?

I became concerned after attending the Great Lakes Grazing Conference in Akron, Ohio, in February. At a session on raising poultry on pasture it was clear that many people were aware of the model pioneered and refined by Joel Salatin at Polyface Farm in the Shenandoah Valley of Virginia.

The audience gave the impression they had been mulling it over for a long time. You could feel the wheels turning. Was it right for them? Should they take the plunge?

At the break I spoke with a man who seemed to have collected his ideas equally from a couple of methods and was planning to raise some chickens this year. Another fellow told me of a vast farmers market on the west side of Cleveland where women dressed in mink coats paid outrageous prices for "Amish grown" broilers. It was obvious that the success of pastured poultry and the burgeoning demand for better chicken, driven by the disgust and fear of consumers, was creating much entrepreneurial interest.

What may be getting lost in the process? As hundreds and thousands more pastured poultry farms spring up, some sincere farmers may think they "get it" when they don't. The waters are muddy, and APPPA members owe beginners an effort at clarity. The public also is hopelessly confused about "free-range" and "organic" chickens and just what such terms mean. Those of us who follow the Polyface model of pastured poultry add a further wrinkle.

"Free-range" birds are being sold both by well-meaning farmers and by agribusiness factories. These factory birds are as bad as conventional birds, for that is what they are. One of the articles in the March 1998 Consumer Reports was an expose of free-range chickens, illustrated by a picture of an Amish farm with some "free range" broilers huddled in the dirt outside a large chicken house. The article explained the USDA's liberal requirements for calling fowl free range (basically access to the outdoors, whether any birds go out or not, and regardless of the quality of the range).

Particularly revolting was the cynical attitude of the large, supposedly alternative producers. One executive, who correctly condemned the "myth" that Rock-Cornish are capable of running around, nevertheless remained on the fake free-range bandwagon. A company president, asked by Consumer Reports about a mere 50 chickens out of 5,000 that were outside pecking in the dirt, said "They move around, and we can get the price we're asking, so obviously there must be something right with the chicken."

What about farmers who place pens on pasture and simply open the door? Maybe they think that their birds are superior because the chickens are in the fresh air and sunshine, even though the range is mud or picked-over grass and the medicated feed ration was formulated by an anonymous corporation or by the local farm cooperative. Their chickens undoubtedly are better than factory birds, but they are not in the same league as chickens raised in the Polyface (Joel Salatin) model-nor does this "free-ranging" improve the land or integrate with other forage-based enterprises.

To practice the Polyface model, let alone to refine and add to it, a person must understand the philosophy and biological principles at its heart. Much of this involves managed grazing. An expert, or even we excited amateurs, could talk for days about animal impact and rest, about the health bonuses-for animals, and ultimately for humans-of succulent grasses and legumes, and about the environmental and economic benefits of grass farming.

Usually we don't have the luxury of days in which to explain ourselves, so we must boil down the ideas to their core. Emphasize why daily or twice-daily moves to fresh pasture accomplish the goals of healthy birds and healthy pastures, as well as offering the potential for multi-species grazing and income. Tell them why keeping the birds moving across clean forage eliminates the need for coccidiostats in the feed.

Point out that free-ranging poultry over one area is nothing but continuous grazing. Explain why this is at least as bad for poultry and pastures as when hoofed stock are involved: unrelenting animal impact leads to a steady deterioration in the diversity, quality, and quantity of forage-and that leads to weeds, bare spots, and the risk of parasite and disease buildup.

Raising pastured poultry, Joel Salatin says, that is fit to eat takes the commitment and labor of individual farmers and families. To a large degree, this method substitutes labor and creativity for capital. Many producers are needed in many communities. The public is concerned and confused and we must take the trouble to educate them. Tell your customers what you do and why. Show them.

And keep in mind that we must continually help well-meaning beginners understand and practice the underlying principles of pastured poultry. Rather than more opportunists working an angle, we need more strong allies to help spread this exciting, challenging, and inspiring revolution.

It's all in the Vernacular
Shelia Hamilton Fall 2000

When marketing product to the general public it is always important to consider the impression you are giving by the words you use. As you describe your method of raising pastured poultry (or free-range for that matter), be it over the phone or in person, your potential customer is forming a picture in their mind of living conditions that you are placing your birds in.

We raise our birds in moveable SHELTERS not PENS. This conjures up a much more favorable picture for the customer. The word "shelter" denotes a place that is safe and will keep the birds from harm, whereas "pen" depicts an enclosed space or barrier type atmosphere.

We take our birds to the PROCESSING FACILITY not the SLAUGHTERING PLANT. Even comparing the words "processing" and "slaughtering" can make a difference. "Processing" is a much gentler word, whereas the word "slaughter" can sound very gruesome.

Remember that you know how your creatures are raised and handled but your prospective customer only has their imagination to draw from.

Definitions:
SHELTER - To provide protection for; cover from harm or danger; shield. To place under cover for safety.
PEN - Any small place of confinement; a small enclosure.
PROCESS - Produced by a special method. A course or method of operations in the production of something.
SLAUGHTER - The act of killing; specifically the butchering of animals for market. Wanton or savage killing; massacre; carnage.

Which Would You Choose?
Kalista Kaufmann   Fall 1999

"*Woke up, got out of bed; dragged a comb across my head. Found my way downstairs and drank a cup, looking up I noticed I was late. Found my coat and grabbed my hat, made the bus in seconds flat. Found my way upstairs and had a smoke and somebody spoke and I went into a dream. Ack! Cough! Blah!*"

Whew, I have got to remember to stop singing. And speaking of smoke, I really wish they'd put in better ventilation.

I live in a huge dark single purpose building. And what is that purpose, you might ask? To raise chickens, just like me. The building is split in half, by a walkway down the middle. I live on the right half. The floor is a mixture of sawdust, feces and moldy feed. The walls are constructed out of sheet metal, which most of the other chickens like to peck at. Can you imagine the racket 8,000 chickens can make? Now I know what it would be like to live in a drum. The ceiling is also sheet metal with automatic waterers hanging from it. Unfortunately, they are rarely cleaned out and are always murky.

This is my home. Oh yes, and did I mention that I share my side with five thousand other chickens? No? Well, I do, and it's not exactly comfy living with that many other chickens. Automatic feeders are spread out on the floor. I live on the far end near the back. I pretend that it is less crowded back here, but I know in reality it's not. I have about a foot of my own "personal" space. The challenge is to travel another five feet to reach the water and feed. I do have to occasionally crawl across my neighbors' backs to get to it. I tend to just sit around and get fat. You could say that I don't exactly get my daily allotment of exercise. I suppose I could run across the backs of my fellow kind, but I don't think they would like that too much.

I like to start my day off with a song. My favorite band is the Beatles. The workers who make sure everything is running smoothly often listen to the Beatles. That's how I came to like them so much. My favorite album is Sergeant Peppers Lonely Hearts Club Band. Unfortunately, it is so dusty and the air smells terribly of feces that I usually break down in a coughing fit while attempting to sing. As you probably noticed that is exactly what happened to me again this morning.

The air is so bad that many of us develop a chronic respiratory illness and die prematurely. The workers are encouraged to wear masks. Some of the older ones refuse it. You can hear them coming from way down on the far end because they are coughing so much. Guess they never thought it could harm them.

I can't argue, though. Chickens are meant for one purpose, food. We are mass produced, and that's that. There is no other life for a chicken.

"It's a beautiful night tonight, isn't it?"
"Yes, it is. It has been awhile since there were no clouds…"

In fact, it has been almost a week since the sky was clear. I live out in the pasture. At night I like to gaze up through the chicken wire and view the stars. I know almost all of the constellations. My house is made out of wood, chicken wire, and metal. It's twelve feet long and ten feet wide. The back half is covered by metal. The front half is divided in two, with one half covered by used aluminum newspaper plates and the other by chicken wire. When it's not raining, most of us chickens like to lie in the open half and take in the stars. If it's raining, we head for shelter in the back. I live with fifty other chickens. We all know each other. Every morning the workers come down to move our house to fresh grass. That's the best part of the day; when they move the pen there is always a bunch of insects. The best part of waking up is a fresh juicy grasshopper in your beak. When it's moving time, all of us rush to the front to catch all of these insects.

After we have been moved, another worker comes along and fills our feeders. Our feeders are three PVC pipes cut in half which hang in the front of our house. They feed us a mixture of organic corn, fishmeal, soybeans, and sea kelp. Occasionally they sprinkle some grit on top to aid in our digestive process. Then a worker will come along and fill the bucket that holds our fresh water. The water is always cool and crisp in the morning. The bucket is connected to our water dispenser that hangs in the front. In the evening the workers come back and fill our feeders and water again.

We all have plenty of space and it's nice to lie in the fresh grass and smell the newly cut hay drifting by. I prefer the evening when the sun goes down. The pasture we live on is a hill that slants toward the west. Each night brings a new sunset that is better than the last, with colors of orange, red, purple and yellow. It's one of the most amazing things to see. I can't imagine life any other way.

Kalista Kaufmann, Chippewa Falls WI wrote this piece for a senior English paper.

Her teacher's comments: "I wonder what differences in the chickens might exist because of the way they're raised?"   Hmmm...

Alternative Poultry Production in France
Skip Polson Summer 2000

About 10 years ago I first learned of a special type of range poultry being raised in France. I saw photos of red-feathered birds, and it sounded like people were calling then " La Belle Rouge" chickens. Now, I know that "rouge" means "red" because I've been to Baton Rouge-("red stick") Louisiana. And I know "belle" means "woman" (as in "Southern belle"). So I cleverly used those facts (nearly my entire knowledge of French, by the way) to interpret this as "the red lady." I thought that sounded like a pretty nice name for a productive and highly appreciated red chicken.

Little by little over the years I heard more about these French birds and the system of producing them for market. When I finally discussed this in depth with someone who knew more about it, I realized that I'd been misinterpreting the French words all along. They weren't saying "red lady" at all. They were saying "Red Label" (Label Rouge in French, which sounds exactly the same to me). What a revelation! They weren't talking about a breed of chickens! They were talking about a certification and labeling system which is used in France to denote a high standard of quality and wholesomeness. The red label also assures consumers that specific approved production practices have been used to produce the product and that its stated place of origin is true.

The Red Label system began in France in the early 1960s, as French consumers objected to the increasing industrialization of their food system. Many people did not want to lose their ties to traditional food production practices, and the sources of local and regional foods which they preferred. The Red Label system is now used extensively throughout the country. Specific Red Label standards and production requirements have been set for many different agricultural products. If they are produced according to the established requirements, they can be certified and marketed with the Red Label attached. Poultry, eggs, ham, salmon, fresh garlic and traditional sausages are a few examples of products for which Red Label certification is possible.

Poultry producers were among the earliest and most vigorous adopters of the Label Rouge concept, beginning in 1965. Since then they have refined and developed the system in conjunction with the relevant national government agencies (the Ministries of Agriculture and Commerce). The Label Rouge system has earned a very positive reputation over the past 35 years, and is well accepted by French consumers. Now approximately half of the chickens purchased by French households are sold with the Red Label certification, at twice the price of the conventional industrial birds sitting beside them in grocery store meat coolers!

Knowing that pastured poultry producers in North America might learn a lot from the experience and success of this French system, I was very pleased to have the opportunity to be in southwestern France in May of 2000. There, near the town of Pau, I was able to view on-farm production first-hand and discuss the system in detail with members and employees of Groupe Euralis, a large agricultural cooperative.

What did I see? I observed a production system which looks a lot like a standard confinement system on the surface, but which differs from it in many significant ways. The French Label Rouge producers typically use permanent housing, but smaller buildings than the conventional confinement industry. Compared to the conventional industry, they raise smaller batches of birds, with much lower stocking densities. Their birds are encouraged to be outside quite a bit, but they also spend a considerable amount of time indoors. These producers share our concern for feed quality and humane treatment of the birds. They want their birds to live a "normal" chicken life, with the ability to move around freely and have normal interactions with other birds and the environment.

Label Rouge birds can be produced and marketed by a single company (or co-op), or by a consortium of separate companies. In either case, very specific criteria and standards are identified and must be met at every stage of production, processing and marketing. For example:

1. Genetics: the birds must be one of a few approved crosses. Some of the crosses have red feathers. Some black, some have naked necks. All of them are considered by the French Label Rouge producers to have the preferred "normal" rate of growth, as opposed to the Cornish Cross types used by the French confinement industry which grow "abnormally" fast. Label Rouge birds must be raised at least 81 days, compared to 41 to 45 days for the confinement industry birds.

2. Hatcheries must meet strict sanitation requirements and pass inspections and bacteriological tests twice per year.

3. Building sizes are limited to 400 square meters with a maximum width of nine meters. Thus, the largest chicken houses allowed are a little narrower and only about one-third as long as the largest confinement houses currently used in both France and the USA. The houses I visited had straw bedding, side curtains for ventilation, solid lower walls, and chicken doors at floor level along the length of the building. They also had several interior chicken-wire partitions across the width of the building to keep the birds in groups of about 800 within the house.

4. The stocking density is limited to a maximum of 11 birds per square meter, while the confinement industry keeps about twice as many birds in the same area.

5. Outdoor access: all Label Rouge birds must have access to the outdoors daily from 9 a.m. to dusk after they are six weeks old. Weather permitting, the birds go outside at a younger age based on the discretion of the producer. This access is provided by floor-level doors, about two feet tall and four feet wide, which are spaced along the length of the building. All buildings are required to have a certain amount of door space, based on the length of the building. The required amount of door space is sufficient for all of the birds to move readily in and out if they choose. If the yard around a chicken house is fenced, there must be at least one hectare (2.47 acres) of yard for each building. Conventional industry birds, on the other hand, are fully confined for their entire life.

6. Starter feed must contain at least 50% cereal grains, while grower feed must contain at least 75% cereals. Neither one can include any animal by-products, growth stimulants, or other additives. Antibiotics may only be used when prescribed by a veterinarian. The conventional industry in France does not have a feed standard.

7. A sanitation period of at least three weeks is required between flocks.

8. Each farm is inspected at least once per flock.

9. Each processing plant is inspected at least once per month, including bacteriological tests. They strive for a very high level of cleanliness in butchering and processing. No water baths are allowed for Label Rouge birds; they are placed separately on stainless steel racks and air chilled.

These are just a few of the requirements which must be met. Additional specific criteria are listed for each segment of the industry. Detailed records of compliance with these requirements are kept by every participating hatchery, farm, feed mill, slaughter house, and distributor. This system enables every Label Rouge bird sold at retail throughout the country to be completely traceable. Each product package is given a code number which allows that product to be identified back to the processing plant, farm, hatchery, and breeder flock that it came from.

Based on all of these requirements the producers of Label Rouge birds are able to characterize them as a farm chicken raised in a natural manner; with no animal by-products in their feed; and processed, packaged and distributed in a rigorously clean and sanitary way. Label Rouge producers also work hard to capitalize on the image of the region of the country where they are located, whether this is in the foothills of the mountains, the coastal forests, or the open countryside in central France. This is the equivalent of me promoting my birds because they have received all the benefits of the "ocean breezes, gentle showers and tropical sunshine in our Aloha State."

The quality guarantees of the Label Rouge-system are appreciated by many French consumers. Label Rouge birds now occupy half of the retail market for chicken in France, and they typically sell for twice the price (or more) of the conventional confinement birds sitting right beside them on the shelf. In the grocery stores I visited, all of the chickens were attractively packaged in clear plastic wrap and bright colorful labels. The Red Label seal was prominently displayed on the Label Rouge birds. The whole-carcass Label Rouge birds ranged from 3.5 to 3.75 pounds in weight and were priced from $2.14 to $3.11 per pound. The conventional birds beside them weighed three pounds and cost $1.10 per pound.

The producer I spent the most time with was Mr. Maysonnave Christian, a former baker, blacksmith and tobacco farmer. Mr. Christian is now focusing on Label Rouge poultry production, and this enables him to farm full-time on his family's 10 acres. In addition to two permanent chicken houses, he also uses an innovative system of portable houses which he built himself after studying the design used by another farmer. Each of his eight portable houses is floorless, five meters wide and 12 meters long, with side walls two meters tall, and a peaked roof. They are designed to sit in one place for the life of one flock of chickens, and then be moved with a tractor to a new location. Thus they were built as lightweight as possible: the framing was done with angle-iron and the walls are light-weight plywood, with light-weight corrugated metal roofing overhead. Like all Label Rouge housing, they have ample floor-level doors to allow the birds easy access to the grassy yard surrounding them.

Mr. Christian also built a special dolly to help him move these houses. It is simply a single axle with a trailer tire on each end. It is a little more than five meters long, just long enough to fit under the width of the houses and stick out a little bit on each side. Thus, after the chickens have been taken to the butcher, he can remove the internal equipment (feeders, waterers, and brooder hovers), lift one end of the house with his tractor's front-end loader, roll the dolly axle under that end, set the house down on the dolly, and secure the house to the dolly. Then he drives the tractor around to the other end of the house, lifts it up, and pulls it to its new location on fresh grass for the next batch of birds.

In the new location he puts in new chopped-straw bedding, and cleans and re-installs the internal equipment (brooder hovers, feeders and waterers). Then he lets it all sit empty for the required three-week sanitation break between batches of birds. Mr. Christian typically sets four of these houses in a line, end-to-end with a separation of about 30 feet between them. This enables him to run one moveable water line and one electric supply cable to all the houses.

In the old location he scoops up the old bedding, puts it on the compost pile, and then tills that spot for a new planting of corn, wheat, grass or a vegetable crop, This system seems to work very well for him.

At the time I visited in early May the portable houses were occupied by two-week-old birds who were just beginning to venture outside into the six-inch-tall green grass surrounding the buildings. Twenty yards away was a small cornfield with three-inch sprouts quickly changing the field from brown to green. Mr. Christian said he would soon put up portable fencing to keep the young chickens out of the corn until it was mature enough that they would not hurt it. At that time he would remove the portable fencing and let the chickens into the cornfield to forage for bugs and enjoy its shade through the remainder of their life (most of June and July).

With his two permanent and eight portable buildings Mr. Christian is able to raise flocks of 16,000 birds three times per year, for a total of 48,000 birds per year. Mr. Christian is one of 18,000 farmer members of Groupe Euralis coop, 400 of whom raise chickens. All together, they raise seven million birds per year. He enters into a contract to raise birds for the coop, and the coop provides the chicks, the feed, and technical advice and assistance to him through its employees. This arrangement is very much like the contract grower systems we are familiar with in the USA.

While I was not privy to precise financial information about Mr. Christian's farm enterprises, he told me that he is paid more per Label Rouge bird than conventional producers receive for their birds. It is important to keep in mind, however, that not all of his birds can be marketed with the Label Rouge. If individual bird have any physical defects or do not weigh the required minimum at slaughter they have to be marketed without the Red Label at the lower prices paid for conventional poultry.

It also appeared to me that, as a member of the coop, Mr. Christian had a much greater sense of genuine ownership, partnership and effective participation in the whole enterprise than contract growers typically do here in the USA. Of course, this is a very subjective observation, and it may be incorrect. The Label Rouge system itself, however, encourages a sense of partnership among the various entities involved in producing the final product. They all have to work together, each one doing their part toward achieving the top-quality Label Rouge goal and its higher payment in the market place. If one of the partners fails to satisfy the Label Rouge requirements, thus causing the birds to be marketed at the conventional price without the Red Label, that partner must compensate the other partners for the loss.

Lessons for all APPPA members:
1. The Label Rouge producers in France have succeeded in clearly differentiating their product from that of the conventional confinement industry. Here in North America, we have not accomplished this yet on a broad scale, despite the fact that some individual producers are doing a very good job of educating their own customers. We do not have clear definitions which distinguish pastured poultry from birds raised in other production systems. For example, the term "free range" does not have a meaningful and precise technical definition, even though it is a term commonly used by the general public. And, in fact, the term is used misleadingly by some poultry producers.

2. The French government and the Label Rouge poultry producers have kept quality, cleanliness, traceability, and truth in labeling as top priorities in the Label Rouge certification system. We would always be wise to do the same in any marketing or certification efforts we undertake, whether they are individual efforts or joint ones.

3. A very large number of French families are willing to pay double (or even more) for Label Rouge birds. Clearly it is possible to compete effectively in the marketplace with the conventional confinement industry. We should never forget that.

4. We need to realize that we (APPPA) can take labeling as far as we wish in the future. We can even go as far as something similar to the French Label Rouge system if we think that would be useful for us. Certainly we need to establish a clear definition as in lesson #1 above. And at some point in time, we may want to establish certification standards for part or all of the pastured poultry production system. These standards might deal with things such as stocking densities, housing requirements, pasture quality and rotation methods, feed composition, butchering and processing methods, marketing practices (such as the maximum allowable shelf life of birds in retail stores), and other components of our system. Many of our producer members will always want to market directly to their customers they may never need detailed labeling and certification criteria. Some others, however, may wish to pursue enterprise partnerships and indirect marketing strategies which would benefit from such standards. We need to recognize this and be open to consideration of this possibility. Do we now want (or need) a greater degree of structure and formal organization in the North American Pastured Poultry industry? Will we in the future? These are questions for all of us to consider. And they are probably going to remain before us for some time to come!

Merci beaucoup to Mr. Francois Paybou for teaching me about the Label Rouge system and for helping me make the necessary contacts for farm visits in France. Many thanks also to Mr. Maysonnave Christian, Mr. Yves Bignalet and Mrs. Marie Pozzo di Borgo of Groupe Euralis for their wonderful hospitality during my visit.

I am also very grateful to Heifer Project International (HPI) for enabling me to combine this visit to France with other HPI-related travel. (To learn more about HPI, visit www.heifer.org)

More information about the Label Rouge system can be found in the following sources:
¤ Paybou, Francois, "Technical And Economic Feasibility Study Of Adopting French Label Rouge Poultry Systems To Illinois," Thesis submitted in partial fulfillment of the requirements for the degree of Master of Science in Agricultural and Consumer Economics in the Graduate College of the University of Illinois at Urbana-Champaign, 2000.
¤ Westgren, Randall E. "Delivering Food Safety, Food Quality, and Sustainable Production Practices: The Label Rouge Poultry System in France," American Journal of Agricultural Economics, 81 (Number 5, 1999), pages 1107-1111.

Lessons From A French System of Production- Label Rouge
By Keith Richards Fall 2001

About half of the poultry producers in France raise their birds under a production system known as Label Rouge (Red Label). They provide their flocks access to large shaded open spaces; feed them no animal matter or antibiotics and raise them at least 81 days before slaughter -- as opposed to 40-50 days in industrial systems. By using breeds that have been selected for slow growth potential and free-range conditions and by following production and processing standards that maximize quality, Label Rouge products are proven to be superior in taste to other poultry on the market.

Although it is more expensive to grow birds in the Label Rouge system, farmers receive more than double the price of industrial-raised poultry by promoting their high quality and marketing under a coordinated labeling system. And French consumers are responding positively to their products. Label Rouge has been able to capture 30% of the overall market for poultry in France and 60% of the whole bird market, with increasing sales every year.

The Label Rouge farmers coordinate production, supplies, processing, and marketing within organized filieres or regional "supply chains," allowing them to hold down production costs and maintain control of pricing decisions. To ensure consistently high quality and the trust of consumers, the whole system is guided by strict standards designed by each filiere, certified by independent organizations, and regulated by the government.

Developing a Strategy for Family Farm Survival
The Label Rouge system was started in the 1960s by a group of traditional poultry farmers in response to the spread of factory farming after World War II. Their primary concern was to preserve the economic viability of small family farms. While they knew they couldn't compete with the new industrial poultry producers on price, they felt they could beat them on the basis of quality. So they developed a production system that would produce superior tasting products, and a business structure that would keep control in the hands of family farmers to ensure a fair economic return.

After years of refining their systems, the Label Rouge production standards create high quality meat through numerous steps. For example, houses and flocks are smaller -- about one-fourth the size of conventional poultry houses -- to reduce stress. Natural feed rations are used with no animal by-products, and routine medications are not permitted. Broilers require access to the outdoors, and there is a minimum sanitation period in poultry houses of three weeks between flocks.

Farmers are required to raise specialty "rustic" breeds that grow at a slower rate. Meat from these birds has a firmer, more flavorful texture. According to researchers at the National Institute for Agricultural Research in France, the special breeds and slow growth account for 80% of the difference in taste.

Transport to processing must be less than two hours. Label Rouge poultry are processed using air chilling instead of immersion chilling, which is the standard in the U.S. Air chilling is considered superior in reducing microbial cross-contamination, along with preventing water uptake from a chlorinated chill tank.

Food safety is also a crucial part of the program. A pathogen reduction program is applied not only at the processing level (as required in the U.S.), but also throughout the whole system. For example, breeding flocks, hatcheries, feed mills, farms, processing plants, and transportation are all monitored for salmonella and other pathogens. Because of this, only three percent of final Label Rouge products are contaminated with salmonella compared to 69% of other French poultry products.

Coordinated Management
Each group of Label Rouge farmers works collaboratively, not only with each other, but also with breeders, hatcheries, feed mills, processors, and information providers. This whole unit makes up a *filiere*, with their products identified by a unique label.

While they are required by Label Rouge guidelines to develop standards and operating procedures cooperatively, each filiere has the discretion to choose a business model that works for them. Some groups of Label Rouge farmers retain ownership of a hatchery, feed mill or processing plant, while others enter into operating agreements with independently owned businesses.

By coordinating with the processors within their filiere, producers control their own distribution, promotion and advertising. This gives them more leverage in negotiating with the large distributor/retailers that dominate the French food system.

The whole arrangement is designed to create a more equitable balance of power between producers and other players in the food system. The producers receive many of the benefits of a vertically integrated system— such as cost savings on inputs and supply management —but unlike corporate vertical integration, control says in the hands of local family farmers and businesses.

Regional Brands Create Marketing Advantage

While all Label Rouge products have a distinctive red label, each filiere has its own brand name that is strongly tied to a regional area. In France, certain regions are associated with certain tastes, so the regional brand builds on this identity for consumers. Each brand has its own image — such as a small farm, birds roaming in a forest setting, or a highlighted regional name—that dominates the product label.

By marketing Label Rouge overall as a symbol of quality, yet tying each Label Rouge product to a specific regional group of farmers, a unique niche in the marketplace is created for each product—a niche that cannot be replaced by cheaper, mass-produced foods.

To make sure that consumers have confidence in their claims, private third-party certifiers ensure that Label Rouge standards are being followed. Inspections are carried out rigorously on each farm, feed mill, processing plant, and hatchery. Taste tests are also done five times a year to ensure products are actually superior to standard poultry.

Each group must meet the Label Rouge baseline standards, but are also encouraged to develop unique standards based on their regional situation. For example, one group uses small portable poultry houses while another requires tree plantings to make their houses blend into the countryside.

Support from Government, Consumers, and Trade Associations

The Label Rouge system has strong support from both government and private sectors. Since the French government is committed to strengthening rural development, federal agencies are responsible for maintaining certification standards, accrediting the certifiers, and protecting against label infringement. The National Institute for Agricultural Research provides research support.

Strong consumer organizations are involved in the development of standards and certification for each filiere. This results in standards that are responsive to consumer interests, such as a recent ban on the use of genetically modified organisms (GMOs). A non-governmental trade association collects a check-off from the sale of each bird to do national consumer education campaigns about the benefits of Label Rouge poultry. Coordinated public education is now also cited as one of the keys to high premiums paid for Label Rouge.

What started as an attempt to save small family farms in one region has turned into an important tool for maintaining the rural landscape and providing economic development throughout France. Small farms and family owned businesses such as processors and hatcheries have developed thanks to the partnerships created around Label Rouge. The Label Rouge system is being applied to other products such as pork, eggs, and butter.

Practical Applications for the U.S.

There is a need to take successful models of sustainable agriculture from across borders and see if systems or practices can be applied in our own backyards.

Anne Fanatico, a technical specialist at the National Center for Appropriate Technology's ATTRA program, believes the Label Rouge system holds promise for U.S. producers. She says, "If the range poultry industry in the U.S. doesn't coordinate, then it will lose the opportunity to further common goals. Growth will be happenstance and a proliferation of terms and brands may continue, contributing to consumer confusion about organic, natural, green, eco-label, free-range, free-roaming, pastured, etc."

Fanatico and Holly Born are heading up efforts to analyze the feasibility of employing key components of the Label Rouge system in the U.S. over the next few years. In partnership with the American Pastured Poultry Producers Association (APPPA), farmers in Wisconsin and Arkansas, Southern SAWG and others they will look at the availability of specialty breeds, the possibility of creating coordinated supply chains, and issues around quality certification in a project titled "Greener Fields."

Their work will feed into the growing demand for eco-labels that go beyond protecting the environment to include concepts of farm economic viability, local food security, and quality in product standards. The Greener Fields project supports this type of work by bringing organizations and individuals together to collaborate on eco-label development, create shared principles, and explain the benefits of eco-labeling to the public.

Poultry Research Review
By Jody Padgham Fall 2003

The modern concept of raising birds on pasture is about ten years old now (Joel Salatin's first book *Pastured Poultry Profit$* was published in 1993.) Though the movement is still very young, interest in pastured poultry continues to grow. One way this is evident is an increasing amount of research dedicated to learning about pastured poultry systems and markets. We thought it would be timely to get you up to date on some recent research, some of which is still ongoing. I will just give you a short list here, and refer you to websites or phone contacts so you can follow up on any research that interests you. We have listed some of these for you before, but I will include them again in case you missed the first time.

The Importance of Genetics: Biological fitness and productivity in range-based systems comparing standard turkey varieties and industrial stocks. Donald Bixby, DVM, American Livestock Breed Conservancy. (919/545-0022) two-year project funded by SARE. (2002-03) Web link to 6 pp first year report: http://www.sare.org/reporting/report_viewer.asp?pn=LS01-122

This research is looking at Bourbon Red, Black Spanish and Blue Slate standard (heritage) turkeys and comparing them to commercial strains in a pasture-based system using DNA analysis and immune system response. Although the commercial variety reached market weight in fewer days and grew to a larger size in the first year of the project, the standard varieties had lower mortality and better immune response. DNA micro-satellite analysis showed some standard varieties are only distantly related to the narrowly bred commercial strains, providing valuable genetic diversity essential for the long-term sustainability of turkeys. Data analysis and interpretation is still under way in this two-year project.

Pasture Raised Products Message and Strategy by Kim Shelquist, Market Research Consultant for the Food Routes Network/ Midwest Collaborators. (Oct/Nov 2002)
On the web at http://www.leopold.iastate.edu/pdfs/pasture_focus_group.pdf or available from the Leopold Center at 515/294-3711.

This research held six focus groups of consumers across the Midwest to learn about customer attitudes so as to better market pastured poultry products. A total of 67 people were involved in six focus group sessions in six locations in MN, WI and IA. NOTE: This was not a randomly selected group, but fit the criteria of "high school diploma or higher, tendency to recycle and see some impact of food buying habits on the environment and minimum of $40,000 household income (adjusted down in one lower cost of living area)"

The summary of the results concludes:
1. In general, pasture-raised is the term these participants favor. Though there were some positive comments about free-range, grassfed and natural.
2. Regardless of the term, a definition and standards must be set. There was much confusion as to what the terms mean and who is verifying them.
3. Opportunities and barriers exist to marketing pasture raised products.
a. Respondents were open to trying new things
b. They care about the health benefits they see from pasture raised
c. They want to support local farmers and have concerns about animal welfare
d. They are not comfortable trusting products without regulation
e. They expect the products to cost more, and that means they must be noticeably

higher quality
f. They question how convenient these products are to find and use.
4. Respondents must know what's in it for them. While they care about the environment and animal welfare, they really care about how the product benefits them and their families. In the end their purchase decisions are based on convenient access to healthy, good tasting food at a reasonable price. Any slogans used to market pasture raised products must emphasize these characteristics.
5. Respondents ask for traditional communications. They get their information about food issues from traditional media (television, radio, print, etc.) and they do pay attention to advertising. Word of mouth from friends and family is a potentially powerful tool for combating the lack of trust in untried products.

There is a lot more information in this paper about these consumer's attitudes. Please reference the paper itself to learn more.

Exploring Market Opportunities for Local and Sustainable Poultry Production and Providing Education in Small Scale Poultry Production for Idaho Growers. Mandi Thompson, October 2002.
This study started with consumer research on interest and knowledge of alternative poultry production. It was found that there are five important characteristics valued when buying poultry products: freshness of product (71%), food safety (69%), quality (67% and USDA inspected (62%). At least 50% of the respondents also said that things such as "humanely raised," Idaho grown,""organic," and "grass-fed," were at least somewhat important. For more detail and other conclusions of the research, go to http://www.ams.usda.gov/tmd/FSMIP/FY2001/ID0334.pdf

---

Consumer Preferences for Organic/Free Range Chicken. Liz Neufeld, August 2002, Kansas State University
This paper offers a lot of good background information, including a comprehensive summary of research that has been done on pastured poultry with costs of production and profitability tables.
A survey sent to 1,000 consumers in Kansas City in September 2001 shows that
1. There was a general lack of knowledge about the existence or availability of free-range chicken.
2. These customers were not price-sensitive (a $0.50/lb price difference had no noticeable affect on customer buying)
3. Consumers reported that taste, appearance and USDA approval were the most important factors to them in purchasing chicken, in that order.
4. Purchasing a healthy product was seen as the greatest motivation in seeking free-range chicken.
5. Free-range chicken rated above regular chicken in all attributes except value, and overall taste was rated very favorably.
(For the purpose of this study, free-range was taken to be "access to outdoor pens and allowed to roam and forage freely.") The full paper can be found at http://www.agmrc.org/poultry/info/ksufreerangech.pdf

---

Impact of Three Pasture Species and Commercial Mash Diets on Pastured and Commercial Caged Layer Hens' Eggs: Omega-3 Fat, Vitamin A and E content H.D. Karsten, P.H. Patterson, G.W. Crews and R.C. Stout; Crop & Soil Sciences and Poultry Science Departments, The Pennsylvania State University (2003).

This research showed that eggs produced by hens on foraged pasture legume mix had an average 18% more omega-3 fat than hens on foraged grass pastures. Omega-3 fat, Vitamin A and E were higher in eggs of hens that foraged pasture (summplemented with commercial mash) than hens fed commercial mash only.

Three Year Study of Large-scale Pastured Poultry Farms Don Schuster, University of Wisconsin Center for Integrated Agricultural Systems (2002). (See the full report on page 7.) Key findings include:
1. Pastured poultry can be an excellent supplementary enterprise (under 1,500 birds per year) on diversified farms, particularly if these farms already direct market other farm products.
2. Net incomes from the five farms studied averaged nearly $2.50 per bird.
3. Microbial tests indicated that on-farm processing can be at least as safe as processing at state or federally inspected facilities.
4. When a set of human nutritional variables was tested for, no significant differences were found between poultry raised under pastured and confinement conditions.

5. The success of primary pastured poultry enterprises (over 1,500 birds per year) depends on several important off-farm variables including access to state or federally inspected processing facilities and effective marketing mechanisms.

Label Rouge: Pasture-Based Poultry Production in France. by Anne Fanatico and Holly Born. (2003) You may remember our former APPPA Coordinator Diane Kaufmann's interest in the Label Rouge project from France. A final synopsis of the details of Label Rouge production systems and possibilities in the U.S can be found in this 12 page report from ATTRA. To see the report go to http://attra.ncat.org/attra-pub/PDF/labelrouge.pdf or call ATTRA at 1-800-346-9140 to request a copy.

Chicken is a healthier choice than high-protein food bars (Med Sci Monit, 2003; 9(2) CR 84-90) From EatWild.com In recent years, many people have been turning away from meat and eating more soy and whey products. The hope is that eating vegetable proteins will be better for their health than eating meat.

Now it appears that two of the most popular high-protein food bars---Atkins Advantage Bar® and the Balance Bar® ---may be less desirable for your health than chicken. In a new study, healthy volunteers were given one of the two high-protein bars or an equal weight of chicken or white bread. After each test meal, researchers measured the volunteers' blood sugar and insulin levels. In general, low and stable levels are associated with better health.

Both the Atkins Advantage Bar® and the Balance Bar® resulted in higher glucose levels than the chicken. The big surprise was that the Balance Bar® raised insulin levels even higher than white bread, a food known for its elevated insulin response.

The chicken proved superior in other ways as well. Weight for weight, it had less fat, fewer calories and only 2 grams less protein than the ultra-high protein Advantage Bar®.

Composting Dead Livestock
Jody Padgham Spring 2002

A publication put out by the Leopold Center for Sustainable Agriculture at Iowa State University in 1999 details the specifics of successfully composting dead livestock. Poultry composting became popular with large poultry producers in warm climates by the late 1980s, but it is actually an old tradition that can be practiced by anyone. We will summarize the paper here. You can download the report in it's entirety as an adobe acrobat file by going to http://www.ag.iastate.edu/centers/leopold/pdfs/SA8.pdf

Why Composting?
Hopefully we won't be loosing many birds in our operations, but if a predator strikes or we have management issues that take us by surprise we may have a disposal issue to contend with. Burial is always an option, but composting takes less labor time, can be better for water tables and will give you the additional benefit of rich compost when it is done.

Composting is a process that speeds the normal decay process by optimizing the environment for naturally occurring bacteria and fungi. The rate of decomposition is governed by environmental conditions and what the organisms have to feed on. To ensure rapid decomposition you must pay attention to:

Moisture. This is crucial. The moisture needs to be between 40-60%, which means the medium will be moist but not soggy. If you can squeeze moisture out, it's probably too soggy.

Co-composting Materials. These provide additional carbon and reduces the attractiveness to insects and rodents. Most commonly used are wood chips, ground corn cobs, or sawdust, which in addition will adsorb liquid run off during the composting process. These materials also keep the mix porous, aiding in oxygenation.

Carbon and Nitrogen. These are key components to track. Without a proper balance of C and N, microbial growth is retarded which reduces the decay rate. Ideal conditions occur at C:N of 25:1, but composting will occur in any range from 10:1 to 50:1. Temperature and odor are a good way of judging the C:N ratio. If you have ammonia odors it means the ratio is low, and can be remedied by adding sawdust. If moisture levels are where they should be and there are no ammonia odors but things are composting very slowly, it is probable that insufficient nitrogen is the cause. This is likely if you use sawdust. To correct, add manure to lower the C:N ratio.

Oxygen. Composting can occur either with air (aerobic) or without (anaerobic). Aerobic is preferred, as its major byproducts are water, carbon dioxide and heat. Anaerobic will produce little heat, but will create smelly by-products such as hydrogen sulfide and organic acids. It is difficult or expensive to maintain very high oxygen levels in a home compost pile, but if you avoid overly wet compost, periodically turn the pile and use relatively coarse co-composting medium you will be better off in avoiding odor problems.

Heat Retention. Heat is an important by-product of bacterial activity. Internal temperatures within properly composting operations often reach 120-150° F. This temperature range stimulates the rapid growth of heat loving bacteria which promote decay. Disease causing micro organisms will also be killed which improves the safety of the finished product. It is important to make on-farm composting piles large enough that internal heat will be generated and retained. Temperatures tend to go down toward the outsides of piles, and so carcasses should be at least 9-12 inches away from the edge and in a bin or pile large enough (6 x 8 ft.) to have a substantial "core" volume.

Composting Facilities
On-farm composting of small carcasses usually is done in simple uninsulated bins. Bins help to keep the material in a compact pile, reduce blowing and scattering of compost materials, and make carcasses less accessible to predators and rodents. Given most of us will have very low volumes of birds to compost, a bin would not be necessary but a nice luxury if you have the time and space to build one.

These systems can be used for birds up to 30 lbs. If a facility is intended for long-term use, you might build it from treated lumber or concrete, but pole-shed construction is fine for the occasional user. A roof or tarp cover is recommended to prevent excess moisture accumulation that can lead to undesirable odors and leachate. You will want some kind of all-weather base to reduce ground and water contamination.

Bin Design: If you plan to use a skid loader, build a bin six feet front to back and eight feet wide. Normal bin height is five to six feet. Wooden bin walls can be constructed using 2 x 6 in. or 2 x 8 in. treated lumber, or treated plywood with 2 x 6 in. stiffeners. Removable drop boards should be used for the front entrance, or doors that open horizontally, so birds can be dropped in.

(The article goes on from here talking about a typical system with 200 lbs. of bird loss a day. Now is the moment to appreciate another of the many benefits of pastured poultry, which is LOW MORTALITY. So I'll happily skip over that part)

Poultry Composting Operations

Poultry composting is begun by placing a 12 in. layer of dry poultry litter (co-composting material) in the bottom of the bin. When the carcasses release excess moisture, this layer will prevent the release of any odorous leachate. Carcasses are placed on top of the base layer at least nine inches from the bin walls. Carcasses should not touch each other, too many in one spot will lead to localized wet spots and poor decay. After the carcasses are positioned in the bin, they are covered with 4 to 6 in. of co-compost material. Incomplete coverage can lead to fly problems. Layering of the co-compost and the carcasses continues until the bin is filled to a depth of five feet. In a properly operating compost process, new material added to the bins reaches a temperature of 120-150° F within 24 to 48 hours. Internal temperatures can be monitored with a long compost thermometer to insure that microbial action is relatively uniform throughout the composting bin. Be sure to poke the probe in at several locations. It is not unusual to find hot and cold spots within the same bin, so a single temperature reading can be misleading. If a bin fails to heat up, moisture excess or deficiency is the most common cause, and you will need to unload the bin and mix in compost from an active bin to stabilize it. During cold weather, warm co-compost material can help to initiate decay.

Heating Cycles

After a bin is filled, it should undergo a primary heating cycle lasting 10-14 days. During this time rapid microbial action depletes the oxygen within the bin, the rate of decay slows and the temperature falls. Following the primary cycle, the partially composted waste should be turned or removed from the primary bin and placed in a secondary bin. The mechanical action of moving the compost breaks up the pile, redistributes excess moisture and introduces a new oxygen supply. Once the pile is moved, a secondary heating cycle occurs causing further decomposition. By the end of the secondary heating cycle, carcasses as large as 15-20 lbs. are normally reduced to bones that are reasonably clean and free of tissues that cause odors and attract insects and predators. Large birds (more than 15 lbs.) many need a third cycle to achieve complete decay.

Land Application

Stabilized compost is a great amendment when applied to land. Though soft tissues will be decayed, bones will still be present but should not be a problem with land application. If you add larger animals to your compost you may want to bury the large bones and skulls. Finished compost should be analyzed for nutrient content and applied to land in rates consistent with crop requirements. (note: if you are organic, be sure you understand the new organic standards for composting, which are EXTREMELY specific) .

CHAPTER FOURTEEN
Resources

Books and Periodicals

Birds of a Feather: Saving Rare Turkeys from Extinction, Carolyn Christman and Robert Hawes, American Livestock Breeds Conservancy, Pittsboro, NC 1999

Chicken Health Handbook, Gail Damerow, Storey Communications, Pownal, VT 1995

Chicken Tractor: The Permaculture Guide to Happy Hens and Healthy Soil, 2nd ed. Andy Lee and Patricia Foreman, Good Earth Pubications, Lexington VA 1994

Commercial Chicken Production Manual, 3rd ed., Mack O. North, AVI Pub., Inc., Westport CT, 1984

Commercial Poultry Nutrition, 3rd ed., S. Leeson and JD Summers, University Books, Guelph, Ontario, 2005

Day Range Poultry, Andy Lee and Patricia Foreman, Good Earth Pubications, Lexington VA 2002

The Dollar Hen: The Classic Guide to American Free-Range Egg Farming, Milo Hastings, Norton Creek Press, Blodgett, OR 2003

Fast Food Nation, Eric Schlosser, Houghton Mifflin Books, New York, NY, 2001

Feeding Poultry: The Classic Guide to Poultry Nutrition, G.F. Heuser, Norton Creek Press, Blodgett, OR 2003

Genetics of the Fowl, F.B. Hutt, Norton Creek Press, Blodgett, OR 2003

Grassfed Gourmet Cookbook, Shannon Hayes, Eating Fresh Publications, 2004

Holy Cows & Hog Heaven, Joel Salatin, Polyface,Inc. Swoope, VA, 2004

Hubbard Classic Breeder Management Guide, Research and Development Department, Hubbard Farm, New Hamshire. www.hubbardbreeders.com

The Legal Guide for Direct Farm Marketing, Neil D. hamilton, Drake University, Des Moines, IA, 1999

Nourishing Traditions, Sally Fallon, New Trends Publishing, Washington D.C., 1999

Pasture Perfect, Jo Robinson, 2004. www.eatwild.com

Pastured Poultry Profit$ Joel Salatin, Polyface, Inc., Swoope, VA, 1998

Poultry Health Handbook, 4th ed., L. Dwight Schwartz, D.V.M., The Penn State University, University Park, PA 1994

Remedies for Health Problems for the Organic Layer, Karma Gloss, Northeast Region SARE, Burlington, VT, 2004

Sucess With Baby Chicks, Robert Plamondon, Norton Creek Press, Blodgett, OR 2003

The Stockman Grassfarmer. Monthly publication. 800-748-9808

Organizations

APPPA (American Pastured Poultry Producers Association). Newsletter, annual meeting, web resources, electronic listserve. www.apppa.org for updated contact info.

PASA, (Pennsylvania Association for Sustainable Agriculture) Membership organization dedicated to education about sustainable food and farming. Resources, newsletter, annual conference. www.pasafarming.org 814-349-9856

MOSES, (Midwest Organic and Sustainable Education Service) Education and resources for those interested in sustainable organic production. Books, annual conference, newsletter. www.mosesorganic.org, 715-772-3153

SSAWG (Southern Sustainable Ag Working Group) Annual conference, resources, networks. www.ssawg.org

ACRES USA ,Books, and annual conference and newsletter. www.acresusa.com 800-355-5313

SARE (Sustainable Ag Research and Education) Grants and grant reports. www.sare.org

ATTRA, (National Sustainable Agricultre Information Service.) Online and printed resources for sustainable farming. www.attra.org, 800-346-9140

Index

A

Acadian Seaplants 7, 133
accidents 215–215
accounting 129, 202, 204–205
advertising
    business plan 208-209
    in France 237-239
    role of 165
    success with 21,45
    survey 8
    turkey 102
aerial predation 52
agritourism liability 223
air quality
    in brooder 100
    and mortality 130-132
alfalfa meal 75, 89-91, 121–125, 128
Alzheimer's, and omega-3s 196
ammonia
    and hens 75,89
    and mortality 130
    composting and 245
    in brooder 14
    problems with confinement 154, 194
animal protein in feed 119–121
antibiotic free
    in brooders 18
    in France 240
    on large pastured farms 7
    marketing 167-168, 198
    organic 189
antibiotics
    in commercial flocks 27,31,196
antibodies 137
ascites 6, 11, 28, 129–130
asset protection 219–223
ATTRA
    Label Rouge 241, 244
    organic resources 189-190
    poultry toolbox 208
attractive display 170
avian disease
    danger of 105
    preventing 144
Avian flu 139-142

B

bacteria
    beneficial 135
    brooder 18-19
    cleanliness and 131
    confinement 27
    E-coli 130, 197
    eggs 82-83
    food safety 171
    fowl cholera 113
    in compost 241
Barred Rock 6
battery brooders 20
bedding
    brooder 13, 15, 18-19, 100, 232
    deep 34, 60-61, 74-76, 91, 99
    floored pasture pens 10-11, 20, 60-61
    hen houses 71, 74-76, 82, 85, 88
    in France 237-238
    keeping dry 66
    problems with 18-19, 134-135, 224
beef liver 16–17, 79
bell-type waters 20, 56, 64-66
Bennett, Dan
    articles about 18-19, 64-66, 79-80, 200-201
    articles by 15-17, 96-99, 204-205
    bio-security 137
biological extracts 119
bird flu 141, 142
Blackhead disease 7, 138
bleach 43, 82-84, 145–147
bleed out 147–148
bloody manure 135
Blue Slate 192, 242
Bourbon Red 98, 107, 192, 242
brand identity 170
    in France 236-242
break-even analysis 209, 212
breeding stock 23, 26-29, 34
    turkeys 105-106
Brewers Yeast 132
Broad Breasted White (BBW) 107–108
brochure
    developing 178-179
    for marketing 77, 166, 168, 175
brome 9, 112
brooder
    basics 2-3, 12-13, 16-17
    coccidia 134-135
    examples 43-44, 54-56, 60, 99, 128, 226
    feed 122
    geese 112
    mortality in 130-132
    problems 14-15, 18-19, 224
    pullets 79
    record-keeping 202-205
    turkey 96, 105
    water for 66
broody hen 34
Brower
    eqiuipment 58, 65, 149, 159
    feeders 71, 80, 119, 206
    pen 40, 96
    profile 42
budgets 204-205
business plan 167, 212–213
business structure 217–222

C

calcium
    feed supplement 91
    in soil 104
    feed mixes 121-124, 136
    turkey feed 101
candling eggs 83–84
cannibalism 127, 133
capital costs 207–207
capons 32
cash-flow 209
cattle panels 37, 43-47
certified organic poultry 192
certified organic processor 44
chef
    marketing eggs to 77-78
    selling chickens to 166
chemical free food 167
Cheyenne River Sioux Tribe 164
chicken tractor 38–40, 57–58, 103, 229-230
chicks
    brooding 12-21
    feed 122-123
    genetics 22-35
    health 127, 130-131, 224
    loss 224, 229-230
    organic 190
    pheasant 109-110
    pullets 72-73, 79-80
    record-keeping 200-209
    use in marketing 170, 174
children
    helping 37, 43-45, 110, 129
    and marketing 174
    and turkeys 92
Christmas
    egg market 87
    geese 114
    turkeys 94-95, 102-106
citric acid 100
CLA (conjugated linoleic acid) 196–198
cleanliness
    brooder 15-19

eggs 88
    disease prevention 134, 137
    home processing 145-147
    and mobile processing 162
    watering systems 64-66
co-composting materials 245
coccidia
    in brooder 12-14, 220
    raw milk treatment 17
    symptoms 131, 134-135
    treatment 131, 134-135
    vinegar treatment 44
collecting feathers 113
compost
    from bedding 10, 19, 60
    offal 158
    producer examples 54-56, 62, 71, 99, 117
confinement poultry systems
    brooding 19
    comparison with 228-230, 233-234
    disease 130, 135, 144, 194
    feed conversion 118
    French comparison 236-242
    genetics and 22-24, 26-30, 35
    pp advantages over 35, 57, 77, 196-199
    turkeys 105-107
    water systems 65
conjugated linoleic acid (CLA) 196–198
consumer demand 167–168
cooking demonstrations 213
copper sulfate 131
Corndel 28–29
Cornish Cross
    alternatives to 22-35, 237
    genetics 22-35, 229-230
    hatching 27-29
corrugated steel for pens 36–39
coupon 179
cover crop 115-118
cows, grazing with 79
coyote 143
Creative Growers 51–53, 116–117
cryovac 174
CSA
    liability and 214-215
    selling meat through 166, 173
    vegetable 103
cup drinkers 16, 18
curly toe 130, 132
customer information, collecting 200-205, 212, 213
customers
    developing 129, 165-169, 173-176, 178-179, 193-194, 212-213, 225
    education of 166, 174-176, 178-179, 196, 198-199
    expectations of 27, 30-32
    farmers market 170-172, 184-186
    for eggs 73, 75, 77-78, 80, 87
    for geese 111-114
    in france 239
    liability and 214-216, 222-223
    multiple purchases 63
    on-farm pick-up 68-70, 207, 227
    satisfying 187, 211, 228, 231-232, 234
    studies of 243
    turkey 95, 97-99, 103-104, 106-108
cut up poultry 184-187

D

database 21, 202–205
day-range systems 7-8, 9-11, 20-21, 47-48, 51-63, 79-80, 98-99
death loss 7, 15-16, 20, 28, 127, 130-132, 138-143, 224, 228-230
    tracking 200-205
deep litter 14, 34, 60-61, 74-76, 91, 99
Delaware Chicken 26-32
Delehanty, Tom 50, 188, 200
delivery 87, 104
demographics 212
deposit 98, 106, 129, 201-203
Diane Kaufmann 127, 200, 244
diarrhea 18, 131, 138
direct marketing 165-179
    assessing 184-187
    in business plan 212-213
    insurance 214-216, 222-223
    legal issues 217-222
    on web 181-183
    tracking 200-204, 208-211
disease
    avian influenza 139-142
    blackhead 7, 138
    coccidia 12, 14, 17, 44, 131, 134-135
    enteritis 130
    fowl cholera 113, 139
    human 140-142, 196
    preventing 18-19, 79-80, 105-106, 137, 144, 234
employees 213, 216, 220
enteritis 130
entertainment
    and insurance 218-219
    marketing as 161-162
equipment 64-70
    Tenosynovitis (viral arthritis) 137
disease-free 106, 111, 144
disease resistant 28-29, 105, 111
display
    farm market 165, 170-172
    packaging 174
docking station for MPU 162–163
dolly, pen moving 36, 43, 57, 71, 103
domain name 181–183
Dot Posters 172, 184-186
Dotson Farm and Feed 7, 136
dual purpose breed 26
ducks
    and disease 144
    hatching eggs 34
    on pasture 109-110
    processing challenges 4
    wild, and disease 139-142

E

educating customers 166, 174-176, 178-179, 196, 198-199
egg cartons 73, 81, 183
EggCartons.com 81
egg fertility 23
egg handling 23-24, 72-73, 79-80, 82-84, 87-88
eggmobile 23, 54, 71-73, 85, 129, 187
egg processing equipment 83-84
eggs
    blood spots on 138
    breeding 22-24, 27-29, 33-35, 106
    candling 80, 83
    cleanliness 75, 88, 137
    economics 72-73, 80
    fed to turkeys 92
    markets 80
    and omega-3 196-197, 243
    production levels 26, 71-72, 87, 90-91, 109
    shelters for 57-58, 71-74
    turkey breeders 106
egg washing 82-84, 87-88
egg yolks 22, 75-78, 124-125
electric fence 2, 7, 31, 226
    electronet 23, 79-80
    feathernet 23
    for breeders 29
    for geese 112
    for turkeys 100, 106, 107
electric heat
    for hens 89-91
    in brooder 7, 12, 16, 60, 99
Elmwood Stock Farm 7, 191-192
Embden Geese 112
Entertainment
    marketing as 161-162

Index

capital costs 209
cleanliness 137, 144
for egg processing 83-84
insurance for 216-223
investment in 80
processing 1456-164
specialized for pp 42
scavaging 66, 82, 99
estate planning 222
Embden geese 112
employees 231, 216, 220
enteritis 130
entertainment
equipment, cont.
    invenstment in 80
    processing 145-164
    scavenging 66, 82, 99
    specialized for pp 42
estate planning 222
evisceration 147, 152, 154, 156
expenses
    estimating 208-211
    tracking 200-205

F

Family Limited Partnerships 220
Farm Bureau 223
farmers market
    displays 165, 170-172
    examples 56, 99, 191-192
    and insurance 220-223
    and processing laws 145
    samples 78
    surveys at 184-186, 212
    tips for 165-166, 170-172, 174-176, 187, 211
farm liability insurance 214-223
Featherman 149, 155
feathers
    clipping wings 110
    goose 111-114
    pecking 89
    plucking 147-149, 153-164
    predator signs 143
    preening and disease 134
    ruffled 139
    scalding 146-153
feed
    alfalfa 124
    for breeding stock 23, 27-29
    in brooder 13, 16-17
    consumption in alternate breeds 28, 30-32, 35
    ducks and pheasants 109
    French system 233
    geese 111-112
    hens 72, 80, 88-91
    and heat stress 88-89
    kelp 133
    limiting 130-132
    mixing own 35, 121, 128, 231-232
    organic 35, 44, 55, 103-104, 188-191, 231-232
    pasture 115-119
    rations 121-125
    raw milk 16-17, 19. 135
    research 243
    size grind 119, 122
    soybeans 91, 101, 121-125
    in system comparison 9-11
    tracking costs 200-210
    turkey 92, 97, 100-101, 106
    witholding 146
feed conversion 10, 28, 118-119, 229
    turkey 107-108
feeders
    in brooder 13, 18-19
    companies 67
    raft feeder 60
    styles 44, 49, 52, 54, 57-58, 60-62, 75, 80, 101, 119, 206, 226
feed mills 125, 232
    in France 237, 240-241
Fertrell 15, 20, 126, 129, 133, 135, 137, 185
    Broiler-Grower Ration 9, 97, 202, 222
    Nutri-Balancer 55, 80, 120-121, 128, 136
    ration recommendations 101, 123-125
fescue 9, 116-118
    killing 60
fiber in diet 91, 121
Filemaker 202-203
filiere 240–241
financial planning 169
financing 210
fishmeal 55, 97, 101, 121-126, 128-129, 136, 197
fixed costs 11–260, 45–260, 209-213
flavor
    compromised 30-33
    duck, pheasant 110
    enhancing egg 91
    from feed ingredients 122
    superior 170, 192, 229
    turkey 105, 107
flax 122, 197
flock size 36
food safety 145, 158, 171-172, 193-194, 197
forage
    for chicks 122
    composition 9, 55, 116-121
    compensating for in winter 125
    and geese 112
    grasses 35, 55, 60-61, 98, 116-119, 233
    quality 36, 57, 76, 192, 196, 233-234, 238, 243
    turkeys 92, 107
foraging
    challenged 22
    enhanced 28-31, 73
fowl cholera 113
fox 58, 143
freezer 68–69, 163, 192
French system 236–242
frozen product 171, 184-186, 192

G

galvanized roofing 39
garden
    chickens in 23
    market garden 55, 60-62, 103, 115
gas brooders 7, 43
geese 34, 66, 111-114
    processing 146, 149, 159
genetically modified organisms(GMOs)
    in feed 188-190, 231
    in France 241
genetics
    breeds used 6, 229-230
    challenges with 14, 26, 30-32
    grass-bred 22-24, 118
    refinements in 22-35, 56, 232
    turkey 105-108
giblets 142-143, 231
Gillis 67
Glass, Kip
    articles about 49, 226-227
    articles by 9-11, 107-108, 187, 210-211, 224-225
global market 27
goals
    for brochures 175
    French production 241
    identifying 167-169, 212-213, 233
goose feathers 113
Goosemobile 111–114
goslings 111–113
GQF 16–17, 60, 65, 206
grassfed 174-176, 229
    benefits 196-199
    geese 111
grazing
    benefits 198-199
    improved 23, 57, 60, 233-234
    multi-species 191

pasture 115-118
rotation 73, 118
turkeys 106-107
Greener Pastures Poultry 54-56, 170-172, 184-185
grit 16, 19, 49, 54, 93, 119, 217, 235
ground turkey 102
Grower's Discount Labels 180
growth hormones 7, 31

H

Hansen, Mike and Debra 43-48
hatcheries 13, 26-28
disease 137, 144
hatcheries, cont.
French 237, 240-241
genetic stock 26-28, 119, 230, 232
Moyer's 105-106
techniques 24
variations 118
Walters 105-106
hatching
cleanliness 137
eggs for breeders 22-26, 34-35
hawks 62
heart attacks
breeding 28
and heat 88
preventing 44, 120, 122-123, 131-132
heat stress
affect of 88-89, 225
aleviating 132
genetics and 22
Heifer Project International 5, 13, 22, 161-164, 239
heirloom (see heritage)
hen house 34, 57, 85, 91
heritage breed
chickens 28, 33
turkeys 98-99, 105-109, 192, 242
Blue Slate 192, 242
Bourbon Red 98, 107, 192, 242
Naragansett 105
Royal Palm 105-107, 192
historical turkey (see heritage)
hog panels 40-48
Holistic Management 213
home-school 20-22, 128–129
homeowner's insurance 215
hoophouse 23, 48, 74-76, 79-80, 191
hot weather
hens 88-89, 91
and pens 39-40
water 64-66
HTML 181-183

humidity
brooder 100, 130
eggs 24, 82
hydrogen pyroxide 120

I

immune system 123-124, 131, 134, 242
income
eggs 73, 80
protecting 217-218
tracking 202-203, 208-211, 227-228
turkey 95, 105-108
income statement 208
income tax 217
incorporation as business 217-222
incubator
custom incubation 24
equipment 25, 42, 67, 81
injury, insurance and 213-216, 220
insects, consumption of 119, 235
inspected processing 4-6
and consumers 243
geese 111
in France 237
and MPUs 159-164
organic 190
producer examples 44, 56, 128, 154, 192, 231
turkeys 104, 106
insurance 214-223
internet 181
Internet Service Provider 181
inventory 171, 207
Island Grown Farmers Cooperative 164

J

K

kelp
in feed rations 44, 80, 121, 235
sources 133, 136
Kencove 3, 23, 59–60, 100
killing cones 21, 150
knives 146, 156, 158

L

labels 180
Label Rouge 8, 208, 236-244
labor
in budget 209-211
comparison 9-11
producer examples 20, 44, 128, 154, 206-207, 226-227
watering systems and 65

lamb 63, 70, 98, 111, 155, 163–164, 166, 170–171, 176, 191
latex gloves 145
lawsuits 215–216, 219-222
laying rate 72
Lee, Andy   day-range 31-32, 57-63
pen design 37-38
on Zero Mortality 228
leg problems 11, 30, 44, 132, 229, 232
lesions 138, 194
liability
cost of insurance 63
MPUs and 158-164
need for 213-223
licensing
legal 218
marketing 172
MPU 162
lime 14, 55, 138
Limited Liability Company 217, 218
Limited Life Partnerships 220
linebreeding 27, 31
listless 135
listserve 33, 181
LLC
examples 55-56
incorporating 217-222
local restaurants 45
loss of appetite 138
loyal customers, creating 77, 167

M

Malathion 138
manganese deficiency 132
manure
benefits of 117
in brooder 14, 18
comparisons 9-11, 48, 51, 57-58, 60-62, 73-75
in compost 54
negative implications in 193, 228-230
in nests 82, 88
as symptom 87, 122, 130-132, 134-135
turkeys and soil benefit 103-104
margin
assessing 209
eggs 80
goals 207
turkey 110
marigold petal dust 75
marketing plan 167, 213
market research 212-213
Mattocks, Jeff
articles about 15, 20, 97, 126, 133, 135

Index

articles by 88, 90, 121-125, 130-132
medicated chick starter 14
medications
    alternatives to 119-120, 135, 240
    using 14
Microsoft Excel 202-204, 223
Microsoft Outlook 204
minerals 120-121, 133
mini-barn 20, 60
mink 143
mission statement 178-179, 212
mixing your own feed 35, 121
mobile processing (MPU) 7, 157-164
molasses 128

mortality
    causes of 11, 14, 18-19, 130-132, 138, 229-230
    reducing 12, 18-19, 28, 120-122
    stories 54-56
    turkeys 94
motor oil 138
Moyer's Chicks 25, 87

N

Narragansett 105
National Organic Program 188
nest boxes 23–24
    for egg production 71-76, 82, 85, 88
    and health 138
    isolating genetics 34
newsletters 8, 21, 166, 199
next-day air 105
nipple waterers/drinkers 18-20, 43, 64-66, 99, 135
nitrogen
    in compost 245
    in pasture 116-118
NOFA 103
non medicated feed 112
Nutri-Balancer 80, 119-121, 123-126, 129, 136

O

oats
    in feed 55, 89, 122, 125, 128
    geese 112
    in pasture 115-116
offal, composting 128, 158, 160, 162
oil gland 147
omega-3 fatty acids 78, 196-199, 243
on-farm processing 4, 146
    examples 111, 128
    liability and 221

MPUs 161-164
online sales 192
operating costs 208
opossum 143
orchardgrass 117
organic
    allowed materials 80, 102-104, 120, 126, 133
    consumers 243
    definition 188-190
    in France 241
    in market 167-168, 136, 137, 173, 178, 200, 208, 213
    operations 43-45, 191-192, 231-232, 235
organic certification 6, 190, 192
Organic Seal 190
overfeeding 124
owls 48, 57, 62, 143, 230
oyster shell 54, 191

P

packaging 171, 174, 187
paddock
    and day-range 57, 63
    egg production and 71, 79, 85-86
    geese 112
    turkeys 100-105
palatability 118
pancake brooders 20
parasites 58, 138, 234
PASA 195
pasture
    benefits of 22-24
    comparison of pen systems 9-11, 51, 57-58, 60-63
    composition of 55, 115-118
    genetics for 22-35
    turkeys 92, 96
    quality of 73, 79, 115-118
pastured marketing 170-176, 178-179, 184-186, 191-199
    challenges of 233-234
Pastured Peepers 3, 22–24, 27-29
Pastured Poultry Profit$ 1, 15, 38, 70, 194
pathogens 18, 27, 58, 76, 134, 240
pecking 89
permits 68, 157-164, 171–260, 213
pheasant 109–110
Phenolic Acids 124
photos
    and liability 219
    in marketing 165-166
    on websites 182-183
Pickwick-Zesco 149, 156
pigs, and hoophouses 75

Plamondon, Robert 12, 48, 64–65, 83–84, 138, 182, 224
plan 2-6
    biosecurity 137-144
    business plan 212-213
    liability plan 217-222
    marketing 165-172
    MPU 159-164
    pen 46-49, 51-53
planning
    for egg production 87, 90-91
    for profitability 201-205, 208-211, 224-225
Plasson waterers 9, 56, 65, 206
plucking 145-149, 153-164
pneumonia 103
Polyface Farm 9, 22–23, 77, 233
poultry netting (see electric fencing)
    examples 20, 34, 37, 47, 52, 57-58, 60-63, 79-80, 85-87, 93, 106
poults 92-94, 100-108
pre-order 4, 106, 111
predators 2-4, 7, 143
    in brooder 14, 17
    and day-range 58, 62, 230
    pen design and 36-49, 57
Premier Fence 23, 34, 58, 93, 100
pressure regulator 64
price
    customer expectations 242-243
    eggs 72-73, 77-78
    equipment 66, 68, 80, 85, 153, 157, 161-164
    examples 5, 45, 56, 80, 85-87, 129, 173, 207, 231, 238
    geese 113
    of pens 38 pheasants and ducks 109-111
    setting 4, 166-172, 177, 184-187, 200-205, 208-213
    turkeys 94-95, 105-108
probiotic 17, 19, 120
processing
    cost of 63
    ducks 110
    eggs 82-84
    equipment 42, 67, 83, 146, 148-164
    french 237, 240
    MPUs 157-163, 192
    on-farm 145-147
    organic 190, 192
    producer examples 44, 56, 99, 128, 192, 206, 227, 231
    record-keeping 201
    turkeys 94, 102-104
Producer Profiles

Alexander 98
Bennett 202
Coulimore 128
Glass 222
Hansen 43
Neuberger 111
Ritch 20
Silverman 54
Smith 85
Stone (Elmwood) 187
Tschoepe 227
Ussery 33
production records 129, 201–202
Production Reds 72
product liability 214-217, 223

profit
    estimating 4, 200-205, 208-211
    examples 8, 45, 63, 98, 108-110, 129, 187, 203, 227-230, 243
protein 121-125
    ducks 109
    inhibitor in soy 127
    layer feed 72, 89-91
    pheasants 109
    turkey feed 92, 97, 100-101, 106
pullet 25
Purdue 26, 35
PVC
    feeders 3, 23, 54, 60
    pens 36-37, 40, 42, 50, 57, 61-63, 96-97
    waterers 98, 101

Q

Quickbooks 197-203

R

raccoons 4, 52, 109
Rainbow 191
ration
    hens 90-91
    ingredients, recipes 55, 121-125
    supplements 119-121
    turkey 92, 97, 100-101, 106
rats
    in brooder 3, 14, 17
    in pens 52, 54
raw milk, for coccidiosis 16-17, 19, 131
record-keeping 21, 99, 129, 200-211
Red Label 236–240
refrigerated trailer 68
regional brand 241
regulations 5-8
    for direct marketing 209
    for eggs 83
    for farmers markets 170-172
    for MPU 158-164
Reiff, Eli 7, 149-154
relationship marketing 173
research
    cost and time 45
    French system 236-242
    large pp systems 7
    marketing meat 170-172, 184-186
    on-farm feed supplements 119-121
    pp nutrition analysis 197
    review 242-244
    system comparisons 7
    turkeys 103-104
reservoir, water 18, 47–48, 64, 66, 191
respiratory problems 130, 137, 194, 235
restaurants
    eggs 77-78
    examples 6-8, 43-45, 54-56, 85-87, 109-110, 192, 207
    marketing to 165-169, 172-173
Rhode Island Red 6, 25–26, 191
riboflavin deficiency 132
roosters, for breeding stock 23-24, 29, 34
roost mites 138
roosts
    cleanliness 138
    for hens 72, 75, 85
    turkeys 107
Rouen ducks 109
Royal Palm 105–107, 192
rye in pasture 115–116

S

Salatin, Joel
    articles about 1, 7. 15, 28, 38, 41-42, 49, 51, 57-58, 73, 85, 96, 106, 109, 154, 173, 175, 194, 197, 201, 232-234, 242
    articles by 36-37, 71, 74-78, 92-95, 118-121
salmonella
    in conventional poultry 177
    danger of 145
    prevention in brooder 105
    testing for 197, 240
SARE grants 103, 127, 161, 171, 184, 197, 242
sawdust
    bedding 13, 75
    in compost 245
scalder 94, 99, 145-149, 153-164
scalding
    ducks 110
    geese 113
Schafer, David 149, 153, 155, 174-176
school bus, for processing equip 159–161
seaweed 133
shackles 21, 152
shade
    pens 10, 30, 38-39, 48, 63, 226
    turkeys 93, 105
Shell, Tim 18-19, 22-24, 27-31, 58, 64-66, 118, 134-135, 230
sheep and poultry 73, 112, 117, 192
shipping
    chicks 2, 13, 25-26
    finished product 105
signs 165-166, 169, 190
    and marketing 213
Silverman, Aaron 20, 51, 54-55, 115-116, 128, 148, 166, 180, 197, 199, 202, 222
sinusitis 130
skunks 4, 109 232
Slow Foods 98, 105
Small Business Administration 218
soil test 104
South Dakota Goose Association 114
soybeans
    in ration 91, 101, 121-125
    problems with 122, 127
spraddle leg 130
sprouted grain 35, 75
state-inspected 5, 128
stocking densities 36, 52, 56, 91, 132, 226
straight-run 55
straw 40, 75, 225, 237–238
stunted growth 138
sudden death syndrome 131
sugar 12, 17, 124, 244
surveys 72, 184-185, 213
swollen joints 137

T

tannins 122
taste
    eggs 78
    improved 27-35, 57
    marketing 175, 194, 236-242
    turkeys 92, 102, 105-108
taxes 217–222
temperature
    in brooder 12-13, 24, 60, 100, 105, 130-131, 228-230
    cooler 68069
    for compost 245-246
    and food safety 171, 179
    for hens 88-91
    in pens 38, 55, 74-76, 85
    processing 145-147, 158

turkeys 93
    for washing eggs 84
Thanksgiving
    and egg demand 87
    turkeys 93-110, 192
The Stockman Grass Farmer 15
transport crate 44, 146-147, 151
trusts 220
Trypsin inhibitor 122, 127
Tschoepe, Karla 231
turkey fryers 146
turkeys
    brooding 92, 96, 100, 105
    diseases 138, 140
    economics 103-104, 107-108
    feed 92, 97, 100-101, 106, 123
    freezing 68-70
    heirloom, heritage 105-108, 242
turkeys, cont.
    marketing 94-95, 97, 102, 106-107
    on pasture 106-107
    pens and shelters 37, 40, 49, 93
    production 92-108
    processing equip 149-152, 159-164
turmeric 44
Tyson 26, 30, 32, 66, 177, 188-189

U

umbrella insurance policy 215
uninsured 214
URL 181–183

USDA inspected processing 44, 56, 104, 106, 154, 161-164, 172, 192, 243
used equipment 66, 157

V

vaccination 137, 140-142
value-added products 170, 180, 206, 216
variable expenses 45
ventilation
    in brooder 19, 130
    pens and shelters 36-40, 74-75, 222, 235, 237
vinegar
    in brooder 16-17
    to help stress 122, 128, 132
Viral Arthritis 7, 137
virus 137-142
vitamins
    to combat stress 89
    in eggs 243
    in feed 120-121, 124, 229
    in pastured meat 196-197

W

walk-in cooler 68
water
    in brooder 12-19, 100, 224-227
    cleanliness 113, 134-137, 193, 224, 228-230
    day-range 54-63, 79-80
    in pens 36-41, 43-45, 48-49, 74-76, 101, 206

    in processing 145-147, 157-164
    systems 9-10, 64-67, 85-87, 100, 224-227
    washing eggs 134-137, 193, 224, 228-230
waterfowl 3, 109, 146
wax 113, 146
weasel 143
Web hosting service 182
web page 181–183
weight gain
    breed differences 28, 30
    on pasture 118-120
    problems 127, 224, 230
    tracking 200
wet litter 18, 122
White Pekin ducks 109
white turkeys 92, 99
wholesale 7-8, 171-172, 177, 184-187
wild birds 14, 43, 110, 134
Willard's Water 120
windbreak 50
winter ration 91
withdraw food 146
wood shavings 34, 85
word-of-mouth marketing 2, 45, 114, 129, 227
worm eggs, and blackhead disease 138

X Y Z

Printed in the United States
204535BV00001B/1-16/A